Methods in Molecular Biology™

Series Editor
John M. Walker
School of Life Sciences
University of Hertfordshire
Hatfield, Hertfordshire, AL10 9AB, UK

For further volumes:
http://www.springer.com/series/7651

Methods in Molecular Biology

miRNomics: MicroRNA Biology and Computational Analysis

Edited by

Malik Yousef

Dabburiya Village, Israel

Jens Allmer

Izmir Institute of Technology, Izmir, Turkey

Humana Press

Editors
Malik Yousef
Dabburiya Village, Israel

Jens Allmer
Izmir Institute of Technology
Izmir, Turkey

ISSN 1064-3745 ISSN 1940-6029 (electronic)
ISBN 978-1-62703-747-1 ISBN 978-1-62703-748-8 (eBook)
DOI 10.1007/978-1-62703-748-8
Springer New York Heidelberg Dordrecht London

Library of Congress Control Number: 2013953868

Printed on acid-free paper

Humana Press is a brand of Springer
Springer is part of Springer Science+Business Media (www.springer.com)

Dedication

JA dedicates the book to his wife Açalya and his son Lukas Aren who was born during the editing of this book.

MY dedicates the book to his parents Mustafa and Haji and his wife Sahar and his daughters Miar and Umaymah and his sons Anas and Mohammed.

Preface

Micro RNAs (miRNAs) have just recently been discovered, but the scientific interest is enormous. When we started becoming interested in miRNAs around 2006, there was already a significant body of knowledge to draw from. Of course we don't work on our own and it is often necessary to introduce students, colleagues, or new collaborators to the current state of miRNA research. Sending out large amounts of review articles is one option but we wanted to improve the situation and decided to put together a book which touches on all the topics relevant to miRNA research. The outcome is what you are currently holding in your hands. We aimed to encompass relevant information from biogenesis of miRNAs, their biological function, computational analyses, as well as medical implications. Each of these topics may be a book in its own right, and this book neither can nor intends to be exhaustive for any of the subjects that it touches upon.

This book is rather intended to present an overview of the current state of the art and aims to put the respective areas of research into a larger perspective. One of the editors is a computer scientist and the other is a biologist by training, and this hopefully ensured that both sides, the computational and the biologic component, are adequately treated in this book but both of us also wanted to highlight medical implications in respect to miRNAs. So if you are coming from the computer science side and want to learn more about the biology of miRNAs, this book will provide the information you need. Likewise, if you come with a biology perspective, you will find computational, statistic, and medical information. The book can be read from the beginning to the end for newcomers to the field but those that find themselves already knowledgeable in parts of miRNA research, statistics, or any other topic tackled in the book can safely skip such chapters.

Scientists today are usually overworked and it is difficult to find colleagues who are able to spend time to write or review book chapters and no wonder we received many replies to our initial invitations along these lines: "Thank you for the invitation and I really support the idea of writing such a book, but unfortunately I cannot participate due to …". Undeterred by such setbacks we gathered an initial group of authors willing to contribute and started the process. After about 6 months we had to realize that not all of the authors were holding up their end of the deal and we had to find new contributors for the affected chapters. As you can see, we were successful, but the overall process from initiation to copy-editing was stretched out to almost 2 years.

Any endeavor, and especially editing a book, is not possible without help and we would like to thank the chapter authors for their contributions. We further appreciate that some of the authors also pitched in with reviewing other chapters in the book. We would also like to extend our gratitude to our external reviewers: Mesut Muyan, Yusuf Tutar, Ekrem Varoğlu, Sreeparna Banerjee, Hasan Otu, and Gerhard Wilhelm Weber. Working together with professional editors at Methods in Molecular Biology was a great pleasure and especially David Casey and John Walker solved many a puzzling question.

Dabburiya, Israel — *Malik Yousef*
Izmir, Turkey — *Jens Allmer*

Contents

Contributors

BÜNYAMIN AKGÜL • *Molecular Biology and Genetics, Izmir Institute of Technology, Izmir, Turkey*

JENS ALLMER • *Molecular Biology and Genetics, Izmir Institute of Technology, Izmir, Turkey*

PATRIZIO ARRIGO • *CNR ISMAC U.O.S of Genoa, Genova, Italy*

ÇIĞIR BIRAY AVCI • *Faculty of Medicine, Department of Medical Biology, Ege University, Izmir, Turkey*

YUSUF BARAN • *Molecular Biology and Genetics, Izmir Institute of Technology, Izmir, Turkey*

DEBORA BARONI • *CNR ISMAC U.O.S of Genoa, Genova, Italy*

YALIN BAŞTANLAR • *Department of Computer Engineering, Izmir Institute of Technology, Izmir, Turkey*

HANS BINDER • *Interdisciplinary Centre for Bioinformatics, University of Leipzig, Leipzig, Germany; LIFE Leipzig Research Centre for Civilization Diseases, University of Leipzig, Leipzig, Germany*

MEHMET VOLKAN ÇAKIR • *Interdisciplinary Centre for Bioinformatics, University of Leipzig, Leipzig, Germany*

TOLGA CAN • *Department of Computer Engineering, Middle East Technical University, Ankara, Turkey*

ERDAL COŞGUN • *Faculty of Medicine, Department of Biostatistics, Hacettepe University, Ankara, Turkey*

BALA GÜR DEDEOĞLU • *Biotechnology Institute, Ankara University, Ankara, Turkey*

AYSE ELIF ERSON-BENSAN • *Department of Biological Sciences, Middle East Technical University, Ankara, Turkey*

ŞERMIN GENÇ • *Department of Neuroscience, Institute of Health Science, University of Dokuz Eylul, Izmir, Turkey*

ÇAĞDAŞ GÖKTAŞ • *Molecular Biology and Genetics, Izmir Institute of Technology, Izmir, Turkey*

SYED MUHAMMAD HAMID • *Molecular Biology and Genetics, Izmir Institute of Technology, Izmir, Turkey*

HAMID HAMZEIY • *Molecular Biology and Genetics, Izmir Institute of Technology, Izmir, Turkey*

LYDIA HOPP • *Interdisciplinary Centre for Bioinformatics, University of Leipzig, Leipzig, Germany; LIFE Leipzig Research Centre for Civilization Diseases, University of Leipzig, Leipzig, Germany*

KALYANI KORLA • *Department of Biochemistry, University of Hyderabad, Hyderabad, India*

RALPH LEO JOHAN MEUWISSEN • *Department of Internal Medicine, School of Medicine, University of Dokuz Eylul, Izmir, Turkey*

CHANCHAL K. MITRA • *Department of Biochemistry, University of Hyderabad, Hyderabad, India*

HASAN OĞUL • *Department of Computer Engineering, Baskent University, Ankara, Turkey*

MERYEM GÜLFEM ÖNER • *Department of Neuroscience, Institute of Health Science, University of Dokuz Eylul, Izmir, Turkey*

MUSTAFA ÖZUYSAL • *Department of Computer Engineering, Izmir Institute of Technology, Izmir, Turkey*
MÜŞERREF DUYGU SAÇAR • *Molecular Biology and Genetics, Izmir Institute of Technology, Izmir, Turkey*
KEMAL UĞUR TÜFEKCI • *Department of Neuroscience, Institute of Health Science, University of Dokuz Eylul, Izmir, Turkey*
HENRY WIRTH • *Interdisciplinary Centre for Bioinformatics, University of Leipzig, Leipzig, Germany; LIFE Leipzig Research Centre for Civilization Diseases, University of Leipzig, Leipzig, Germany*
MALIK YOUSEF • *The College of Sakhnin, Sakhnin, Israel; The Galilee Society Institute of Applied Research, Shefa Amr, Israel*
GÖKMEN ZARARSIZ • *Faculty of Medicine, Department of Biostatistics, Hacettepe University, Ankara, Turkey*

Chapter 1

Introduction to MicroRNAs in Biological Systems

Ayse Elif Erson-Bensan

Abstract

MicroRNAs are 20–24-nucleotide-long noncoding RNAs that bind to the 3′ UTR (untranslated region) of target mRNAs. Since their discovery, microRNAs have been gaining attention for their ability to contribute to gene expression regulation under various physiological conditions. Consequently, deregulated expression of microRNAs has been linked to different disease states. Here, a brief overview of the canonical and alternative microRNA biogenesis pathways and microRNA functions in biological systems is given based on recent developments. In addition, newly emerging regulatory mechanisms, such as alternative polyadenylation, in connection with microRNA-dependent gene expression regulation are discussed.

Key words MicroRNA, Genome, RNAi, Biogenesis

1 Introduction

MicroRNAs are hairpin-derived 20–24 nucleotide long noncoding RNAs that bind to the 3′ UTR of target mRNAs, leading to either translational delay or mRNA degradation. Followed by their initial discovery, microRNAs are now accepted to play pivotal roles in development, differentiation, and other normal physiological functions, while deregulated expression of microRNAs is important in a wide variety of different pathologies [1]. Given that all known or predicted protein-coding genes are thought to be expressed only from a small percentage (i.e., 1.5–2 %) of the whole genome, it is quite remarkable that microRNAs constitute about 1–2 % of all genes in worms, flies, and mammals. Moreover, each microRNA is predicted to control a vast number of (up to a few hundred) target genes [2]. Hence, microRNA regulated gene expression is a novel and important mechanism regulating the expression of a significantly large portion of the genome.

The first microRNAs discovered as noncoding small RNAs were lin-4 and let-7 as developmental regulators of *Caenorhabditis elegans* [3, 4]. Initially considered as an unusual worm specific

Malik Yousef and Jens Allmer (eds.), *miRNomics: MicroRNA Biology and Computational Analysis*, Methods in Molecular Biology, vol. 1107, DOI 10.1007/978-1-62703-748-8_1, © Springer Science+Business Media New York 2014

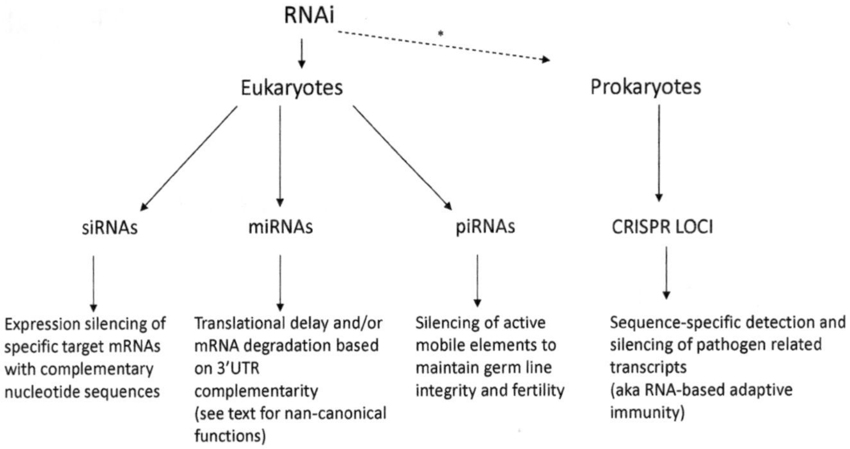

Fig. 1 RNAi in eukaryotes and a similar mechanism in prokaryotes. RNAi consists of siRNAs (small interfering RNAs), microRNAs (microRNAs), and piRNAs (piwi-interacting RNAs) in eukaryotes. CRISPR loci in prokaryotes and archaeal genomes harbor short invader-derived sequences that are transcribed to target complementary RNAs. (*This RNA-based adaptive immunity seems to protect prokaryotes from pathogens and is distinct from eukaryotic RNAi machinery)

gene expression regulation mechanism, today microRNAs are accepted as important regulators of gene expression in eukaryotes. In fact, microRNA-mediated gene regulation is part of a larger mechanism known as RNA interference (RNAi) which also involves a group of small regulatory RNAs (such as siRNAs (small interfering RNAs) and piRNAs (piwi-interacting RNAs)) (Fig. 1). While siRNAs generally function to interfere with the expression of specific target genes with complementary nucleotide sequences, piRNAs function against active mobile elements to maintain germ line integrity and fertility.

2 MicroRNAs in Biological Systems

Above mentioned RNAi and similar mechanisms seem to be present and functional in all biological systems, emphasizing the importance of RNA during evolution. When compared to eukaryotic RNAi, prokaryotes have a functionally analogous RNAi-like defense system that is likely to have evolved independently and hence is not homologous to that of eukaryotes [5] (Fig. 1). "Clustered regularly interspaced short palindromic repeats" (CRISPRs) based system found in prokaryotic and archaeal genomes is similar to RNAi in eukaryotes, in the sense that they both rely on small RNAs for sequence-specific detection and silencing of transcripts. CRISPR loci harbor short invader-derived sequences and are transcribed as long RNAs that are processed into smaller invader-targeting RNAs (also known as prokaryotic silencing, psiRNAs). The psiRNA and associated proteins target complementary RNAs for degradation or translational delay. This RNA-based adaptive immunity seems to protect prokaryotes from potential genome invaders such as viruses [6]. However, the mechanism of this RNA silencing is distinct from eukaryotic RNAi and is largely dependent on CRISPR-associated (Cas) or RNase III family nucleases [7]. Nevertheless, recently, a large number of 20 nucleotide long noncoding RNAs has been discovered by deep sequencing in hyperthermophilic archaeon *Sulfolobus solfataricus* [8]. Identity and function of these small RNAs and their potential to function as microRNAs remain to be investigated.

When plants and animals are considered, although key proteins of RNAi are conserved, there are major differences in the biogenesis, function, and evolution of microRNAs. To begin to evaluate these differences, understanding microRNA gene structures and how they evolved in plants and animals is crucial. For plants, while there are translational repression cases [9], because there is generally extensive complementarity between a microRNA and its target sequence, inverted duplication of genes is thought to have given rise to proto-microRNA regions. Such regions, when transcribed and processed by the siRNA machinery, are complementary to the original gene's mRNA sequence [10]. This may explain the very few, if not single, number of plant microRNA target mRNAs compared to numerous mRNA targets of animal microRNAs. Moreover, in plants, coevolution of microRNA–mRNA pairs is thought to be a common mechanism whereas animal microRNAs seem to have emerged in genomes first and then acquired many mRNA targets [10]. Another possibility to explain microRNA gene evolution both in plants and in animals is via transcribed repetitive sequences and inverted repeat transposable elements [11, 12].

There are also differences in the genome distributions of microRNA genes in plant and animal genomes. The majority of plant microRNAs are transcribed from individual transcription units that are in intergenic regions [13]. Some plant microRNAs with identical mature sequences can be found as tandem clusters in the genome whereas in animals, clustering of unrelated microRNAs is more common [10]. In animal genomes, a significant number of microRNA genes are embedded within the sense strands of introns of protein-coding genes and are expressed together with the host gene [14]. Such microRNAs, named as "mirtrons," are originated from spliced introns and are generally processed into pre-microRNAs without Drosha activity [15]. There are other examples of microRNAs located in large introns with independent expression patterns from the host because they have their own regulatory units [16]. MicroRNAs can also be found in the exons of noncoding RNAs (e.g., miR-155 is encoded within an exon of the noncoding RNA known as B-cell integration cluster) [17, 18]. In essence, microRNA genes appear to map to different parts of the genome, and hence, their transcriptional regulation and expression patterns would be expected to vary greatly.

2.1 Regulation of MicroRNA Genes

Given the high number of microRNA genes in the genome and their ability to contribute to gene expression regulation mechanisms, it is important to understand how microRNAs themselves are transcribed and regulated (*see* also Chapter 18). In general, the pri-microRNAs with hairpin RNA secondary structures are transcribed by RNA polymerase II both in animals and plants [19, 20]. Identifying where RNA polymerase II binds on DNA, i.e., on microRNA promoters, and where microRNA transcription starts, are essential to understand how microRNAs themselves are regulated. However, identification of promoter sequences and transcription factors that control microRNA expression can be challenging due to lack of a predictable and conserved microRNA gene structure as in protein coding genes. Currently, around 200 human transcription factor and microRNA relationships have been gathered and integrated into a database: TransmiR [21].

Given their roles in development and differentiation, microRNA promoters are likely to be very tightly regulated by specific transcription factors in a spatial and temporal manner. For example, experimental evidence suggests embryonic stem cell transcription factors such as Nanog and Oct3/4 to be associated with the miR-302-367 promoter which is regulated in a development specific manner [22]. In addition to specific transcription factor-dependent regulation, further complicated mechanisms are likely to exist such as feedback loops and hormonally controlled microRNAs. For example, Estradiol (E2) induces expression of several microRNAs in hormone responsive tissues in human, rat and mouse and some of these microRNAs in turn down-regulate ERα (estrogen receptor α) [23]. Thus, E2 responsive microRNAs may have a switch like function for hormone induced changes in cells. Hence, not surprisingly, aberrant microRNA expression has already been associated with breast and endometrial cancers [23].

Another transcriptional regulator of microRNA expression is p53. As part of a positive feedback loop, p53 induces expression of miR-34a which then suppresses SIRT1 (silent mating type information regulation 2 homolog). SIRT1 normally deacetylates p53 and causes a decrease in p53-mediated transcriptional activation of cell cycle arrest and apoptotic genes (e.g., p21, PUMA, BAX). Eventually, SIRT1 mediates the survival of cells through inhibiting p53-dependent apoptosis. Hence, downregulation of SIRT1 by p53 activated miR-34a leads to increased p53 activity due to decreased deacetylation of p53 [24] (Fig. 2). Such networks and pathways may also crosstalk with each other via microRNAs as well.

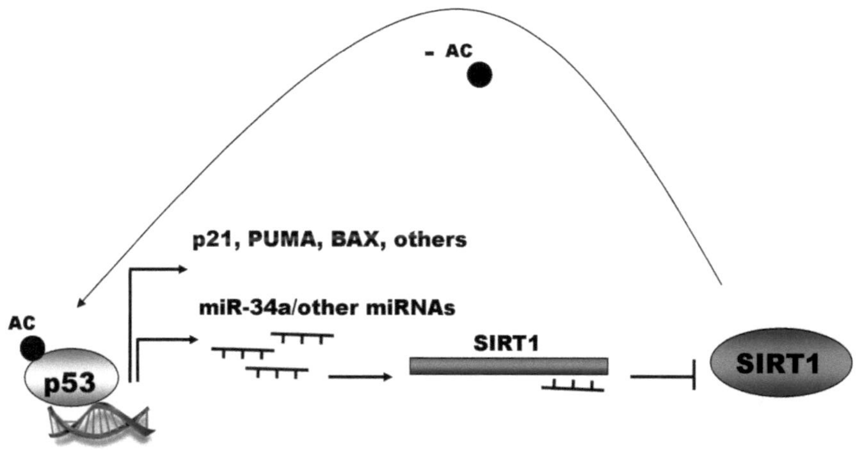

Fig. 2 p53 induces expression of miR-34a which then binds to *SIRT1* mRNA and suppresses SIRT1 translation. SIRT1 normally mediates the survival of cells through inhibiting p53-dependent apoptosis by deacetylating p53 and causing a decrease in p53-mediated transcriptional activation of cell cycle arrest and apoptotic genes (e.g., p21, PUMA, BAX). Hence, as part of a positive feedback mechanism, downregulation of SIRT1 by p53 activated miR-34a leads to increased p53 activity due to decreased deacetylation of p53

For example, miR-145 as another direct target of p53 negatively regulates c-Myc and this contributes to the miR-145-mediated inhibition of tumor cell growth both in vitro and in vivo [25].

In addition to such experimental findings that identify regulators of microRNA transcription, new supportive computational approaches are also emerging to identify microRNA promoter regions and transcription factors [26]. MiRStart is such an integrative prediction tool that combines relevant experimental datasets to identify transcription start sites of microRNAs [27]. It should be emphasized that identification of microRNA promoter regions will also contribute to the understanding of the role of epigenetics in microRNA transcription, especially in neoplasms. Genome-wide profiling already provided valuable data on chronic lymphocytic leukemia where a clear difference in global DNA methylation patterns upstream of microRNA sequences was detected compared to healthy samples [28]. Considering that some microRNA genes are located in other protein coding genes, abnormal epigenetic regulation of these host genes is also expected to cause deregulated microRNA expression. Hence, the role of epigenetics in microRNA transcription is likely to be an expanding field of research in the near future.

2.2 MicroRNA Biogenesis

Discovery of microRNA-dependent gene regulation is a relatively new area of interest and major effort in the field is currently focused on target mRNA identification. However, as mentioned above, transcriptional control of microRNA genes themselves or how they are processed into mature and functional microRNAs are also appealing concepts. There are interesting new studies focusing on

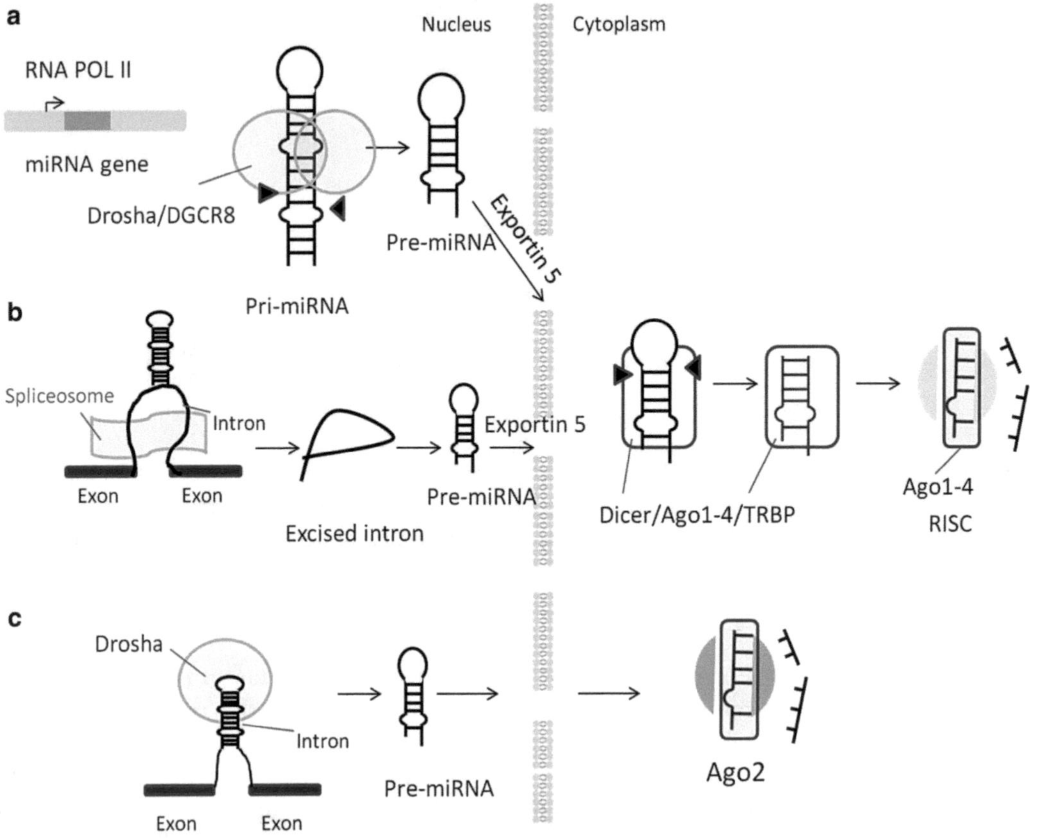

Fig. 3 MicroRNA Biogenesis Pathways. (**a**) RNA polymerase II transcribed pri-microRNA is processed by the microprocessor complex (Drosha (RNase III) and DGCR8) into pre-microRNA with a stem-loop structure of 50–70 nucleotides and a 3′ overhang of a few nucleotides. Pre-microRNAs are then recognized and transported to the cytoplasm by Exportin 5. In the cytosol, pre-microRNAs are then cleaved by Dicer/Argonaute (AGO1–4), assisted by TRBP (transactivation-responsive RNA binding protein), to form the microRNA/microRNA* duplexes which are later separated to be incorporated into the microRNA-(AGO) protein complex called RISC (RNA-induced silencing complex). (**b**) Intronic microRNAs, mirtrons, are processed by the spliceosome without Drosha activity. Following debranching of the lariat of the excised intron, pre-microRNA-like hairpins form. Following their recognition by Exportin-5, pre-microRNAs also become direct Dicer substrates. (**c**) Simtrons, in contrast to mirtrons, may require Drosha but not splicing, DGCR8, or Dicer activity

unraveling the details of microRNA biogenesis via canonical and alternative pathways. According to the canonical pathway (Fig. 3a), in animals, following their transcription by RNA polymerase II, a microprocessor complex containing Drosha (RNase III), and its RNA binding partner DGCR8 (also known as Pasha in invertebrates) cleaves the pri-microRNA (primary microRNA). The cleavage site is approximately 11 bp away from the stem and ssRNA junction of the hairpin structure [29]. The newly cleaved pre-microRNA (precursor microRNA) has a stem-loop structure of 50–70 nucleotides and has a 3′ overhang of a few nucleotides. This overhang is recognized by Exportin 5 during transport of the pre-microRNA to the cytoplasm [30]. Following the nuclear export, pre-microRNAs are then cleaved from their terminal loops by the cytoplasmic RNase III Dicer [31]. The newly formed microRNA/microRNA* duplexes are then separated and the mature microRNA is incorporated into a microRNA-Argonaute (AGO) protein complex called RISC (RNA-induced silencing complex). AGO proteins share PAZ (Piwi-Argonaute-Zwille), MID (located between the PAZ and the PIWI domain), and PIWI (P element-induced wimpy testes) domains. The PAZ domain has a docking function for the 3′ end of the RNA and the MID domain anchors the 5′ phosphate of the terminal nucleotide of the small RNA (for a review, *see* ref. 32). The PIWI domain has a structure that is similar to RNase H and is associated with endonucleolytic activity [33]. Hence, AGO proteins with PIWI domains can cleave the target RNAs bound to the small RNAs. Of the four human AGO proteins (AGO1–4), only AGO2 has this so-called "slicer" activity [32]. Although the significance of alternating AGO protein use remains to be investigated, recent studies suggest that AGO identity may contribute to length determination of bound microRNAs as in the case of shortened microRNA 3′ ends during mammalian brain development linked to different use of AGO proteins [34]. Consistent with these findings, miR-451 was shown to only associate with AGO2, suggesting mature microRNAs to interact with specific AGO proteins, which may also have a possible influence on the length of some microRNAs [35]. Incidence and consequences of this specificity remains to be further investigated while Dicer is known to be responsible for the length diversity of microRNAs with asymmetrical structural motifs in their precursor structures [36].

In addition to the above mentioned core processing enzymes, there are other RNA-binding proteins such as FXR1P (fragile X-related protein 1), FMRP (fragile X mental retardation protein), FXR2P (fragile X-related protein 2) that either enhance or inhibit microRNA maturation through interacting with the terminal loop structures. Pri-miR-18a, pri-let7, pri-miR-1, pri-miR-21 are some of the microRNAs that are identified to be controlled to a certain extend by RNA binding proteins (hnRNP A1, KSRP, MBNL1, and MCPIP1, respectively). Other examples of microRNA-associated proteins are; DEAD box RNA helicases p68 (DDX5) and p72 (DDX17) which are involved in the maturation step [37]. Furthermore, the well-known tumor suppressor, p53 interacts with the Drosha processing complex through its association with DEAD-box RNA helicase p68. p53 then facilitates the processing of several primary microRNAs into precursor microRNAs (miR-16-1, miR-143, miR-145, and miR-206). This interaction somewhat causes increased pri-microRNA processing activity in doxorubicin-treated cells [38]. While their exact mode of action remains to be delineated, such RNA binding proteins are thought to bind to the terminal loops of microRNA progenitors and alter structure and interactions of the microRNAs [39].

In spite of the shared components of the microRNA biogenesis pathway among organisms, variations do exist at different steps. For example, because plants lack an ortholog of Drosha in the nucleus, Dicer-like 1 (DCL1) cleaves the pri-microRNA to form the pre-microRNA which is further processed by the same enzyme to form the microRNA/microRNA* [40, 41]. As microRNA/microRNA* biogenesis proteins are located in the nucleus and the process is completed within specialized nuclear dicing bodies (D-bodies) [42–44]. HASTY (HST), the Arabidopsis ortholog of Exportin-5, is then thought to transport the microRNA/microRNA* duplexes into the cytoplasm to be loaded into the AGO protein complexes. In *C. elegans* and vertebrates, there is a single Dicer protein whereas in *Drosophila* there are two, one of which is (Dcr-1) involved in microRNA cleavage and the other (Dcr-2) takes part in the siRNA biogenesis pathway [45]. In Drosophila, in contrast to human, all AGO proteins possess the above mentioned slicer activity.

Based on what we know about the complexity of microRNA regulation, a better understanding of microRNA biogenesis mechanisms is needed to better evaluate the microRNA effect in cells. As a major determinant of a cell's microRNA pool, we do not know how the microprocessor complex itself is regulated in detail. To answer that question, a recent study suggested clues on how the microprocessor complex can regulate its own levels. The microprocessor complex can cleave hairpins at the 5′ UTR of *DGCR8* mRNA and cause a decrease in the protein levels which is further reflected on the complex formation [46]. Moreover, DGCR8 stabilizes the Drosha protein via protein–protein interactions. This regulatory feedback loop between Drosha and DGCR8 may contribute to the general homeostatic control of microRNA biogenesis [46]. DGCR8, with other endonucleases, seems to control the fate of different classes of RNAs such as mRNAs, small nucleolar RNAs (snoRNAs), and long noncoding RNAs [47]. Dicer, as well, is likely to have other functions in development and differentiation as various mouse dicer knockout studies suggested [48]. Unraveling such non-microRNA-dependent functions of microRNA biogenesis regulators will be interesting as part of a larger regulatory network.

2.3 Alternative MicroRNA Biogenesis

There is also intriguing evidence pointing out the existence of alternative microRNA biogenesis pathways that can processes microRNAs without Drosha activity from introns that bear hairpin structures similar to Drosha processed pre-microRNAs (Fig. 3b). Such microRNAs (also known as mirtrons), with splice acceptor and donor sequences at their termini were first described in *D. melanogaster* and *C. elegans* [49, 50] and later in higher organisms. Interestingly, while some mirtrons are evolutionarily conserved across species, some seem to have emerged recently in small introns [51].

Discovery of these mirtrons led to the finding that splicing takes over the role of Drosha in the biogenesis of certain microRNAs and spliceosome-excised introns effectively bypass the Drosha cleavage step. This mirtron specific biogenesis mechanism merges with the canonical microRNA pathway at the nucleus to cytoplasm transfer step via Exportin-5.

Remarkably, a new group of microRNAs have also been defined as splicing-independent mirtron-like microRNAs, "simtrons" (Fig. 3c). Simtrons, in contrast to mirtrons, may require Drosha but not splicing, DGCR8, or Dicer. While the involvement of DGCR8 and Dicer in this pathway needs further clarification, processing of the miR-451 is an example of Dicer-independent simtron. Drosha cleaves the pri-miR-451 into a 42 nucleotide hairpin with a short 19 nucleotide stem structure which is directly loaded into the vertebrate "Slicer" Argonaute (AGO2). AGO2, then, cleaves the passenger strand of the pre-microRNA at positions 10–11 across from the hairpin 5′ end [52]. Thus, AGO2-mediated cleavage of the hairpin structure yields a 30 nucleotide intermediate, whose 3′ end is released to generate the mature miR-451 [53].

Some small nucleolar RNAs (snoRNAs), and transfer RNAs (tRNAs) are also processed into microRNA-like molecules independently of the Microprocessor complex [54, 55].

Considering these recent discoveries of such alternative biogenesis mechanisms, it appears that our understanding of microRNA biogenesis pathways is far from complete. The evolutionary significance and possible functional variations in these pathways may await further studies.

3 Targets and Functions of MicroRNAs

It has been well established that microRNAs participate in gene expression control and act as fine tuners, generally by binding to 3′ UTRs of target mRNAs. As stated above, in spite of commonalities, there are pivotal differences in microRNA–mRNA interactions among plant and animal cells. Plant microRNAs bind to their few or only targets generally via perfect complementarity, which suggests that plant microRNAs may be acting similarly to siRNAs and cause direct mRNA cleavage [56]. On the other hand, animal microRNAs bind to their many targets via less than perfect complementarity at the 5′ end seed sequence [2, 57]. Hence, predicting and confirming the mRNA targets of a single microRNA is a comparatively difficult task in animal cells. To further complicate the microRNA target recognition mechanisms, deep sequencing studies showed expression of isomiRs, individual microRNA variants that are heterogeneous in length, in a cell type specific manner. IsomiRs are generally reported to have 3′ end variations owing to the fact that 3′ microRNA ends extend from within the PAZ domain of the AGO proteins and are therefore prone to exonucleolytic cleavage and modifications, possibly by exoribonucleases and nucleotidyl transferases [58–60]. Functional significance of isomiRs with 3′ or 5′ end variations remains to be investigated in terms of target specificity.

3.1 Noncanonical Functions of MicroRNAs

In addition to the vast number of microRNA–mRNA interactions defined in the literature, we are just beginning to understand novel and noncanonical functions of microRNAs. For example, the star strands have long been overlooked as they were generally thought to be expressed at much lower levels and/or degraded while the other strand whose 5′ end is less stably base-paired is chosen as the guide strand to interact with the RISC complex [32]. Increasing evidence now suggests that both strands of the microRNA duplex can indeed be functional and have significant impact on cells. For example, ectopic expression of miR-24-2 star strand in MCF-7 breast cancer cells results in in vivo and in vitro suppression of cellular survival [61]. To further signify their potential roles, bioinformatics analyses demonstrate that some microRNA star seed sequences are conserved among vertebrates [48]. Hence, there are new experimental approaches to express the star form of the microRNA duplex, either by transient transfection of microRNA mimics or introducing artificially designed stem-loop sequences into short hairpin RNA (shRNA) overexpression vectors [62]. This way, functions and roles of star strands can be investigated in detail to better evaluate the consequences of deregulated microRNA expression cases in cancer cells.

Another controversial role suggested for microRNAs is "RNA activation (RNAa)." It turns out microRNAs can positively regulate gene expression by targeting promoter elements of protein coding genes. For example, miR-373 transfections were able to induce transcription of E-cadherin and cold-shock domain-containing protein C2 (CSDC2) by targeting the complementary regions in corresponding promoters where RNA polymerase II binding was enriched [63]. Mouse Cyclin B1 transcription was also induced via several microRNAs and miR-466d-3p/AGO1 complex was found to be selectively associated with the Cyclin B1 promoter [64]. In addition to promoter binding microRNAs, earlier, miR-10a was shown to bind to 5′ UTR of ribosomal protein mRNAs and enhance their translation [65].

MicroRNAs have also been implicated in the so-called senescence-associated transcriptional gene silencing. Cellular senescence triggered by neoplastic events involves repression of proliferation-promoting genes generally regulated by the retinoblastoma protein (RB). Interestingly, in pre-malignant cells, AGO2, let-7, and RB1 were reported to physically interact to repress RB target genes in case of senescence [66]. In essence, AGO2 was described as the effector protein for let-7 to execute a silent-state chromatin state at target promoters [66].

4 Alternative Polyadenylation

Based on above mentioned findings, microRNAs appear as strong trans-regulators of gene expression in a wide perspective. Interestingly, recent studies showed that microRNA-dependent mRNA targeting patterns can be altered under different physiological states via changes in the 3′ UTR lengths of some target mRNAs. Normally, the polyadenylation signal position on precursor mRNAs determines where the poly(A) tail should be added followed by endonucleolytic cleavage during mRNA maturation. Hence, the position of the added poly(A) tail is one of the determinants of 3′ UTR lengths of mRNAs. It appears there are more than one proximal or distal alternative polyadenylation (APA) signals on mRNAs. Therefore, preferential use of these proximal or distal poly(A) signals can potentially change the length of the 3′ UTRs. Surprisingly, cells seem to use this method to add an extra level of control to microRNA-dependent regulation by choosing different APA signals in a context-dependent manner. For example, in response to an activation signal, the CD4+ T lymphocyte transcription profile switches to transcripts with shorter 3′ UTRs by using more proximal poly(A) sites [67]. In cancer cells, altered 3′ UTR lengths due to APA were also reported to be more abundant compared to normal counterparts [68, 69]. Moreover, while we know deregulated microRNA expression profiles to correlate with disease phenotypes [70], such global changes in 3′ UTR lengths apparently also correlate with clinically distinct cancer subtypes [71]. These findings suggest APA to be a mechanism to evade microRNA-dependent negative regulation. Our group also demonstrated 3′ UTR shortening of *CDC6*, a regulator of DNA replication, in response to estrogen in breast cancer cells as a way to evade microRNA-dependent regulation [72]. Another recent study showed APA in a tissue-specific manner, which also caused changes in microRNA-mediated regulation [73]. It appears that cells may be using APA and shorten or lengthen their mRNA 3′ UTRs to alter microRNA binding sites as a way to more robustly upregulate or downregulate protein levels due to proliferative signals and/or to provide tissue specific microRNA regulation. Therefore, while microRNAs are generally accepted as fine tuners of gene expression regulation, there appears to be other mechanisms in the microRNA–mRNA interaction dynamics. Another such regulation may be via pseudogenes where a surprising function has been proposed for pseudogenes in accord with microRNAs. It seems that pseudogenes do contribute to gene expression regulation by acting as microRNA decoys and hence regulate the levels of their functional ancestral tumor suppressor and oncogene copies [74]. For example, *PTENP1*, a pseudogene of the *PTEN* (Phosphatase and tensin homolog) tumor suppressor, is transcribed and a number of microRNAs are capable of targeting the common 3′ UTR on both the gene and pseudogene transcripts. Because *PTENP1* acts as a microRNA decoy, knockdown of this pseudogene transcript leads to reduced levels of *PTEN* mRNA and protein [75]. This suggests existence of a pseudogene transcript to reduce the cellular concentration of microRNAs, and therefore allowing the original gene transcript to escape microRNA-mediated repression.

5 Conclusion

In conclusion, our insight on microRNAs is expanding with an impressive pace. In addition to major efforts to understand microRNA function via predicting and validating mRNA targets, alternative biogenesis pathways and noncanonical functions of microRNAs appear as interesting new avenues and are gaining attention. These new findings will provide a more comprehensive view of the importance of microRNAs in plant and animal genomes and how microRNAs may be incorporated into other gene expression related mechanisms such as APA of 3′ UTRs and as of yet unidentified others.

References

1. Erson A, Petty E (2008) MicroRNAs in development and disease. Clin Genet 74:296–306
2. Bartel DP (2009) MicroRNAs: target recognition and regulatory functions. Cell 136:215–233
3. Lee RC, Feinbaum RL, Ambros V (1993) The C. elegans heterochronic gene lin-4 encodes small RNAs with antisense complementarity to lin-14. Cell 75:843–854
4. Wightman B, Ha I, Ruvkun G (1993) Posttranscriptional regulation of the heterochronic gene lin-14 by lin-4 mediates temporal pattern formation in C. elegans. Cell 75: 855–862
5. Shabalina SA, Koonin EV (2008) Origins and evolution of eukaryotic RNA interference. Trends Ecol Evol 23:578–587
6. Hale CR, Zhao P, Olson S, Duff MO, Graveley BR, Wells L, Terns RM, Terns MP (2009) RNA-guided RNA cleavage by a CRISPR RNA-Cas protein complex. Cell 139:945–956
7. Wiedenheft B, Sternberg SH, Doudna JA (2012) RNA-guided genetic silencing systems in bacteria and archaea. Nature 482:331–338
8. Xu N, Li Y, Zhao YT, Guo L, Fang YY, Zhao JH, Wang XJ, Huang L, Guo HS (2012) Identification and characterization of small RNAs in the hyperthermophilic archaeon Sulfolobus solfataricus. PLoS One 7:e35306
9. Naqvi AR, Sarwat M, Hasan S, Roychodhury N (2012) Biogenesis, functions and fate of plant microRNAs. J Cell Physiol 227: 3163–3168
10. Axtell MJ, Westholm JO, Lai EC (2011) Vive la difference: biogenesis and evolution of microRNAs in plants and animals. Genome Biol 12:221
11. Piriyapongsa J, Mariño-Ramírez L, Jordan I (2007) Origin and evolution of human microRNAs from transposable elements. Genetics 176:1323–1337
12. Piriyapongsa J, Jordan IK (2008) Dual coding of siRNAs and microRNAs by plant transposable elements. RNA 14:814–821
13. Reinhart BJ, Bartel DP (2002) Small RNAs correspond to centromere heterochromatic repeats. Science 297:1831
14. Rodriguez A, Griffiths-Jones S, Ashurst J, Bradley A (2004) Identification of mammalian microRNA host genes and transcription units. Genome Res 14:1902–1910
15. Berezikov E, Chung W, Willis J, Cuppen E, Lai E (2007) Mammalian mirtron genes. Mol Cell 28:328–336
16. Isik M, Korswagen HC, Berezikov E (2010) Expression patterns of intronic microRNAs in Caenorhabditis elegans. Silence 1:5
17. Tam W (2001) Identification and characterization of human BIC, a gene on chromosome 21 that encodes a noncoding RNA. Gene 274: 157–167
18. Eis P, Tam W, Sun L, Chadburn A, Li Z, Gomez M, Lund E, Dahlberg J (2005) Accumulation of miR-155 and BIC RNA in human B cell lymphomas. Proc Natl Acad Sci U S A 102:3627–3632
19. Lee Y, Kim M, Han J, Yeom KH, Lee S, Baek SH, Kim VN (2004) MicroRNA genes are transcribed by RNA polymerase II. EMBO J 23:4051–4060

20. Xie Z, Allen E, Fahlgren N, Calamar A, Givan SA, Carrington JC (2005) Expression of Arabidopsis MICRORNA genes. Plant Physiol 138:2145–2154
21. Wang J, Lu M, Qiu C, Cui Q (2010) TransmiR: a transcription factor-microRNA regulation database. Nucleic Acids Res 38:D119–D122
22. Barroso-del Jesus A, Lucena-Aguilar G, Menendez P (2009) The miR-302-367 cluster as a potential stemness regulator in ESCs. Cell Cycle 8:394–398
23. Klinge CM (2012) microRNAs and estrogen action. Trends Endocrinol Metab 23:223–233
24. Yamakuchi M, Lowenstein CJ (2009) MiR-34, SIRT1 and p53: the feedback loop. Cell Cycle 8:712–715
25. Sachdeva M, Zhu S, Wu F, Wu H, Walia V, Kumar S, Elble R, Watabe K, Mo YY (2009) p53 represses c-Myc through induction of the tumor suppressor miR-145. Proc Natl Acad Sci U S A 106:3207–3212
26. Wu JH, Sun YJ, Hsieh PH, Shieh GS (2012) Inferring coregulation of transcription factors and microRNAs in breast cancer. Gene 518(1): 139–144. doi:10.1016/j.gene.2012.11.056
27. Chien CH, Sun YM, Chang WC, Chiang-Hsieh PY, Lee TY, Tsai WC, Horng JT, Tsou AP, Huang HD (2011) Identifying transcriptional start sites of human microRNAs based on high-throughput sequencing data. Nucleic Acids Res 39:9345–9356
28. Baer C, Claus R, Frenzel LP, Zucknick M, Park YJ, Gu L, Weichenhan D, Fischer M, Pallasch CP, Herpel E, Rehli M, Byrd JC, Wendtner CM, Plass C (2012) Extensive promoter DNA hypermethylation and hypomethylation is associated with aberrant MicroRNA expression in chronic lymphocytic leukemia. Cancer Res 72:3775–3785
29. Han J, Lee Y, Yeom KH, Nam JW, Heo I, Rhee JK, Sohn SY, Cho Y, Zhang BT, Kim VN (2006) Molecular basis for the recognition of primary microRNAs by the Drosha-DGCR8 complex. Cell 125:887–901
30. Okada C, Yamashita E, Lee SJ, Shibata S, Katahira J, Nakagawa A, Yoneda Y, Tsukihara T (2009) A high-resolution structure of the pre-microRNA nuclear export machinery. Science 326:1275–1279
31. Ketting RF, Fischer SE, Bernstein E, Sijen T, Hannon GJ, Plasterk RH (2001) Dicer functions in RNA interference and in synthesis of small RNA involved in developmental timing in C. elegans. Genes Dev 15:2654–2659
32. Kim VN, Han J, Siomi MC (2009) Biogenesis of small RNAs in animals. Nat Rev Mol Cell Biol 10:126–139
33. Jinek M, Doudna JA (2009) A three-dimensional view of the molecular machinery of RNA interference. Nature 457:405–412
34. Juvvuna PK, Khandelia P, Lee LM, Makeyev EV (2012) Argonaute identity defines the length of mature mammalian microRNAs. Nucleic Acids Res 40:6808–6820
35. Dueck A, Ziegler C, Eichner A, Berezikov E, Meister G (2012) microRNAs associated with the different human Argonaute proteins. Nucleic Acids Res 40:9850–9862
36. Starega-Roslan J, Krol J, Koscianska E, Kozlowski P, Szlachcic WJ, Sobczak K, Krzyzosiak WJ (2011) Structural basis of microRNA length variety. Nucleic Acids Res 39:257–268
37. Gregory RI, Yan KP, Amuthan G, Chendrimada T, Doratotaj B, Cooch N, Shiekhattar R (2004) The Microprocessor complex mediates the genesis of microRNAs. Nature 432:235–240
38. Suzuki HI, Yamagata K, Sugimoto K, Iwamoto T, Kato S, Miyazono K (2009) Modulation of microRNA processing by p53. Nature 460: 529–533
39. Choudhury NR, Michlewski G (2012) Terminal loop-mediated control of microRNA biogenesis. Biochem Soc Trans 40:789–793
40. Kurihara Y, Watanabe Y (2004) Arabidopsis micro-RNA biogenesis through Dicer-like 1 protein functions. Proc Natl Acad Sci U S A 101:12753–12758
41. Fahlgren N, Howell MD, Kasschau KD, Chapman EJ, Sullivan CM, Cumbie JS, Givan SA, Law TF, Grant SR, Dangl JL, Carrington JC (2007) High-throughput sequencing of Arabidopsis microRNAs: evidence for frequent birth and death of MICRORNA genes. PLoS One 2:e219
42. Park MY, Wu G, Gonzalez-Sulser A, Vaucheret H, Poethig RS (2005) Nuclear processing and export of microRNAs in Arabidopsis. Proc Natl Acad Sci U S A 102:3691–3696
43. Fang W, Fang W, Lin C, Lin C, Zhang H, Zhang H, Qian J, Zhong L, Xu N (2007) Detection of let-7a microRNA by real-time PCR in colorectal cancer: a single-centre experience from China. J Int Med Res 35:716–723
44. Song L, Han MH, Lesicka J, Fedoroff N (2007) Arabidopsis primary microRNA processing proteins HYL1 and DCL1 define a nuclear body distinct from the Cajal body. Proc Natl Acad Sci U S A 104:5437–5442
45. Miyoshi K, Miyoshi T, Hartig JV, Siomi H, Siomi MC (2010) Molecular mechanisms that funnel RNA precursors into endogenous small-interfering RNA and microRNA biogenesis pathways in Drosophila. RNA 16: 506–515
46. Han J, Pedersen JS, Kwon SC, Belair CD, Kim YK, Yeom KH, Yang WY, Haussler D, Blelloch R, Kim VN (2009) Posttranscriptional crossregulation between Drosha and DGCR8. Cell 136:75–84
47. Macias S, Plass M, Stajuda A, Michlewski G, Eyras E, Caceres JF (2012) DGCR8 HITS-CLIP reveals novel functions for the Microprocessor. Nat Struct Mol Biol 19:760–766

48. Yang JS, Phillips MD, Betel D, Mu P, Ventura A, Siepel AC, Chen KC, Lai EC (2011) Widespread regulatory activity of vertebrate microRNA* species. RNA 17:312–326
49. Ruby JG, Jan CH, Bartel DP (2007) Intronic microRNA precursors that bypass Drosha processing. Nature 448:83–86
50. Okamura K, Hagen JW, Duan H, Tyler DM, Lai EC (2007) The mirtron pathway generates microRNA-class regulatory RNAs in Drosophila. Cell 130:89–100
51. Curtis HJ, Sibley CR, Wood MJ (2012) Mirtrons, an emerging class of atypical microRNA. Wiley Interdiscip Rev RNA 3:617–632
52. Diederichs S, Haber DA (2007) Dual role for argonautes in microRNA processing and post-transcriptional regulation of microRNA expression. Cell 131:1097–1108
53. Yang JS, Lai EC (2010) Dicer-independent, Ago2-mediated microRNA biogenesis in vertebrates. Cell Cycle 9:4455–4460
54. Ender C, Krek A, Friedländer MR, Beitzinger M, Weinmann L, Chen W, Pfeffer S, Rajewsky N, Meister G (2008) A human snoRNA with microRNA-like functions. Mol Cell 32:519–528
55. Cole C, Sobala A, Lu C, Thatcher SR, Bowman A, Brown JW, Green PJ, Barton GJ, Hutvagner G (2009) Filtering of deep sequencing data reveals the existence of abundant Dicer-dependent small RNAs derived from tRNAs. RNA 15:2147–2160
56. Rhoades MW, Reinhart BJ, Lim LP, Burge CB, Bartel B, Bartel DP (2002) Prediction of plant microRNA targets. Cell 110:513–520
57. Doench JG, Sharp PA (2004) Specificity of microRNA target selection in translational repression. Genes Dev 18:504–511
58. Schirle NT, MacRae IJ (2012) The crystal structure of human Argonaute2. Science 336:1037–1040
59. Elkayam E, Kuhn CD, Tocilj A, Haase AD, Greene EM, Hannon GJ, Joshua-Tor L (2012) The structure of human argonaute-2 in complex with miR-20a. Cell 150:100–110
60. Neilsen CT, Goodall GJ, Bracken CP (2012) IsomiRs—the overlooked repertoire in the dynamic microRNAome. Trends Genet 28: 544–549
61. Martin EC, Elliott S, Rhodes LV, Antoon JW, Fewell C, Zhu Y, Driver JL, Jodari-Karimi M, Taylor CW, Flemington EK, Beckman BS, Collins-Burow BM, Burow ME (2012) Preferential star strand biogenesis of pre-miR-24-2 targets PKC-alpha and suppresses cell survival in MCF-7 breast cancer cells. Mol Carcinog. doi:10.1002/mc.21946 [Epub ahead of print]
62. Qu B, Han X, Tang Y, Shen N (2012) A novel vector-based method for exclusive overexpression of star-form microRNAs. PLoS One 7:e41504
63. Place RF, Li LC, Pookot D, Noonan EJ, Dahiya R (2008) MicroRNA-373 induces expression of genes with complementary promoter sequences. Proc Natl Acad Sci U S A 105:1608–1613
64. Huang V, Place RF, Portnoy V, Wang J, Qi Z, Jia Z, Yu A, Shuman M, Yu J, Li LC (2012) Upregulation of Cyclin B1 by microRNA and its implications in cancer. Nucleic Acids Res 40:1695–1707
65. Orom UA, Nielsen FC, Lund AH (2008) MicroRNA-10a binds the 5′UTR of ribosomal protein mRNAs and enhances their translation. Mol Cell 30:460–471
66. Benhamed M, Herbig U, Ye T, Dejean A, Bischof O (2012) Senescence is an endogenous trigger for microRNA-directed transcriptional gene silencing in human cells. Nat Cell Biol 14:266–275
67. Sandberg R, Neilson JR, Sarma A, Sharp PA, Burge CB (2008) Proliferating cells express mRNAs with shortened 3′ untranslated regions and fewer microRNA target sites. Science 320:1643–1647
68. Mayr C, Bartel DP (2009) Widespread shortening of 3′UTRs by alternative cleavage and polyadenylation activates oncogenes in cancer cells. Cell 138:673–684
69. Fu Y, Sun Y, Li Y, Li J, Rao X, Chen C, Xu A (2011) Differential genome-wide profiling of tandem 3′ UTRs among human breast cancer and normal cells by high-throughput sequencing. Genome Res 21:741–747
70. Erson AE, Petty EM (2009) microRNAs and cancer: new research developments and potential clinical applications. Cancer Biol Ther 8:2317–2322
71. Singh P, Alley TL, Wright SM, Kamdar S, Schott W, Wilpan RY, Mills KD, Graber JH (2009) Global changes in processing of mRNA 3′ untranslated regions characterize clinically distinct cancer subtypes. Cancer Res 69: 9422–9430
72. Akman BH, Can T, Erson-Bensan AE (2012) Estrogen-induced upregulation and 3′-UTR shortening of CDC6. Nucleic Acids Res 40: 10679–10688
73. Ghosh T, Soni K, Scaria V, Halimani M, Bhattacharjee C, Pillai B (2008) MicroRNA-mediated up-regulation of an alternatively polyadenylated variant of the mouse cytoplasmic {beta}-actin gene. Nucleic Acids Res 36:6318–6332
74. Pink RC, Wicks K, Caley DP, Punch EK, Jacobs L, Carter DR (2011) Pseudogenes: pseudo-functional or key regulators in health and disease? RNA 17:792–798
75. Poliseno L, Salmena L, Zhang J, Carver B, Haveman WJ, Pandolfi PP (2010) A coding-independent function of gene and pseudogene mRNAs regulates tumour biology. Nature 465:1033–1038

Chapter 2

The Role of MicroRNAs in Biological Processes

Kemal Uğur Tüfekci, Ralph Leo Johan Meuwissen, and Şermin Genç

Abstract

MicroRNAs (miRNAs) are tiny regulators of gene expression on the posttranscriptional level. Since the discovery of the first miRNA 20 years ago, thousands of them have been described. The discovered miRNAs have regulatory functions in biological and pathological processes. Biologically, miRNAs have been implicated in development, differentiation, proliferation, apoptosis, and immune responses. In this work, we summarize the role of miRNA in biological processes taking into account the various areas named above.

Key words Apoptosis, Immune responses, MicroRNA, Cell signaling, Development, Differentiation, Autophagy

1 Introduction

The discovery of miRNAs caused the beginning of a new era in molecular biology. In eukaryotic organisms, more than 15,000 miRNAs have been defined so far and the number is increasing everyday [1]. MicroRNAs regulate gene expression by degradation or translational repression of their mRNA target genes [2]. In addition to these mechanisms, miRNAs can alter histone modifications and DNA methylation [3, 4]. It has been proposed that approximately 30 % of the human protein coding genes are controlled by miRNAs [5]. MicroRNAs are involved in several biological processes such as proliferation, differentiation, apoptosis, development, angiogenesis, and immune response via regulating their target genes [6]. Here, we present an overview of our current understanding the role of miRNA in biological processes. In the following sections we will first focus on development and differentiation.

Malik Yousef and Jens Allmer (eds.), *miRNomics: MicroRNA Biology and Computational Analysis*, Methods in Molecular Biology, vol. 1107, DOI 10.1007/978-1-62703-748-8_2, © Springer Science+Business Media New York 2014

2 Development and Differentiation

Recent studies indicated that miRNAs have regulatory roles in most developmental processes. In invertebrates and zebra fish embryos dicer deficiency leads to lethality by developmental arrest [7].

Studies of miRNA function in mammalian development started with the generation of Dicer-1 and Argonaute2 (Ago-2) null mice in 2003. According to these studies, lethality occurs on embryonic day (ED) 7.5 and therefore neither embryonic stem (ES) cells nor pluripotent stem cells could be isolated [8]. In case of Ago-2 null mice, other members of the Ago family proteins were preferentially expressed in different regions of the embryo [8]. Thus, it was concluded that different Ago family members have different impacts on gene regulatory networks during development. In embryonic development, miR-196 and miR-10 were discovered to target homeobox clusters (HOX) [9]. MiR-196 regulation of HOXb8 mRNA was confirmed by luciferase assay. Using massive parallel signature sequencing technology, Mineno et al. profiled miRNA expression of a single entire mouse embryo [10]. As a result of their study, miR-2 and miR-193 showed ED 10.5 specific expression and are therefore thought to be regulators of developmental transitions.

So far, miRNA expression analyses in human embryonic stem cells have been performed using different approaches [11]. Transcription factor analyses were performed by Boyer et al. in an ES cell line [12]. According to this study, developmental transcription factors POU class 5 homeobox 1 (OCT4), sex determining region Y(SRY)-box 2 (SOX2), and Nanog Homeobox (NANOG) were found to be interacting with 14 miRNAs. Moreover, these transcription factors were found to bind the promoters of miR-137 and miR-301 genes. On the other hand, Strauss et al. profiled 248 miRNAs in 13 different ES cell lines and examined their correspondence to embryoid bodies (EB) [13]. In the aforementioned study, ES cell-specific miRNAs were defined. However, according to the comparison of ES and EB cell lines, no differences in the miRNA profiles were detected among experimental groups. Furthermore, in that study, miRNA signatures of somatic cells were analyzed and compared to those of ES and EB cells. Results suggested that miRNA expression signatures of somatic cells are much more complex than ES cells.

In hematopoiesis, hematopoietic stem cells (HSC) are subjected to proliferation, commitment, maturation and apoptosis through a network of transcription and growth factors [14]. It was found that in these processes, three miRNAs, namely, miR223, miR-181, and miR-142 are preferentially expressed in hematopoietic tissues [14]. Presence of the miRNA machinery enzymes during organogenesis and tissue morphogenesis is a proof of miRNA activity in morphogenesis. In situ hybridization studies suggest that Ago and Dicer transcripts in distal mesenchyme and epithelium overlap in ED 11.5–14.5 mouse embryos [7]. Moreover, these transcripts are highly expressed in lung parts that undergo branching morphogenesis [15]. In another study, using *Dicer1* deficient mouse lungs, extreme defects in morphology and increased rate of apoptosis in lung epithelium were observed [16]. In skin development, skin-specific miRNAs were identified by Yi et al. using Dicer inactivation [17]. They suggested that miR-200 and miR-19/20 families are specific to epidermis and miR-199 is particularly expressed in hair follicles.

In hematopoiesis, hematopoietic stem cells (HSC) are subjected to proliferation, commitment, maturation, and apoptosis through a network of transcription and growth factors [14]. In these processes, three miRNAs, namely, miR223, miR-181, and miR-142 are preferentially expressed in hematopoietic tissues [14]. Presence of the miRNA machinery enzymes during organogenesis and tissue morphogenesis is a proof of miRNA activity in morphogenesis. In situ hybridization studies suggest that Ago and Dicer transcripts in the distal mesenchyme and epithelium overlap in ED 11.5–14.5 mouse embryos [7]. Moreover, these transcripts are also highly expressed in lung parts that undergo branching morphogenesis [15]. In another study, using *Dicer1* deficient mouse lungs, extreme defects in morphology and increased rate of apoptosis in lung epithelium were observed [16]. In skin development, skin-specific miRNAs were identified by Yi et al. using Dicer inactivation [17]. They have suggested that miR-200 and miR-19/20 families are specific to epidermis and miR-199 is specifically expressed in hair follicles.

2.1 Development of Nervous System

In nervous system development, localized transcriptional and translational controls are programmed in different neuronal regions. As a result of microscopic examinations, particular mRNAs and polysomes were found to exist in dendrites and axons of a neuron [18]. Studies in Dicer knockout zebra fish initially showed the involvement of miRNA in nervous system development and neural cell differentiation [19]. When it comes to mammals, many miRNAs enriched in brain and neurons were characterized using short RNA cloning, sequencing, and northern blotting [20]. Moreover, analysis of miRNA expression signatures in fetal and neonatal mouse brains revealed specific timing of sudden expression alterations in the developing brain [21]. Furthermore, using polysome profiling methods, miRNAs were also shown to regulate translation in interaction with polysomes [21]. When it comes to specific miRNAs in the developing brain, embryonic neurons were found to be enriched with miR-124 and miR-128. On the other hand, in astrocytes, miR-23, miR-26, and miR-29 were abundantly detected [7]. These findings suggest that divergent miRNA distributions can make a contribution to protein composition in many kinds of specialized cells in the developing nervous system.

For instance, of the brain specific miRNAs, miR-124a is the most abundant in about 25–50 % of the total brain [21]. MiR-124a is implicated in switching neural progenitor cells into neuron lineage by repressing non-neuronal gene activities [22]. After maturation, miR-124a expression can still be detected in mature neurons of the cerebral cortex. Finally, during synaptogenesis, neurotransmitters regulate miRNA-mediated synaptic plasticity by affecting local production of proteins [23]. In this process, dendrite enriched miRNA miR-134 was found to regulate size of the dendritic spine by silencing LIM-domain containing protein kinase 1 (LimK1) [23]. MiR-134 inhibition can be inactivated by synaptic stimuli and brain-derived neurotrophic factor (BDNF) to rescue translation of LimK and spine extension. MicroRNAs not only do play a role in nervous development but are also important in muscle development, as we will outline in the following sections.

2.2 Muscle Development

During development of skeletal and cardiac muscles, miR-1 and miR-133 co-regulate events without exception in a variety of organisms [7]. MiR-1 is abundantly expressed in muscle progenitor cells and differentiating muscle after the formation of mesoderm [24]. During heart development between ED 8.5–11.5, miR-1 is strongly expressed and represses expression of heart and neural crest derivatives expressed 2 (Hand2) transcription factors, which is responsible for the differentiation of ventricular cardiomyocytes [24]. However, the expression of miR-1 and miR-133 declines at ED 13.5 [7]. In addition, miR-1 is also capable of facilitating myotube formation via promoting differentiation and inhibiting myoblast proliferation [7]. According to promoter analysis of miR-1, some muscle differentiation factors, such as serum response factor (SRF), myogenic regulatory factors (MRFs), and myocyte enhancer factor-2 (Mef-2), were suggested to bind to its enhancer region [24]. Nevertheless, miR-133 was also found to repress SRF in proliferating myoblast, forming a negative feedback loop regulating myotube formation, indirectly. Finally, another miRNA regulating muscle development is miR-181, which is upregulated before all differentiating factors are expressed [14].

Conclusively, miRNAs have mandatory and extensive roles in embryonal and tissue development. This is because the developmental process needs to be spatially and temporally well regulated. Aberrant miRNA expressions may result in developmental defects. Some defects may be lethal, while some may be tolerable as developmental disorders such as Down syndrome, mental retardation, autism spectrum disorders, Rett syndrome, and fragile X syndrome. The complexity of all developmental events therefore demands a dynamic miRNA expression profile. In all stages of development, based on the cell lineage, differentiation and specification in a developing embryo has its own characteristic miRNA signature.

3 MicroRNAs Involved in the Cell Cycle

The cell cycle is defined as series of events involving the growth, DNA replication and nuclear and cytoplasmatic division; in short the life cycle of a dividing eukaryotic cell. The cell cycle consist of four main phases, which are G1 (gap 1), S (synthesis), G2 (gap 2), and M (mitosis) [25]. Multiple checkpoints regulate cell-entrance to the cell cycle and the transition between different phases during a cell cycle. Also miRNAs do participate in many of these crucial cell cycle control pathways.

In G1 phase the *mir-15a-16-1* cluster miRNAs were found to cause cell cycle arrest by inhibiting important cell cycle regulators, such as CDK1, CDK2, CDK6, and cyclins [25]. Moreover, CDK4 and CDK6 are negatively regulated by miR-24, miR-34, miR-124, miR-125b, miR-129, miR-137, miR-195, miR-449, and let-7 [25]. CDK4 and CDK6 are on the other hand positively regulated by other factors, such as D-type cyclins and CDC25A by dephosphorylation of CDKs. But D-type cyclins are negatively regulated by let-7, miR-15 family, miR-17, miR-19a, miR-20a, and miR-3 [25]. These miRs are targeting the genes that induce proliferation, thus causing anti-proliferative effects. In some tumors, miR-124 and miR-137 were found to be hypermethylated, resulting in overexpression of CDK6 [26]. However, miRNAs do target negative key regulators of CDK4/pRB pathway, too, which results in cell cycle entry. A recent study of miR-106a overexpression in cancer cells showed a decrease in pRB expression [27]. Other phosphorylated retinoblastoma protein (pRB) family members, p107/RBL1 and p130/RBL2 are regulated by the *mir-290* and the *mir-17-92* clusters' miRNAs. Moreover, Cyclin-dependent kinase 4 inhibitor (INK4) and other Cip/Kip family proteins, which are all negatively regulating cell proliferation and cell cycle progression, are controlled by several microRNAs. For instance CDK4/6 specific inhibitor $p16^{INK4a}$ is regulated by miR-24 and miR-31 [28]. Other Cip/Kip proteins $p21^{Cip1}$, $p27^{Kip1}$, and $p57^{Kip2}$ are controlled by miR-17-92, miR-106b, miR-221/222, and miR-181 [25]. The ectopic expression of the miR-221-22 cluster has been found to cause tumor growth through inhibiting $p27^{Kip1}$ and $p57^{Kip2}$ and consequently upregulating CDK2 activity [29].

Most of the cell cycle regulating miRNAs act during G1–S transition. However, there are a few examples of miRNAs that take part in later phases of the cell cycle. After duplicating the genome in S phase, cell cycle progression is mostly controlled by the CDK1 complex composed of cyclin A or cyclin B. Both cyclin A and cyclin B are regulated by miR-125, miR-24, and let-7 [25]. During the G2–M transition CDK1–Cyclin B complex is regulated by WEE1 which is also regulated by some miRs, namely miR-195, miR-516-3p, and miR-128a [30]. Furthermore, Cip/Kip CDK inhibitors affect mitotic entry by directly modulating activity of CDK1, or indirectly by miRNAs controlling them. During mitosis, the cell division process is controlled by polo-like kinase 1 (PLK1) which phosphorylates CDC25C, thus activating a CDK1–Cyclin B1 complex [31]. At this M checkpoint miRNAs are believed to play a regulating role, too. A study using naso-pharyngeal cancer cells showed that a decrease in miR-100 expression led to a subsequent PLK1 mRNA expression increase [32]. Aurora B kinase, which is involved in the attachment of the mitotic spindle to centromeres during mitosis, is regulated by miR-24 [25]. However, the exact relationship between mitotic proteins and miRNAs remains unexplored.

In contrast to cell cycle regulation by miRNAs, miRNA expression can itself be regulated by cell cycle controlling factors. According to various studies, some cell cycle proteins, such as the Myelocytomatosis oncogene (MYC) and the E2F transcription factors, were shown to regulate several miRNAs. The transcription factor c-MYC is regulating genes that have roles in development, cell proliferation, differentiation, and apoptosis, and control the cell cycle, regulating G1–S transition. In some cancers, c-MYC was found to be deregulating miRNA expression [25]. According to the first study on this topic in 2005 by Mendell et al., c-MYC induces expression of the mir-17-92 cluster which was considered as an oncogene since it accelerates tumor development in a B-cell lymphoma animal model [33]. Moreover, mir-106a-92-2 is the paralogous cluster of mir-17-92 and its transcription is induced by MYCN in neuroblastoma cells [34]. Furthermore, both c-MYC and MYCN were found to induce miR-9 transcription that regulates coding mRNA of E-cadherin (CDH1) [35]. In addition to these miRNAs, many other miRNA clusters were found to be regulated by MYC: mir-15a-16-1, miR-22, miR-23a/b, miR-26, miR-29, and several let-7 clusters [36].

The E2F transcription factors control cell cycle progression through regulating timely expression of genes required for S and M phases. Some recent studies showed that the E2F1–3 transcription factors directly bind to the promoter of the mir-17-92 cluster [37]. From this cluster, miR-17-5p and miR-20a are able to repress the E2F factor, and thus an auto-regulatory loop exists to regulate proliferation, apoptosis, and accumulation of aberrant E2F1–3 [37]. Moreover, the mir-106b-25 cluster can be activated by E2F1. And then again, miR-106b and miR-93 also regulate E2F1, thus forming another negative feedback loop [38]. In addition, mir-449c-b-a, which is encoded in the first intron of the cell division cycle 20 homolog B (CDC20B) gene, forms a direct target for the E2F1 transcription factor. Of the latter cluster, miR-449a/b targets and inhibits CDK6 and CDC25A, which then results in dephosphorylation of pRB and a G1 cell cycle arrest. These results were shown as an example of negative feedback regulation of the E2F-pRB pathway [39].

It is obvious that miRNAs participate in many crucial cell cycle control pathways. Most of them have anti-proliferative properties due to regulating mitogenic pathways via the activation of CDKs. On the other hand, some miRNAs induce proliferation by targeting CDK inhibitors or pRB family proteins. Finally, cell cycle dependent transcription factors even increase the complexity of the miRNA network during cell cycle progression. They regulate the cell cycle in both a positive and negative manner.

4 MicroRNAs Involved in Cell Signaling

MiRNAs play major roles in many cellular events such as proliferation and differentiation that are controlled by intracellular signaling. The mitogen-activated protein kinase (MAPK), Phosphatidylinositol 3-kinases (PI3K), nuclear factor kappa-light-chain-enhancer of activated B cells (NFkB), Notch, transforming growth factor β (TGF-β), and Hedgehog (Hh) pathways are major examples of these signaling cascades. MicroRNAs may control cell signaling via downregulating their target genes which encode signal transduction molecules participating in these major pathways [40, 41]. For instance Let-7 directly targets the Ras protein which is a member of the MAPK signaling pathway. Phosphatase and tensin homolog (PTEN), a signaling molecule in the PI3K pathway and contributing to cell survival, is targeted by several miRNAs such as miR-21, miR-26a, miR-221, and miR-222 [40]. MiR-21 induces cell proliferation, migration and inhibits apoptosis in various cancers via targeting PTEN [42]. NF-κB is a transcription factor that regulates several genes that are essential for immune response and tumorigenesis. The newly identified miR-301 activates NF-κB via downregulating the NF-κB repressing factor (NKRF) [43]. Several other miRNAs such as miR-146, miR-155, and miR-9 regulate NF-κB pathways by targeting NF-κB regulators and effectors [44]. The Notch and Hh signaling pathways are important gene regulating mechanisms that control cell differentiation during embryogenesis as well as adult homeostasis. MicroRNAs also contribute to the regulation of these pathways. MiR-181 regulates Natural Killer cell development via inhibiting Nemo-like kinase (NLK), an inhibitor of Notch signaling [45]. Downregulation of miR-324-5p increases Hh-dependent gene expression and leads to increased cell proliferation [46].

In addition, miRNA expression is regulated by various signaling pathways. For instance, an activated MAPK pathway upregulates miR-31 expression in rat vascular smooth muscle cells (VSMC) [47]. The nerve growth factor (NGF) induces expression of miR-221, miR-222 whereby NGF induction is itself dependent on activation of the extracellular signal-regulated kinase 1 and 2 (ERK1/2) pathways in PC12 cells, a rat pheochromocytoma cell line of adrenal medulla [48]. MicroRNAs may also create regulatory feedback loops in signaling cascades. There are several regulatory loops between miRNAs and their targets in various cellular processes. A regulatory loop between miR-146 and NF-κB was reported in THP-1 cells, a human acute monocytic leukemia cell line [49]. MicroRNAs may connect different signaling pathways through acting as a mediator of inter-pathway crosstalk [41]. For example, miR-15 and miR-16 link Wnt and TGF-β signaling pathway in early embryonic patterning [50].

In conclusion, miRNAs contribute to the regulation of intracellular signaling pathways. They also establish complicated connections between different signaling pathways. Alternations of intracellular pathways are related with several human diseases.

5 MicroRNAs Implicated in Immune Responses

The immune system provides protection against invading pathogens and is composed of two components: the innate immune system and the adaptive immune system. The innate immune system provides generalized protection, while the adaptive immunity yields more specialized immune responses. A critical role in the immune system regulation has been reserved for miRNAs. Genetic ablation of the miRNA machinery may result in severe defects in immune cell development and may lead to autoimmune disorders and even cancer.

The innate immune system is activated through Toll-like receptors on monocytes and macrophages. In these cells, miR-132, miR-146, and miR-132 were found to regulate immune responses after pathogen recognition [49, 51]. Over-expression of MiR-155 was induced by both bacterial and viral stimuli as well as via exposure to pro-inflammatory cytokines, such as Interferon (IFN) and tumor necrosis factor (TNF). As opposed to miR-155, miR-146 induction may occur only in response to bacterially derived ligands or IL-1 and TNF [49, 51]. Both miR-146 and miR-155 form a negative feedback loop with the Toll-like receptors (TLR) signaling pathway regulating Interleukin-1 receptor-associated kinase 1 (IRAK1), TNF receptor associated factor (TRAF6), Fas-Associated protein with Death Domain (FADD), ribosome inactivating protein (RIP), and inhibitor of kappa B (IKK) [51, 52]. On the other hand, miR-125 downregulation can be seen after stimulation with Lipopolysaccharide (LPS) which is a cell wall component of gram-negative bacteria. According to several reports, miR-125 targets the pro-inflammatory cytokine TNF and, after downregulation of miR-125, an increase in TNF expression takes place under infectious conditions [52].

In the adaptive immune system, pathogenic growth is eliminated or prevented by lymphocytes. During the adaptive immune system response, antigens are presented by antigen-presenting cells and specific antibodies against antigens are produced. According to in vitro studies on T lymphocyte activation, expressions of miR-21, miR-22, miR-24, miR-103, miR-155, and miR-204 were upregulated, while miR-16, miR-26, miR-30, miR-150, and miR-181 were suppressed [53]. However, most functions of these miRNAs in T-cell activation remain elusive. A recent study by Li et al. suggested a role for miR-181 in T-cell selection in the thymus by negative alteration of the TCR-dependent T cell responses to antigens [54]. Moreover, the contribution to T cell function of miR-150, which is suppressed by a regulatory T cell specific transcription factor Foxp3, is still unclear, whereas in B cells, miR-150 expression is stage specific, suggesting that mir-150 has a role in B cell development [55].

Until recently, there was no evidence that miRNAs might have roles in antiviral immune response. However, Jopling et al. reported that hepatocyte-expressed miR-122 is required for Hepatitis C virus (HCV) replication through interaction with 50-noncoding regions of its viral genome [56]. Besides, more recent studies supported the idea that cellular miRNAs do contribute to the response against viral infections. For instance, miR-32 was characterized to prevent accumulation of primate foamy virus type 1 (PFV-1) in human cells. Genome of the virus encodes a protein, called TAS, which suppresses miRNA directed functions [57]. Moreover, miR-24 and miR-93 were also shown to regulate replication of vesicular stomatitis virus (VSV) [58]. According to a study reported by Triboulet et al., reduction in expression of Dicer or Drosha, which are the main enzymes of the miRNA processing pathway, results in enhanced replication of HIV-1 in T lymphocytes via upregulation of histone acetyl-transferase and attenuation of the miR-17-92 cluster [59]. In addition to this, data from another study surprisingly showed that HIV-1 Tat inhibits Dicer function and these data therefore support the idea of miRNA based antiviral effects [60]. Finally, according to a previous study on hepatocytes in which several miRNA expression levels were deregulated after treatment with antiviral cytokine IFN [61]. Precise expression of miR-122, miR-196, miR-296, miR-351, miR-431, and miR-448, which have seed sequence matches with HCV genome, were downregulated. When miRNAs, miR-196 and miR-448 were mutated so that they could not bind to their predicted targets in the HCV genome this then resulted in obliteration of any inhibitory effect on replication [61].

6 MicroRNAs in Angiogenesis

Angiogenesis is the process of blood vessel formation. During this process, endothelial cells (EC) become activated, proliferate, and migrate. In the course of embryogenesis, blood vessels are formed by vasculogenesis in which vascular precursor cells mature and

Table 1
AngiomiRs: miRNAs participate in regulation of angiogenesis in either angiogenic or anti-angiogenic manner

miRNA Name	Type	Target
miR-126	Angiogenic	SPRED1 PIK3R2
miR-23/27	Angiogenic	Sprouty2 SEMA6A
miR-221/222	Anti-angiogenic	c-kit
miR-17-92 cluster	Both angiogenic and anti-angiogenic	Thrombospondin type-1 repeat containing proteins TIMP1 VEGF-A ITGB5

differentiate into a vascular network after which angiogenesis starts due to interaction between accessory cells, pericytes, and vascular smooth muscle cells (VSMC), so that blood vessels can be produced from preexisting vasculature [62]. In adults, angiogenesis becomes quiescent, but it may be activated during wound healing and certain pathological processes. These pathological processes include several human diseases, like cancer.

There is experimental evidence that miRNAs participate in the regulation of vascular development; a first example came from the analysis of mice homozygous for a hypomorphic Dicer allele. In that study, Dicer lacking mice died between days 12.5 and 14.5 of gestation due to absence of angiogenesis [63]. This evidence was further confirmed by Suarez et al. using an EC specific Dicer knockout mice model [64]. According to this model, there were no distinct differences in the vascular phenotypes between both mice models. Moreover, studying the function of Dicer by specific knockdown experiments resulted in reduced proliferation and endothelial sprouts formation. However, knocking down Dicer interestingly resulted in the overexpression of a pro-angiogenic factor via Thrombospondin-1 upregulation [64]. In addition to *Dicer*, *Drosha* was also knocked down which resulted in decreased angiogenesis in human ECs in vitro.

AngiomiRs, the class of miRNAs that regulate genes directing angiogenesis, are frequently detected in ECs [65]. The most important angiomiRs are miR-126, miR-221/222, miR-23/27, and the miR-17-92 cluster, which are summarized in Table 1. Of these miRNAs, miR-126 was characterized as a regulator of sprouty-related, EVH1 domain containing 1 (SPRED1), and phosphoinositide-3-kinase regulatory subunit 2 (PIK3R2), which are negative regulators of the vascular endothelial growth factor (VEGF) signaling, thus impairing the maintenance of vascular integrity [66]. Then again miR-221/222 functions as an anti-angiogenic miRNA by targeting stem cell factor receptor c-kit [67]. Furthermore, miR-23 and miR-27 were found as enhancers of angiogenesis by targeting Sprouty2 and Semaphorin 6A (Sema6A), which are anti-angiogenic. On the other hand, the functions of miR-17-92 cluster in angiogenesis are complex. From this cluster, miR-17-5p, miR-18a, and miR-19a were first analyzed so that the last two were found to target proteins containing thrombospondin type-1 repeats (TSR), and miR-17-5p was found to modulate EC migration and proliferation protein TIMP1 [64]. Contrary to this, the miR-17-92 cluster microRNAs, miR-20a and miR-92a appear to have anti-angiogenic activity by targeting mRNAs of VEGF-A and Integrin β5 (ITGB5), respectively [68]. Finally, the miR-23-27-24 cluster has recently been reported to regulate angiogenesis, as well [69].

7 MicroRNAs and Apoptosis

Apoptosis is a type of active, cell-autonomous induced cell death in which cells that are normally dangerous for the survival of the organism are removed. This death response can be initiated by both intracellular mediators (mitochondria and stressors), through the so-called intrinsic pathway or by death receptors on the cell membrane after activation with a cognate ligand via the extrinsic pathway [70]. In the intrinsic apoptotic pathway, balance between pro-apoptotic and anti-apoptotic proteins of the B-cell lymphoma 2 (BCL2) superfamily define the fate of the cell and regulate mitochondrial membrane permeability [71]. Two members of the anti-apoptotic BCL2 superfamily proteins, Bcl-2 and Bcl extra-large (Bcl-xL) are often found to be overexpressed in human malignancies. Both proteins suppress apoptosis by protecting against permeabilization of the mitochondrial outer membrane after inhibiting pro-apoptotic Bcl-2-associated X (BAX) and Bcl-2 homologous antagonist/killer (BAK) proteins [71]. Other members of the BCL2 family, for instance bcl-2 homology 3 (BH3) domain, BH3-only proteins like p53 upregulated modulator of apoptosis (PUMA) and phorbol-12-myristate-13-acetate-induced protein 1 (PMAIP1 or NOXA) act as cytosolic sensors of cell damage or stress [70].

The intrinsic pathway of apoptosis can be initiated after cell stress triggers BH3-only proteins PUMA and NOXA followed by their activation of multi-domain pro-apoptotic proteins BAX and BAK. Activated BAX and BAK can then translocate into mitochondria, causing its membrane to become permeable [72]. After the mitochondrial membrane is permeable, cytochrome-c and pro-apoptotic SMAC/DIABLO are released to the inter membrane space and cytochrome c can bind to the adaptor protein apoptosis protease-activating factor-1 (APAF1) forming an apoptosome. This apoptosome then recruits and activates caspase-9.

Activated caspase-9, in turn, activates downstream caspases-3, -6, -7, and apoptosis occurs [72]. On the other hand, the extrinsic apoptotic pathway is activated through specific pro-apoptotic receptors upon binding their ligands, such as Apo2L/TRAIL and CD95L/FasL [73]. After activation of these related death receptors, they enable the intracellular death domains to bind to Fas associated death domain (FADD) and trigger the recruitment of initiator caspases, caspase-8 and caspase-10 [70]. Upon their activation, caspase-8 and caspase-10 undergo self-processing into active enzymes and are released into the cytosol to further activate caspases-3, -6 and -7, which then produce an apoptotic signal via the intrinsic pathway.

First publications about regulatory roles of miRNAs in apoptosis appeared in 2003. In these studies on Drosophila, miR-14 and *bantam* were shown to regulate cell death by inhibiting effector caspase Drice and pro-apoptotic high temperature-Induced Dauer formation family member (hid1) genes, respectively [74]. Since then, several miRNA families that regulate apoptosis have been found in both intrinsic and extrinsic pathways. The identified miRNAs have either pro-apoptotic or anti-apoptotic properties.

Some miRNAs are denoted as anti-apoptotic because they suppress pro-apoptotic genes. Interestingly, one of these miRNAs, miR-186*, decreases after stimulation with an apoptosis promoting agent curcumin, thereby causing an increase of Caspase-10 expression [75]. Moreover, miR-24 was shown to regulate the Fas-binding pro-apoptotic protein Fas-associated factor 1 (FAF1) in gastric, cervical, and prostate cell lines [76]. Apoptosome triggering Caspase-9 is suppressed by miR-133 and miR-24a [77, 78]. Apoptosis sensitizers such as BH3-only proteins that are also known as direct activators are also regulated by miRNAs. Of the BH3-only proteins, BIM is regulated by the miR-17-92 cluster, which consists of the miRNAs miR-181a, miR-32, and miR-25, resulting in sensitization to apoptotic stimuli [70]. Another protein from the family of the BH3-only proteins, BAK1, was shown to be regulated by miR-125b in prostate cancer cells [79]. A very recent study with the liver carcinoma cell line HepG2 revealed that PUMA can be targeted by miR-483-3p. Furthermore, ectopic expression of miR-128 in the human embryonic kidney (293T) cell line resulted in induction of mitochondria-mediated apoptosis [80]. Finally, executioner caspases are also regulated by miRNAs. Let-7a has been found to regulate caspase-3 and its ectopic expression also decreases drug induced apoptosis [81].

BCL-2 is an anti-apoptotic protein that inhibits oligomerization of BAX and BAK to prevent mitochondria-mediated apoptosis. Evidence suggests that miR-34 downregulates BCL-2 protein and its expression is correlated with tumor aggressiveness [82]. In addition, ectopic expression of miR-34 in prostate cancer cell lines causes decreased cell growth and proliferation as a result of increased rate of apoptosis [83]. Also, hematopoietic cancer cells obtained from patients showed that miR-15a and miR-16-1 downregulates BCL-2 at the posttranscriptional level [84]. Furthermore, according to studies in glioblastoma, after radiotherapy, miR-181 was found to be downregulated; however, its transient overexpression caused sensitization of cells against radiation by targeting BCL-2 [85]. BCL-W is another BCL-2 family member protein and downregulated by miR-122 in hepatocellular carcinoma cells [86]. Moreover, BCL-W is also downregulated by miR-133b in adenocarcinoma cell lines and target validation studies showed that miR-133b binds to the 3′-UTR of BCL-W [87]. Finally another member of BCL-2 family, BCL-XL is directly targeted by miR-491 in colorectal cancer cell lines [88].

8 MicroRNAs Related to Autophagy

Autophagy is a catabolic process that causes degradation of a cell's own components. The mechanism of this intra cellular degradation is based on the sequestering of cytosolic content via a lysosomal pathway inside double membrane vesicles [89]. The main aim of autophagy is to recycle nutrients by removing damaged proteins and organelles so that homeostatic functions are kept at a basal level [90]. Proper functioning of the autophagy process is a basic requirement for accurate development and defects in autophagy result in numerous human diseases including cancer and neurodegeneration [91]. In cancer pathology, autophagy may be seen as a promoter of cancer cell survival; however, progressive autophagy can result in cell death [92]. Therefore, the role of autophagy is complex and the cellular response depends on genetic composition and environmental cues.

Autophagy is initiated by the beclinl protein, which is the homologue of yeast Atg6. Experimental evidence showed that beclin 1 expression is altered in some diseases due to differences in miRNA expression signatures. According to a study by Zhu et al., beclin 1 expression is inhibited after targeting by miR-30a [93]. Furthermore, as result of their functional experiments with miR-30a agonist and antagonist, miR-30a agonists decreased while antagonists increased autophagic activity, respectively. A recent computational study by Jegga et al. showed that miR-130, miR-98, miR-124, miR-204, and miR-142 are putative posttranscriptional regulators of the autophagy-lysosomal pathway genes [94]. Another study by Xiao et al. showed that miR-204 regulates autophagy in cardiomyocytes [95]. They have found that miR-204 directly targets the LC3-II protein to regulate autophagy as a result of ischemia–reperfusion injury. In a recent study of Frankel et al., autophagy was induced by etoposide and rapamycin in breast cancer cells and they characterized miR-101 as a potent inhibitor of autophagy [96]. According to transcriptome analysis, they identified stathmin 1 (STMN1), RAS-related protein RAB5A and autophagy related 4 homolog D (ATG4D) as miR-101 targets.

In conclusion, the role of miRNAs in autophagy has just started to be explored after a growing number of studies have been published in the last 2 years. There might be much more miRNAs to be discovered in the future that will help unravel the regulatory mechanisms of autophagy.

9 Conclusion

Several studies showed that miRNAs play a major role in many biological processes. The alteration of physiological expression of miRNAs may cause pathogenesis of a variety of diseases. Increasing biological information on miRNAs may help the generation of novel strategies in disease therapies. For example, the use of cell-death-inducing miRNA analogues may have anticancer effects.

References

1. Griffiths-Jones S, Grocock RJ, van Dongen S et al (2006) miRBase: microRNA sequences, targets and gene nomenclature. Nucleic Acids Res 34:D140–D144
2. Almeida MI, Reis RM, Calin GA (2011) MicroRNA history: discovery, recent applications, and next frontiers. Mutat Res 717:1–8
3. Wang Z, Chen Z, Gao Y et al (2011) DNA hypermethylation of microRNA-34b/c has prognostic value for stage non-small cell lung cancer. Cancer Biol Ther 11:490–496
4. Yuan JH, Yang F, Chen BF et al (2011) The histone deacetylase 4/SP1/microrna-200a regulatory network contributes to aberrant histone acetylation in hepatocellular carcinoma. Hepatology 54:2025–2035
5. Ha TY (2011) MicroRNAs in human diseases: from cancer to cardiovascular disease. Immune Netw 11:135–154
6. Huang Y, Shen XJ, Zou Q et al (2010) Biological functions of microRNAs. Bioorg Khim 36:747–752
7. Lee CT, Risom T, Strauss WM (2006) MicroRNAs in mammalian development. Birth Defects Res C Embryo Today 78:129–139
8. Liu J, Carmell MA, Rivas FV et al (2004) Argonaute2 is the catalytic engine of mammalian RNAi. Science 305:1437–1441
9. Lagos-Quintana M, Rauhut R, Meyer J et al (2003) New microRNAs from mouse and human. RNA 9:175–179
10. Mineno J, Okamoto S, Ando T et al (2006) The expression profile of microRNAs in mouse embryos. Nucleic Acids Res 34:1765–1771
11. Suh MR, Lee Y, Kim JY et al (2004) Human embryonic stem cells express a unique set of microRNAs. Dev Biol 270:488–498
12. Boyer LA, Lee TI, Cole MF et al (2005) Core transcriptional regulatory circuitry in human embryonic stem cells. Cell 122:947–956
13. Strauss WM, Chen C, Lee CT et al (2006) Nonrestrictive developmental regulation of microRNA gene expression. Mamm Genome 17:833–840
14. Chen CZ, Lodish HF (2005) MicroRNAs as regulators of mammalian hematopoiesis. Semin Immunol 17:155–165
15. Lu J, Qian J, Chen F et al (2005) Differential expression of components of the microRNA machinery during mouse organogenesis. Biochem Biophys Res Commun 334: 319–323
16. Harris KS, Zhang Z, McManus MT et al (2006) Dicer function is essential for lung epithelium morphogenesis. Proc Natl Acad Sci U S A 103:2208–2213
17. Yi R, O'Carroll D, Pasolli HA et al (2006) Morphogenesis in skin is governed by discrete sets of differentially expressed microRNAs. Nat Genet 38:356–362
18. Brittis PA, Lu Q, Flanagan JG (2002) Axonal protein synthesis provides a mechanism for localized regulation at an intermediate target. Cell 110:223–235
19. Giraldez AJ, Cinalli RM, Glasner ME et al (2005) MicroRNAs regulate brain morphogenesis in zebrafish. Science 308:833–838
20. Lagos-Quintana M, Rauhut R, Yalcin A et al (2002) Identification of tissue-specific microRNAs from mouse. Curr Biol 12:735–739

21. Krichevsky AM, King KS, Donahue CP et al (2003) A microRNA array reveals extensive regulation of microRNAs during brain development. RNA 9:1274–1281
22. Smirnova L, Grafe A, Seiler A et al (2005) Regulation of miRNA expression during neural cell specification. Eur J Neurosci 21:1469–1477
23. Schratt GM, Tuebing F, Nigh EA et al (2006) A brain-specific microRNA regulates dendritic spine development. Nature 439:283–289
24. Sokol NS (2012) The role of microRNAs in muscle development. Curr Top Dev Biol 99:59–78
25. Bueno MJ, Malumbres M (2011) MicroRNAs and the cell cycle. Biochim Biophys Acta 1812:592–601
26. Lujambio A, Ropero S, Ballestar E et al (2007) Genetic unmasking of an epigenetically silenced microRNA in human cancer cells. Cancer Res 67:1424–1429
27. Volinia S, Calin GA, Liu CG et al (2006) A microRNA expression signature of human solid tumors defines cancer gene targets. Proc Natl Acad Sci U S A 103:2257–2261
28. Malhas A, Saunders NJ, Vaux DJ (2010) The nuclear envelope can control gene expression and cell cycle progression via miRNA regulation. Cell Cycle 9:531–539
29. Miller TE, Ghoshal K, Ramaswamy B et al (2008) MicroRNA-221/222 confers tamoxifen resistance in breast cancer by targeting p27Kip1. J Biol Chem 283:29897–29903
30. Butz H, Liko I, Czirjak S et al (2010) Down-regulation of Wee1 kinase by a specific subset of microRNA in human sporadic pituitary adenomas. J Clin Endocrinol Metab 95:E181–E191
31. Glover DM, Hagan IM, Tavares AA (1998) Polo-like kinases: a team that plays throughout mitosis. Genes Dev 12:3777–3787
32. Shi W, Alajez NM, Bastianutto C et al (2010) Significance of Plk1 regulation by miR-100 in human nasopharyngeal cancer. Int J Cancer 126:2036–2048
33. O'Donnell KA, Wentzel EA, Zeller KI et al (2005) c-Myc-regulated microRNAs modulate E2F1 expression. Nature 435:839–843
34. Schulte JH, Horn S, Otto T et al (2008) MYCN regulates oncogenic MicroRNAs in neuroblastoma. Int J Cancer 122:699–704
35. Ma L, Young J, Prabhala H et al (2010) miR-9, a MYC/MYCN-activated microRNA, regulates E-cadherin and cancer metastasis. Nat Cell Biol 12:247–256
36. Chang TC, Yu D, Lee YS et al (2008) Widespread microRNA repression by Myc contributes to tumorigenesis. Nat Genet 40:43–50
37. Woods K, Thomson JM, Hammond SM (2007) Direct regulation of an oncogenic micro-RNA cluster by E2F transcription factors. J Biol Chem 282:2130–2134
38. Petrocca F, Visone R, Onelli MR et al (2008) E2F1-regulated microRNAs impair TGFbeta-dependent cell-cycle arrest and apoptosis in gastric cancer. Cancer Cell 13:272–286
39. Yang X, Feng M, Jiang X et al (2009) miR-449a and miR-449b are direct transcriptional targets of E2F1 and negatively regulate pRb-E2F1 activity through a feedback loop by targeting CDK6 and CDC25A. Genes Dev 23: 2388–2393
40. Ichimura A, Ruike Y, Terasawa K et al (2011) miRNAs and regulation of cell signaling. FEBS J 278:1610–1618
41. Inui M, Martello G, Piccolo S (2010) MicroRNA control of signal transduction. Nat Rev Mol Cell Biol 11:252–263
42. Yang CH, Yue J, Pfeffer SR et al (2011) MicroRNA miR-21 regulates the metastatic behavior of B16 melanoma cells. J Biol Chem 286:39172–39178
43. Lu Z, Li Y, Takwi A et al (2011) miR-301a as an NF-kappaB activator in pancreatic cancer cells. EMBO J 30:57–67
44. Ma X, Becker Buscaglia LE, Barker JR et al (2011) MicroRNAs in NF-kappaB signaling. J Mol Cell Biol 3:159–166
45. Cichocki F, Felices M, McCullar V et al (2011) Cutting edge: microRNA-181 promotes human NK cell development by regulating Notch signaling. J Immunol 187:6171–6175
46. Ferretti E, De Smaele E, Miele E et al (2008) Concerted microRNA control of Hedgehog signalling in cerebellar neuronal progenitor and tumour cells. EMBO J 27:2616–2627
47. Liu X, Cheng Y, Chen X et al (2011) MicroRNA-31 regulated by the extracellular regulated kinase is involved in vascular smooth muscle cell growth via large tumor suppressor homolog 2. J Biol Chem 286:42371–42380
48. Terasawa K, Ichimura A, Sato F et al (2009) Sustained activation of ERK1/2 by NGF induces microRNA-221 and 222 in PC12 cells. FEBS J 276:3269–3276
49. Taganov KD, Boldin MP, Chang KJ et al (2006) NF-kappaB-dependent induction of microRNA miR-146, an inhibitor targeted to signaling proteins of innate immune responses. Proc Natl Acad Sci U S A 103:12481–12486
50. Martello G, Zacchigna L, Inui M et al (2007) MicroRNA control of Nodal signalling. Nature 449:183–188
51. O'Connell RM, Taganov KD, Boldin MP et al (2007) MicroRNA-155 is induced during the macrophage inflammatory response. Proc Natl Acad Sci U S A 104:1604–1609
52. Tili E, Michaille JJ, Cimino A et al (2007) Modulation of miR-155 and miR-125b levels following lipopolysaccharide/TNF-alpha stimulation and their possible roles in regulating the response to endotoxin shock. J Immunol 179:5082–5089

53. Cobb BS, Hertweck A, Smith J et al (2006) A role for Dicer in immune regulation. J Exp Med 203:2519–2527
54. Li QJ, Chau J, Ebert PJ et al (2007) miR-181a is an intrinsic modulator of T cell sensitivity and selection. Cell 129:147–161
55. Xiao C, Calado DP, Galler G et al (2007) MiR-150 controls B cell differentiation by targeting the transcription factor c-Myb. Cell 131:146–159
56. Jopling CL, Yi M, Lancaster AM et al (2005) Modulation of hepatitis C virus RNA abundance by a liver-specific MicroRNA. Science 309:1577–1581
57. Lecellier CH, Dunoyer P, Arar K et al (2005) A cellular microRNA mediates antiviral defense in human cells. Science 308:557–560
58. Otsuka M, Jing Q, Georgel P et al (2007) Hypersusceptibility to vesicular stomatitis virus infection in Dicer1-deficient mice is due to impaired miR24 and miR93 expression. Immunity 27:123–134
59. Triboulet R, Mari B, Lin YL et al (2007) Suppression of microRNA-silencing pathway by HIV-1 during virus replication. Science 315:1579–1582
60. Bennasser Y, Le SY, Benkirane M et al (2005) Evidence that HIV-1 encodes an siRNA and a suppressor of RNA silencing. Immunity 22:607–619
61. Pedersen IM, Cheng G, Wieland S et al (2007) Interferon modulation of cellular microRNAs as an antiviral mechanism. Nature 449:919–922
62. Carmeliet P, Jain RK (2011) Molecular mechanisms and clinical applications of angiogenesis. Nature 473:298–307
63. Yang WJ, Yang DD, Na S et al (2005) Dicer is required for embryonic angiogenesis during mouse development. J Biol Chem 280: 9330–9335
64. Suarez Y, Fernandez-Hernando C, Yu J et al (2008) Dicer-dependent endothelial microRNAs are necessary for postnatal angiogenesis. Proc Natl Acad Sci U S A 105:14082–14087
65. Wang S, Olson EN (2009) AngiomiRs—key regulators of angiogenesis. Curr Opin Genet Dev 19:205–211
66. Fish JE, Santoro MM, Morton SU et al (2008) miR-126 regulates angiogenic signaling and vascular integrity. Dev Cell 15:272–284
67. Poliseno L, Tuccoli A, Mariani L et al (2006) MicroRNAs modulate the angiogenic properties of HUVECs. Blood 108:3068–3071
68. Hua Z, Lv Q, Ye W et al (2006) MiRNA-directed regulation of VEGF and other angiogenic factors under hypoxia. PLoS One 1:e116
69. Zhou Q, Gallagher R, Ufret-Vincenty R et al (2011) Regulation of angiogenesis and choroidal neovascularization by members of microRNA-23 27 24 clusters. Proc Natl Acad Sci U S A 108:8287–8292
70. Vecchione A, Croce CM (2010) Apoptomirs: small molecules have gained the license to kill. Endocr Relat Cancer 17:F37–F50
71. Reed JC (1998) Bcl-2 family proteins. Oncogene 17:3225–3236
72. Henry-Mowatt J, Dive C, Martinou JC et al (2004) Role of mitochondrial membrane permeabilization in apoptosis and cancer. Oncogene 23:2850–2860
73. Ashkenazi A (2002) Targeting death and decoy receptors of the tumour-necrosis factor superfamily. Nat Rev Cancer 2:420–430
74. Brennecke J, Hipfner DR, Stark A et al (2003) bantam encodes a developmentally regulated microRNA that controls cell proliferation and regulates the proapoptotic gene hid in Drosophila. Cell 113:25–36
75. Zhang J, Du Y, Wu C et al (2010) Curcumin promotes apoptosis in human lung adenocarcinoma cells through miR-186* signaling pathway. Oncol Rep 24:1217–1223
76. Qin W, Shi Y, Zhao B et al (2010) miR-24 regulates apoptosis by targeting the open reading frame (ORF) region of FAF1 in cancer cells. PLoS One 5:e9429
77. He B, Xiao J, Ren AJ et al (2011) Role of miR-1 and miR-133a in myocardial ischemic postconditioning. J Biomed Sci 18:22
78. Walker JC, Harland RM (2009) microRNA-24a is required to repress apoptosis in the developing neural retina. Genes Dev 23: 1046–1051
79. Shi XB, Xue L, Yang J et al (2007) An androgen-regulated miRNA suppresses Bak1 expression and induces androgen-independent growth of prostate cancer cells. Proc Natl Acad Sci U S A 104:19983–19988
80. Adlakha YK, Saini N (2011) MicroRNA-128 downregulates Bax and induces apoptosis in human embryonic kidney cells. Cell Mol Life Sci 68:1415–1428
81. Tsang WP, Kwok TT (2008) Let-7a microRNA suppresses therapeutics-induced cancer cell death by targeting caspase-3. Apoptosis 13:1215–1222
82. Bommer GT, Gerin I, Feng Y et al (2007) p53-mediated activation of miRNA34 candidate tumor-suppressor genes. Curr Biol 17:1298–1307
83. Hagman Z, Larne O, Edsjo A et al (2010) miR-34c is downregulated in prostate cancer and exerts tumor suppressive functions. Int J Cancer 127:2768–2776
84. Cimmino A, Calin GA, Fabbri M et al (2005) miR-15 and miR-16 induce apoptosis by targeting BCL2. Proc Natl Acad Sci U S A 102:13944–13949

85. Chen G, Zhu W, Shi D et al (2010) MicroRNA-181a sensitizes human malignant glioma U87MG cells to radiation by targeting Bcl-2. Oncol Rep 23:997–1003
86. Lin CJ, Gong HY, Tseng HC et al (2008) miR-122 targets an anti-apoptotic gene, Bcl-w, in human hepatocellular carcinoma cell lines. Biochem Biophys Res Commun 375: 315–320
87. Crawford M, Batte K, Yu L et al (2009) MicroRNA 133B targets pro-survival molecules MCL-1 and BCL2L2 in lung cancer. Biochem Biophys Res Commun 388: 483–489
88. Nakano H, Miyazawa T, Kinoshita K et al (2010) Functional screening identifies a microRNA, miR-491 that induces apoptosis by targeting Bcl-X(L) in colorectal cancer cells. Int J Cancer 127:1072–1080
89. He C, Klionsky DJ (2009) Regulation mechanisms and signaling pathways of autophagy. Annu Rev Genet 43:67–93
90. Mizushima N, Levine B, Cuervo AM et al (2008) Autophagy fights disease through cellular self-digestion. Nature 451:1069–1075
91. Chen N, Debnath J (2010) Autophagy and tumorigenesis. FEBS Lett 584:1427–1435
92. Chen N, Karantza V (2011) Autophagy as a therapeutic target in cancer. Cancer Biol Ther 11:157–168
93. Zhu H, Wu H, Liu X et al (2009) Regulation of autophagy by a beclin 1-targeted microRNA, miR-30a, in cancer cells. Autophagy 5:816–823
94. Jegga AG, Schneider L, Ouyang X et al (2011) Systems biology of the autophagy-lysosomal pathway. Autophagy 7:477–489
95. Xiao J, Zhu X, He B et al (2011) MiR-204 regulates cardiomyocyte autophagy induced by ischemia-reperfusion through LC3-II. J Biomed Sci 18:35
96. Frankel LB, Wen J, Lees M et al (2011) microRNA-101 is a potent inhibitor of autophagy. EMBO J 30:4628–4641

Chapter 3

The Role of MicroRNAs in Human Diseases

Kemal Uğur Tüfekci, Meryem Gülfem Öner, Ralph Leo Johan Meuwissen, and Şermin Genç

Abstract

About 20 years have passed since the discovery of the first microRNA (miRNA) and by now microRNAs are implicated in a variety of physiological and pathological processes. Since the discovery of the powerful effect miRNAs have on biological processes, it has been suggested that mutations affecting miRNA function may play a role in the pathogenesis of human diseases. Over the past several years microRNAs have been found to play a major role in various human diseases. In addition, many studies aim to apply miRNAs for diagnostic and therapeutic applications in human diseases. In this chapter, we summarize the role of miRNAs in pathological processes and discuss how miRNAs could be used as disease biomarkers.

Key words Biomarker, Human disease, Exosome, Mutation, Circulating miRNA, Single nucleotide polymorphism, Copy number variation

1 Introduction

The discovery of microRNAs (miRNA) resulted in a great revolution in molecular biology. It has been proposed that approximately 30 % of the human protein coding genes are controlled by miRNAs [1]. MicroRNAs are involved in several biological processes. They also participate in the pathogenesis of several human diseases including cancer, immune, cardiovascular and neurological disorders. Genome-wide expression profiles from tissues or body fluids demonstrated that miRNAs could be used as biomarkers in different human diseases [2]. Some miRNAs could be validated as prognostic disease markers and therefore contribute to the estimation of drug responses. Thus, miRNAs could become useful in disease detection, they promise to be new therapeutic targets in human disorders and they may become drugs to treat human disorders. Here, we present an overview of our current understanding the role of miRNA in human diseases.

Malik Yousef and Jens Allmer (eds.), *miRNomics: MicroRNA Biology and Computational Analysis*, Methods in Molecular Biology, vol. 1107, DOI 10.1007/978-1-62703-748-8_3, © Springer Science+Business Media New York 2014

2 MicroRNA Mutations in Human Diseases

Three main genetic changes influence miRNA function: (1) Copy-number variations (CNV) similar to large-scale mutations, (2) Single nucleotide polymorphism (SNP) type mutations, and (3) epigenetic changes. They are localized in miRNA genes, miRNA target genes and miRNA processing genes.

2.1 CNV Like Large-Scale Mutations in MicroRNA Genes

MicroRNA genes can be affected by large-scale mutations such as deletions, duplications and insertions. So far only a limited amount of studies investigated CNV type mutations in miRNA genes [3–5]. An initial study was done in a Han Chinese population [4]. MicroRNA genes were affected by CNV type mutation in 13 of 646 subjects. Results showed that two miRNA genes (miR-338, miR-657) had a gain of function, two miRNA genes (miR-204, miR-383) had a loss of function and one gene (miR-589) had both (loss and gain) of function due to these mutations [4]. In a second study, CNV type mutations were evaluated in 105 DNA samples from different ethnic origins. The authors found 21 human miRNAs that reside within 14 CNV loci [5]. Some loci contained only a single miRNA gene (e. g.: hsa-mir-202 resides in 10q26.3), whereas three loci were found to have more than one miRNA gene (e.g., hsa-let-7g and hsa-mir-135a-1 reside in 3p21.2). A follow-up more detailed bioinformatics analysis showed that 193 pre-miRNAs were located in the regions covered by the 385 CNV markers [3].

To date, there is no clear evidence showing that CNV like mutations in miRNA genes are associated with human Mendelian disorders. A CNV type mutation at 15q was found in ten patients with developmental delays and features of the autism spectrum [6]. Recently, de Pontual et al. have showed the first example of a miRNA gene that is responsible for a developmental defect in humans leading to observable symptoms [7]. They reported a germ line hemizygous deletion of the miR-17∼92 polycistronic miRNA cluster, in a patient with microcephaly, short stature, and digital abnormalities. The relationship between disease and mutation was confirmed with the presence of similar findings in targeted deletion of the miR-17∼92 cluster in mice [7].

There seems to be a causative relation between cancer and CNV type mutations. A small region of chromosome 13q14 is deleted in more than half of chronic lymphocytic leukemia (CLL) patients. Calin et al. showed that miR-15 and miR-16 are located within the missing region in CLL, and that these genes are deleted and down-regulated in most CLL disease cases [8]. Expressions of miR-15a, miR-16, and 13q14 deletion were also evaluated in multiple myeloma (MM) patients. Expression levels of these miRNAs were independent of the chromosome 13 deletion [9]. MiR-15a and miR-16-1 deletions were found in prostate cancer; therefore, this finding suggests that both miR-15a and miR-16-1 could be tumor suppressor [10]. Two large-scale mutations in miR-125-b1 are linked with leukemogenesis. An insertion of miR-125b-1 into the immunoglobulin heavy chain gene was reported in a patient with precursor B cell acute lymphocytic leukemia (B-ALL) [11]. The t(2;11) translocation upregulates miR-125-b expression in nine myelodysplastic syndromes (MDS) and ten acute myeloid leukemia (AML) patients. Alternatively, miR-125b blocks the myelomonocytic differentiation in HL 60 and NB4 leukemic cell lines [12]. Likewise, heterozygous deletion of 2q37 region leads to a loss of miR-562 gene expression encountered in 4 % of sporadic Wilm's tumors [13]. A large germ line deletion in the Neurofibromatosis type 1 (NF1) locus is reportedly connected with a severe NF1 phenotype in a French patient. NF1 is an autosomal disorder connected with central nervous system (CNS) tumors. This patient's deletion included six miRNA genes (hsa-mir-423, -193a, -365-2, and -632) that may contribute to the phenotype of NF1 [14]. A large-scale deletion in chromosome 14 was found in gastrointestinal stromal tumors (GISTs). The loss of 14q in GISTs was correlated with diminished expression of miRNAs located in this region as well as with tumor progression [15]. A CNV type mutation at the miR-218 locus, which causes down-regulation of miR-218 expression, was linked to tobacco smoking and was found in lung cancer [16].

2.2 SNP Type Mutations in MicroRNA Genes

SNPs might cause loss or gain-of-functions of miRNA genes. These functional alternations participate in the pathogenesis of several diseases such as cancer and Mendelian disorders. A germ line miRNA gene mutation linked to familial CLL was reported as a first example [17]. The study showed that mutations in pri-mir-16-1 lead to low expression of the mature miRNA. In another case, an SNP in miR-126 was found in patients with MLL-AF4 ALL [18]. This SNP blocks the miRNA processing.

Results of many association studies were evaluated whether SNPs in miRNA genes increase risk of cancer [19–27]. A polymorphism (rs2910164) in miR-146 predisposes carriers to papillary thyroid carcinoma [19], breast cancer [21], glioma [25], and gastric cancer [20]. However, the same polymorphism could be protective against different cancer type such as prostate, hepatocellular carcinoma, cervical cancer, and esophageal squamous cell carcinoma [20, 28]. There is no explanation yet how one SNP could be a predisposing factor and protective factor for different cancer types. The rs2910164 polymorphism causes reduction of mature miRNAs and loss of inhibitory effect on target genes [19]. GC heterozygotes produce three mature miRNAs that change transcription profiles of cells and alter phenotypic traits. Indeed, the C allele promotes cell proliferation and colony formation in the NIH/3T3 cell line [26]. The rs2910164 polymorphism could be a diagnostic marker for the early onset breast cancer [21], a prognostic marker for glioma [25], and esophageal carcinoma [24].

This polymorphism was also studied in other diseases and it was found that for example the C allele may be a risk factor for dilated cardiomyopathy [20].

SNPs in promoter regions affect transcription factor binding and alter miRNA levels. The rs57095329 (A>G) polymorphism located in the promoter region of miR-146a, has been associated with systemic lupus erythematosus (SLE) [29]. The transcription factor Ets-1 binds near this polymorphism in the miR-146 promoter. Promoter pull-down assays using streptavidin-conjugated agarose beads showed that biotin-labeled A probes bound more Ets-1 protein than the G probes [29]. This result suggested that the rs57095329 polymorphism affects the Ets-1 binding capacity of the miR-146 promoter.

The rs11614913 polymorphism in miR-196a2 was studied in different human cancers as a predisposition factor. The C allele was correlated with poor survival in lung cancer [20] and lymph node metastasis in gastric cancer [30]. The C allele was linked to a significant increase in mature miR-196a expression [20].

MiR-96 is a tissue specific miRNA expressed in the mammalian cochlea during development. Mutations in miR-96 are associated with autosomal dominant, progressive hearing loss in human and mice [28, 31]. MiR-96 mutant diminuendo mice showed that mir-96 is required for development and differentiation of auditory hair cells [31].

2.3 SNPs in MicroRNA Target Genes

Nine thousand six hundred and seventy three human genes are located in a CNV region [32], but CNV type mutations in miRNA target genes are less well studied. Here, we focus on SNP type variations in miRNA target genes. A recent genome-wide study showed that there are thousands of SNPs in miRNA target genes [33]. SNPs in miRNA target genes affect the efficiency of miRNA binding with their target regions via creating or destroying these binding sequences. It has been found that there are 58,977 SNPs in 90,784 miRNA targets, 20,779 of these SNPs destroy target sites while creating 91,711 new miRNA target sites concurrently [33].

The Texel sheep model is the first example of an animal disorder caused by an SNP in a miRNA target gene [34]. Here an SNP is located in the 3′ UTR region (G to A transition) of the myostatin (*GDF8*) gene. This SNP creates a new target site for mir-1 and mir-206, which are both highly expressed in skeletal muscle. This in turn causes translational inhibition of GDF8 which then leads to the muscular hypertrophy of the Texel sheep [34]. Recently, Shao et al. have reported that one SNP 131 C>T in the 3′-UTR of the bone morphogenetic protein 5 (BMP5) gene is associated with fatness in pigs [35]. This C/T transition alters BMP5 interaction with let-7c and miR-184.

There is some evidence that SNPs in miRNA target genes can be associated with human disorders. In an initial human study,

Abelson et al. reported that an SNP in the miR-189 target gene is associated with Tourette's syndrome (TS) [36] which is a developmental neuropsychiatric disorder characterized by persistent vocal and motor tics. They found a frame shift mutation at the miR-189 binding site in the Slit and the Trk-like 1 (SLITRK1) genes [36]. These findings suggested that there is an association between SLITRK1 sequence variants and TS. Using functional studies by mutating target sequences, they have found decreased levels of binding of miRNA to target.

Parkinson Disease (PD) is a movement disorder caused by environmental and genetic factors. One polymorphism, rs1989754, in the fibroblast growth factor 20 (*FGF20*) gene has been associated with an increased risk for PD [37]. Recently, three SNPs have been found in the 3′-UTR of FGF20 in PD [37]. Functional assays showed that the SNP rs12720208 disrupts a binding site for miRNA-433 and causes increased level of FGF20 and this increase in FGF20 is correlated with augmented α-synuclein expression.

SNPs in miRNA target genes were analyzed in different cancer types. *KRAS*, a prominent oncogene in NSCLC, contains a miRNA Let-7 complementary site (LCS) in its 3′-UTR region. The presence of this SNP in the LCS increases the risk for NSCLC since the frequency of the variant allele is fourfold higher in NSCLC patients than compared to controls [38]. There is, however, no connection between LCS polymorphism and disease survival [39]. Other SNPs in miRNA target genes could prove useful as prognostic markers. Indeed, SNP in integrin beta 4 (ITGB4) was correlated with tumor aggressiveness and survival [40]. Furthermore, SNPs in miRNA target genes may participate in drug resistance. For instance an SNP in the 3′ UTR of the dihydrofolate reductase (*DHFR*) gene affects miR-24 binding and leads to DHFR overproduction and consequently methotrexate resistance [41].

2.4 Mutations in MicroRNA Biogenesis Genes

SNPs in genes that encode proteins involved in miRNA biogenesis affect global expression of miRNAs and are thereby linked to human diseases. The Drosha protein plays a role in the initial step of miRNA biogenesis. Overexpression of Drosha was linked to CNV mutations affecting whole chromosomes in cervical cancer [42] and it has been found that a haplotype in Drosha was related with shorter survival in lung cancer [43]. In addition, an SNP within the same haplotype was associated with low mRNA expression of Drosha in adenocarcinoma patients [43]. Eight SNPs (rs7735863, rs6884823, rs3792830, rs669702, rs639174, rs3805500, rs7719666, and rs17410035) were associated with head and neck cancer and they could be prognostic factors for this disease.

Exportin 5 (XPO5) and RAN (Ras-related nuclear protein) mediate the nuclear export of miRNAs. SNPs in these genes were found in esophageal cancer and head-neck cancer [22, 44]. The SNPs rs11077, rs2227301, and rs699937 are located in XPO5 while the SNPs rs14035 and rs11061209 are located in the *RAN* gene [22, 44]. Another frame shift mutation in *XPO5* was found in the colorectal cancer cell lines HCT-15 and DLD-1 [45]. The *XPO5* mutation leads to the retention of pre-miRNAs in the nucleus and diminishes miRNA-target inhibition.

DICER1 and transactivation-responsive RNA-binding protein (TRBP) participate in pre-miRNA processing. The SNPs rs13078 and rs3742330 in DICER were found in oral premalignant lesions (OPLs). One SNP (rs3742330) in DICER increases the risk of OPLs [23]. Melo and colleagues identified truncating mutations in TRBP in sporadic and hereditary carcinomas with microsatellite instability [46]. These mutations diminish TRBP protein expression and thereby cause disturbance in miRNA processing [46]. Furthermore, another SNP (rs784567) in the TRBP gene has been associated with a 20 % risk reduction for developing bladder cancer [47].

The RNA-induced silencing complex (RISC) has a pivotal role in guiding mature miRNAs to their targets. RISC contains Dicer and many other proteins such as Gemin 3, Gemin 4, and Ago. An SNP in the Gemin 3 gene (rs197414) was associated with bladder and esophageal cancer [22, 47]. The rs197412 variant genotypes show reduced OPL risk [23]. Four different SNPs (rs2740348, rs7813, rs2740351, and rs7813) in Gemin 4 are associated with various cancers [48, 49].

Haplotype analyses showed that the H3 haplotype of the Gemin 4 gene is associated with bladder cancer [47], whereas the common haplotype is related to the esophageal cancer risk [22]. SNPs in Ago genes were evaluated in bladder and renal cancer and in OPL, but so far there is no reported association of any SNPs with any of these cancer types [22, 47, 48].

2.5 Epigenetic Regulation of MicroRNA Genes

MicroRNA expression can also be regulated by epigenetic mechanisms especially via methylation of miRNA gene promoter regions [50]. Hypermethylation of miRNA genes have been observed in various cancer types. Hypermethylation of miRNA gene promoters downregulates miRNA expression and decreases the inhibition of target genes [51–53]. On the other hand, the let-7a-3 locus is hypomethylated in ovarian, colon, and lung cancer, which is contrary to let-7a-3's supposed role as a tumor suppressor in these diseases [54, 55]. The methylation status could be a predictive factor for cancer prognosis. Methylation of the miR-137 promoter has been associated to poor survival in squamous carcinoma of the head and neck [56]. The methylation status could also predict metastasis potential of cancer. MiR-148a, miR-34b/c, and miR-9 hypermethylations have been significantly associated with the appearance of lymph node metastases [57].

Many previous studies reported an association between methylation of miRNA genes and cancer [58–66]. Similar relationship may exist between epigenetic regulation of miRNA genes and other disorders.

3 MicroRNA Expression Profiles in Human Diseases

3.1 Cancer

As we discussed before, microRNAs have critical functions in numerous biological processes such as cellular proliferation and cell death. Altered expression of microRNAs might accompany cancer development. The link between miRNAs and cancer was first published by Calin et al. [8] in 2002. MiR-15 and miR-16 were found deleted or down-regulated in the majority (68 %) of chronic lymphocytic leukemia (CLL) patients. Many studies have shown that overexpression of microRNAs contributes to tumor formation by inhibiting tumor suppressor genes and the down-regulation of microRNAs might unblock inhibition of oncogenic genes, thus accelerating tumorigenesis. Oncogenic miRNAs are miR-9, miR-17-92, miR-21, miR-27a, miR-103, miR-106, miR-107, miR-125b, and miR-155 [67]. Tumor suppressor miRNAs are miRNAs are miR-15, miR-16, miR-23b, miR-29a, miR-29b, miR-29c, miR-34a, let-7a, miR-124, miR-133b, miR-137, miR-143, miR-145, miR-192, and miR-215 [67]. Altered miRNA expression patterns could help to classify tumors. For instance, miR-205 may distinguish the two major types of non-small-cell lung cancer with 96 % sensitivity and 90 % specificity [68]. However, these results were not confirmed by a later study [69]. MicroRNA expression changes have prognostic relevance in many cancers. Low miR-135a expression has been correlated with earlier relapse and shorter disease free survival in Hodgkin's lymphoma [70]. Some dysregulated miRNAs have been associated with biopathological features of cancer such as tumor stage, angiogenesis, and metastasis. For example, miR-10b expression is elevated and miR-126, miR-206, and miR-335 decreased in metastatic breast tumors [71, 72]. MicroRNA expression profiles may help predict drug resistance. High expression of miR-21, hsa-miR-23a, hsa-miR-27a, and hsa-let-7g has been associated with chemotherapy resistance in ovarian cancer [73].

In conclusion, miRNA expression profiles may be helpful for cancer diagnosis, prediction of the clinical outcome, and drug resistance. Alternation of miRNA expression might therefore become a new therapeutic approach in cancer.

3.2 Autoimmune Diseases

MicroRNAs have an important role in the development of immune cells and the maintenance of immune system functions. Altered expression of microRNAs has been associated with autoimmune disorders such as rheumatoid arthritis (RA), SLE, and multiple sclerosis (MS) [74]. RA is a systemic autoimmune disorder that is characterized by chronic inflammation within the joint tissue. Initial studies showed that miR-146 and miR-155 were upregulated in synovial fluids, fibroblast, and peripheral blood mononuclear cells (PBMC) of RA patients [74]. Later, elevation of miR-203 and miR-346 and decrease of miR-124a levels were found in RA [75].

Target genes of altered miRNAs are regulator molecules of inflammation such as the TNF receptor associated factor 6 (TRAF6), Interleukin-1 receptor-associated kinase 1 (IRAK1), and tumor necrosis factor-alpha (TNFα) [74].

3.3 Neurological Diseases

Several research groups reported that some miRNAs are only expressed in the central nervous system (CNS) and that this expression is further subject to temporal and spatial differences [76]. MicroRNAs play an essential role in brain development and physiological functions. The necessity of miRNAs in CNS function was confirmed with knockdown studies of miRNA biogenesis pathway genes. Dicer ablation leads to immature neurogenesis [77] and deletion of DiGeorge syndrome chromosome region (DGCR), encoding for another miRNA machinery gene, causes diminished dendritic spine formation [78]. MicroRNAs are involved in CNS disorders through altering neuronal function or regeneration. MicroRNA expression profiles may help to diagnose some CNS disorders or predict CNS disease development. For these purposes, miRNA expression profiles were evaluated in brain tissue, cerebrospinal fluid (CSF), serum, plasma, and PBMC of patients [37, 72].

Alzheimer's disease (AD) is the most common cause of dementia characterized by memory loss and behavioral changes. The accumulation of β amyloid tau phosphorylation and beta secretase (BACE) 1 contribute to AD pathogenesis. Several studies found altered miRNA expression in different brain regions such as cerebrospinal fluid (CSF) and PBMC [76]. The most prominent findings are a decreased expression of miR-29 and miR-107, which target the BACE protein [76].

Parkinson's disease (PD) is a movement disorder characterized by tremor, rigidity, bradykinesia and dopaminergic neuronal loss in the substantia nigra. There is abnormal, mostly containing α-synuclein, protein accumulation in neurons. Genetic pathogenesis is related with mutations in Parkin (PARK2), α-synuclein, PTEN induced putative kinase 1 (PINK1), DJ-1, and leucine rich repeat kinase 2 (LRRK2) genes [79]. MicroRNA expression studies showed that miR-133b expression was higher in brain tissue of PD patients [76]. There is a negative regulatory loop between miR-133b and Pitx3 transcription factor, which is a target of miR-133b. Also, miR-34b/c is downregulated in PD [76].

Stroke is one of the most common causes of human death but if survived it can also lead to disability of patients. Several studies evaluated miRNA expression in rodent brain, serum, and human serum samples [80, 81]. Until now, miRNA expression in brain tissue has only been analyzed in animal studies. MiR-181 is the most evaluated miRNA in animal models for stroke; it was down-, up-regulated, or remained unchanged in ischemic brain tissue [80]. MiR-181 expression levels could be different in selective regions of ischemia. The ischemic region shows increased levels of miR-181, while the penumbra shows decreased levels of miR-181 [82]. MiR-124, miR-497, miR-12, miR-298, miR-50, and mir-672 are upregulated, while miR-155 and miR-362-3p are down-regulated in the ischemic brain [80, 81]. MicroRNA expression changes were found in analyzed serum samples of stroke patients [83]. Results suggested that miR-210 downregulation in blood could be a novel biomarker for clinical diagnosis and prognosis in acute cerebral ischemia [83].

Schizophrenia and bipolar disorder are the two most common psychiatric disorders of which the disease pathogenesis has not been unraveled. Several studies evaluated miRNA expression in various brain regions of schizophrenia patients. MiR-24, miR-26b, miR-29c, and miR-7 expression changes as well as Dicer upregulation were confirmed experimentally [76, 84]. BDNF expression, another major player in schizophrenia, seems to be regulated by miRNAs such as miR-30a and miR-195 [84]. Moreover, it also controls other miRNAs expression. Several miRNA expression alternations were found in brain tissue and plasma samples of patients with bipolar disorder. Interestingly, lithium, which is used as therapeutics in bipolar disorder, treatment changes miR-34a, miR-152, miR-155, and miR-221 expression in 20 lymphoblastoid cell lines [85].

3.4 Cardiovascular Diseases

Cardiovascular diseases are a major cause of human mortality. Many studies suggest that miRNAs have specific roles in cardiac development and disorders. MiR-1 is the first miRNA that has been shown to have numerous functions in the heart, including regulation of cardiac morphogenesis, electrical conduction, and cell-cycle control [86, 87]. Several subsequent studies showed altered expressions of several miRNAs in cardiac tissue of mice and in diseased human myocardium [86]. MicroRNAs that are highly expressed in muscle tissue are called myomiRs. The expression levels of myomiRs such as miRs-208a, miR-208b, and miR-499 are usually altered in heart diseases [86]. Alternation of miRNAs expression could constitute a new treatment approach in cardiovascular disorders. For example, miR-15 inhibition protects mice from cardiac ischemic injury by reducing infarct size and enhancing cardiac function [88].

4 MicroRNA as Disease Biomarkers

4.1 Circulating MicroRNAs

The presence of miRNAs in serum was first reported by Lawrie et al. in 2008. The authors found elevated miRNA levels (miR-155, miR-210 and miR-21) in the sera of patients with diffuse large B-cell lymphoma (DLBCL) [89]. Their finding was confirmed by follow-up studies [90]. Circulating miRNAs are very stable molecules and less susceptible to chemical modification, RNAse degradation, and multiple freeze-thawing cycles [2, 90].

The source of circulating miRNAs could be blood cells in healthy humans. MicroRNAs from healthy individuals were isolated and miRNA expression was evaluated by deep sequencing. The expression of 91 out of 101 miRNAs in the serum was found to be similar to that of PBMCs [90]. Six miRNAs were found only in serum, whereas four miRNAs were exclusively present in PBMCs. This result supports the idea that circulating miRNAs can be released from PBMC cells such as platelets and monocytes. Nevertheless, circulating miRNAs might also originate from a wide variety of cells including inflammatory and cancer cells or resulting from cell senescence. Passive release from PBMC on the other hand does not play an important role in the generation of circulating miRNAs.

A considerable part of circulating miRNAs is contained in micro vesicles (MV) and micro particles (MPs). MPs are large membrane vesicles that originate from plasma membranes; on the other hand, exosomes, which are a kind of MVs, are endosomal originated vesicles [2]. Both types of vesicles contain miRNAs, which regulate several cell biological functions such as protein secretion, immune response, and RNA or protein transfer or cell–cell interaction. Micro vesicles from embryonic stem cells contain abundant miRNAs and mediate transfer of miRNAs into mouse embryonic fibroblasts in vitro [91]. Embryonic stem cells may affect the expression of various genes in neighboring cells by transferred miRNAs.

The mechanism of miRNA secretion is currently unknown. Ceramide dependent secretory machinery might form a release mechanism for miRNAs secretion. Kosaka et al. showed that secretion of miRNA was changed after both overexpression and inhibition of a rate-limiting enzyme that is involved in ceramide biosynthesis [92]. Apoptotic bodies may also mediate miRNAs secretion. Endothelial cell-derived apoptotic bodies are generated during atherosclerosis and increase the production of CXC motif ligand 12 (CXCL12) by transferring miRNAs to recipient vascular cells [93].

Besides MVs and exosomes, high-density lipoprotein (HDL) can transport miRNAs to recipient cells, as well. MicroRNAs may be exported from cells to HDL by neutral sphingomyelinase [94]. Scavenger receptor B1 mediates the cellular uptake of HDL-linked miRNAs into recipient cells. Argonaute 2 (AGO2) is another carrier molecule for circulating miRNAs since a significant amount of circulating miRNA is associated with Ago2 [95]. Additional proteins may also participate in the transport of miRNAs.

Several studies investigate the normal spectrum of circulating miRNAs in healthy individuals by means of different platforms. More than 270 different miRNAs were detected in healthy persons. Twenty miRNAs identified in at least four of these studies are let-7b, miR-16, miR-21, miR-223, miR-24, miR-25, miR-30d, miR-320, miR-106b, miR-142-3p, miR-15a, miR-183, miR-19b, miR-20a, miR-22, miR-26a, miR-451, miR-484, miR-92a [90]. Differences in circulating miRNAs between males and females have not been detected [2, 90]. Age, also, does not alter the level of circulating miRNAs [96].

4.2 Circulating MicroRNAs as Biomarkers

An ideal biomarker has to have some specific features like disease specificity and sensitivity, which can be correlated with clinical evidences and be measured with easy, inexpensive methods. To date, there are no ideal biomarkers for several human diseases so we need new biomarkers for diagnosis. MicroRNAs are stable in various body fluids, they have a tissue or biological stage-specific expression and they can be easily detected by various methods. In addition, expression levels of circulating miRNAs have been associated with different diseases [2, 90]. Thus miRNAs have characteristic features of ideal biomarkers. Consequently they are promising candidate biomarkers for various disorders.

4.3 Circulating MicroRNAs in Human Diseases

The level of circulating miRNAs has been investigated in several human diseases including cancer, cardiovascular disease, and stroke [2, 90]. The below part briefly explains the findings of those studies.

The link between circulating miRNAs and cancer was initially demonstrated by Lawrie et al. in 2008. The authors found a significant rise of the levels of miR-155, miR-21, and miR-210 in serum from DLBCL patients [89]. The level of miR-21 was further correlated with relapse-free survival [89].

Lung cancer is the primary cause of cancer death worldwide. MicroRNAs have functions in lung development and tumorigenesis. The level of circulating miRNAs was evaluated in two groups of lung cancer patients (longer vs. shorter survival). The levels of miR-486 and miR-30d were higher, miR-1 and miR-499 were lower in shorter-survival groups [97]. Alternatively, exosomes were purified from the plasma of patients with NSCLC (adenocarcinoma) and controls [98]. The levels of 12 miRNAs were found at similar levels in tumor tissue and plasma and undetectable in control group. These results support the potential use of circulating miRNA profiles as a biomarker for lung cancer.

Circulating miRNAs were also evaluated as biomarkers in breast and ovarian cancers. Several circulating miRNA's expression profiles were altered in breast cancer including let-7a, miR-10b, miR-21, miR-34a, miR-155, and miR-195 [99–102]. Besides, elevated circulating miR-195 could differentiate breast cancer from other cancer types with a sensitivity of 88 % and a specificity of 91 % [103], whereas the level of miR-34a, miR-10b and miR-155 discriminates clinical stages of breast cancer [101].

Furthermore, circulating levels of miR-195 and let-7a were shown to be decreased in breast cancer patients postoperatively, suggesting a link between circulating miRNA levels and treatment efficiency [100]. When miRNA levels were investigated in serum from ovarian cancer patients, MiR-21, miR-92, miR-93, miR-126, and miR-29a were upregulated whereas miR-155, miR-127, and miR-99b were found to be downregulated [104]. In addition, tumor-derived exosomes were isolated by a modified magnetic activated cell sorting (MACS) using anti-epithelial cell adhesion molecule (EpCAM) and the miRNAs were further analyzed [105]. The resulting miRNA expression profile levels of 175 miRNAs were similar in ovarian tumor cells and exosomes. Only 12 miRNAs were only upregulated in the tumor cells, whereas 31 miRNAs were specifically elevated within the exosomes [105].

Colorectal cancer (CRC) is the third most common cancer worldwide. In several studies different circulating miRNAs have been suggested as disease biomarkers for CRC diagnosis, including miR-17-3p, miR-29a, miR-92a, miR-141, and miR-221 [106]. In addition, circulating miRNAs could be prognostic for CRC. For instance, the occurrence of high levels of miR-29a in serum was linked with liver metastasis of CRC [2].

Several circulating miRNA levels were evaluated in different heart diseases [86]. The level of miR-1, miR-133 miR-208a, miR-328, and miR-499 were altered in acute myocardial infarction [86]. Besides, the levels of miR-208b and miR-499 correlated to serum levels of troponin T and creatine phosphokinase (CPK), two markers of cardiac damage [86]. Patients with other heart disease exhibited a different circulating miRNA expression profile. MiR-17, miR-92a, miR-126, miR-130a, miR-221, and miR-222 were dysregulated in coronary artery disease (CAD), whereas miR423-5p is the primary miRNA, whose level was altered in congestive heart failure [2, 107].

MicroRNA levels in circulation were evaluated in patients with autoimmune disorders such as SLE, systemic sclerosis and plaque psoriasis [108, 109]. Serum levels of miR-200a, miR-200b, miR-200c, miR-429, miR-205, and miR-192 were lower in patients with SLE [109].

Circulating miRNAs have also been associated with inflammatory diseases. For instance, circulating miRNAs were analyzed in patients with sepsis and systemic inflammatory response syndrome (SIRS). MiR-126 and miR-146a were reduced in SIRS patients and miR-126, miR-146a, and miR-223 were significantly reduced in patients with sepsis [2].

4.4 Methods for the Analysis of Circulating MicroRNAs

Expression analyses of circulating miRNAs require effective isolation methods and a robust quantification of miRNA levels. The source of miRNA profiling studies can be plasma, serum, exosomes, or microvesicles. Total RNA purification with Trizol is the most commonly used method for miRNA expression studies. Novel RNA extraction methods were developed for small sized miRNA isolation [110]. But a standardized miRNA purification protocol is needed for cross-comparisons between different studies.

Today, the main miRNA profiling methods are microarrays and quantitative polymerase chain reaction (qPCR). The qPCR method is easy, inexpensive and requires small amount of miRNAs although primer design could become a problem for some miRNAs [2]. The microarray method requires more starting material and needs optimization for probes and hybridization conditions to detect many different miRNAs at the same time [2, 90]. The next-generation sequencing method allows comprehensive and accurate measurement of miRNAs. However, it is a very expensive and complicated method for miRNA profiling. Some new techniques have been developed such as electrochemical sensor technology which directly detects miRNAs without need of PCR and a labeling reaction [111]. New methods may greatly facilitate the application of circulating miRNAs as disease biomarkers.

Another important issue is the normalization of raw data. Since endogenous controls do not exist in serum, different genes or miRNAs have been used for raw data analysis of circulating miRNAs. MiR-16 was used in several studies for normalization [2]. In addition, other biomolecules in body fluids such as creatinine levels were used for normalization [110]. However, spiked-in synthetic nonhuman mature miRNAs could be ideal control for data normalization of circulating miRNAs and would, in addition, allow for absolute quantification [110].

Further research is needed to develop better reagents and standardized protocols for sample preparation and to create more accurate detection and normalization methods for this rapidly expanding field.

5 Conclusion

Several studies showed that miRNAs play a major role in many biological and pathological processes. Alternation of miRNA genes by mutation or other genetic regulatory mechanisms changes miRNA expression. These changes in miRNA expression may contribute to the pathogenesis of a wide variety of human diseases. Furthermore, specific miRNA alternations were correlated with different human disease. Consequently, the levels of miRNAs in diseased tissue or circulating miRNAs were used as diagnostic or prognostic disease biomarker.

References

1. Ha TY (2011) MicroRNAs in human diseases: from cancer to cardiovascular disease. Immune Netw 11:135–154
2. Reid G, Kirschner MB, van Zandwijk N (2011) Circulating microRNAs: association with disease and potential use as biomarkers. Crit Rev Oncol Hematol 80:193–208
3. Duan S, Mi S, Zhang W et al (2009) Comprehensive analysis of the impact of SNPs and CNVs on human microRNAs and their regulatory genes. RNA Biol 6:412–425
4. Lin CH, Li LH, Ho SF et al (2008) A large-scale survey of genetic copy number variations among Han Chinese residing in Taiwan. BMC Genet 9:92
5. Wong KK, deLeeuw RJ, Dosanjh NS et al (2007) A comprehensive analysis of common copy-number variations in the human genome. Am J Hum Genet 80:91–104
6. Miller DT, Shen Y, Weiss LA et al (2009) Microdeletion/duplication at 15q13.2q13.3 among individuals with features of autism and other neuropsychiatric disorders. J Med Genet 46:242–248
7. de Pontual L, Yao E, Callier P et al (2011) Germline deletion of the miR-17 approximately 92 cluster causes skeletal and growth defects in humans. Nat Genet 43:1026–1030
8. Calin GA, Dumitru CD, Shimizu M et al (2002) Frequent deletions and down-regulation of micro-RNA genes miR15 and miR16 at 13q14 in chronic lymphocytic leukemia. Proc Natl Acad Sci U S A 99:15524–15529
9. Corthals SL, Jongen-Lavrencic M, de Knegt Y et al (2010) Micro-RNA-15a and micro-RNA-16 expression and chromosome 13 deletions in multiple myeloma. Leuk Res 34:677–681
10. Porkka KP, Ogg EL, Saramaki OR et al (2011) The miR-15a-miR-16-1 locus is homozygously deleted in a subset of prostate cancers. Genes Chromosomes Cancer 50:499–509
11. Sonoki T, Iwanaga E, Mitsuya H et al (2005) Insertion of microRNA-125b-1, a human homologue of lin-4, into a rearranged immunoglobulin heavy chain gene locus in a patient with precursor B-cell acute lymphoblastic leukemia. Leukemia 19:2009–2010
12. Bousquet M, Quelen C, Rosati R et al (2008) Myeloid cell differentiation arrest by miR-125b-1 in myelodysplastic syndrome and acute myeloid leukemia with the t(2;11)(p21;q23) translocation. J Exp Med 205:2499–2506
13. Drake KM, Ruteshouser EC, Natrajan R et al (2009) Loss of heterozygosity at 2q37 in sporadic Wilms' tumor: putative role for miR-562. Clin Cancer Res 15:5985–5992
14. Pasmant E, de Saint-Trivier A, Laurendeau I et al (2008) Characterization of a 7.6-Mb germline deletion encompassing the NF1 locus and about a hundred genes in an NF1 contiguous gene syndrome patient. Eur J Hum Genet 16:1459–1466
15. Haller F, von Heydebreck A, Zhang JD et al (2010) Localization- and mutation-dependent microRNA (miRNA) expression signatures in gastrointestinal stromal tumours (GISTs), with a cluster of co-expressed miRNAs located at 14q32.31. J Pathol 220:71–86
16. Davidson MR, Larsen JE, Yang IA et al (2010) MicroRNA-218 is deleted and downregulated in lung squamous cell carcinoma. PLoS One 5:e12560
17. Calin GA, Ferracin M, Cimmino A et al (2005) A MicroRNA signature associated with prognosis and progression in chronic lymphocytic leukemia. N Engl J Med 353:1793–1801
18. Harnprasopwat R, Ha D, Toyoshima T et al (2010) Alteration of processing induced by a single nucleotide polymorphism in pri-miR-126. Biochem Biophys Res Commun 399:117–122
19. Jazdzewski K, Murray EL, Franssila K et al (2008) Common SNP in pre-miR-146a decreases mature miR expression and predisposes to papillary thyroid carcinoma. Proc Natl Acad Sci U S A 105:7269–7274
20. Ryan BM, Robles AI, Harris CC (2010) Genetic variation in microRNA networks: the implications for cancer research. Nat Rev Cancer 10:389–402
21. Shen J, Ambrosone CB, DiCioccio RA et al (2008) A functional polymorphism in the miR-146a gene and age of familial breast/ovarian cancer diagnosis. Carcinogenesis 29:1963–1966
22. Ye Y, Wang KK, Gu J et al (2008) Genetic variations in microRNA-related genes are novel susceptibility loci for esophageal cancer risk. Cancer Prev Res (Phila) 1:460–469
23. Clague J, Lippman SM, Yang H et al (2010) Genetic variation in MicroRNA genes and risk of oral premalignant lesions. Mol Carcinog 49:183–189
24. Guo H, Wang K, Xiong G et al (2010) A functional variant in microRNA-146a is associated with risk of esophageal squamous cell carcinoma in Chinese Han. Fam Cancer 9:599–603
25. Permuth-Wey J, Thompson RC, Burton Nabors L et al (2011) A functional polymorphism in the pre-miR-146a gene is associated with risk and prognosis in adult glioma. J Neurooncol 105:639–646

26. Xu B, Feng NH, Li PC et al (2010) A functional polymorphism in Pre-miR-146a gene is associated with prostate cancer risk and mature miR-146a expression in vivo. Prostate 70:467–472
27. Zhan JF, Chen LH, Chen ZX et al (2011) A functional variant in microRNA-196a2 is associated with susceptibility of colorectal cancer in a Chinese population. Arch Med Res 42:144–148
28. Mencia A, Modamio-Hoybjor S, Redshaw N et al (2009) Mutations in the seed region of human miR-96 are responsible for nonsyndromic progressive hearing loss. Nat Genet 41:609–613
29. Luo X, Yang W, Ye DQ et al (2011) A functional variant in microRNA-146a promoter modulates its expression and confers disease risk for systemic lupus erythematosus. PLoS Genet 7:e1002128
30. Peng S, Kuang Z, Sheng C et al (2010) Association of microRNA-196a-2 gene polymorphism with gastric cancer risk in a Chinese population. Dig Dis Sci 55:2288–2293
31. Kuhn S, Johnson SL, Furness DN et al (2011) miR-96 regulates the progression of differentiation in mammalian cochlear inner and outer hair cells. Proc Natl Acad Sci U S A 108:2355–2360
32. Felekkis K, Voskarides K, Dweep H et al (2011) Increased number of microRNA target sites in genes encoded in CNV regions. Evidence for an evolutionary genomic interaction. Mol Biol Evol 28:2421–2424
33. Gong J, Tong Y, Zhang HM et al (2012) Genome-wide identification of SNPs in microRNA genes and the SNP effects on microRNA target binding and biogenesis. Hum Mutat 33:254–263
34. Clop A, Marcq F, Takeda H et al (2006) A mutation creating a potential illegitimate microRNA target site in the myostatin gene affects muscularity in sheep. Nat Genet 38:813–818
35. Shao GC, Luo LF, Jiang SW et al (2011) A C/T mutation in microRNA target sites in BMP5 gene is potentially associated with fatness in pigs. Meat Sci 87:299–303
36. Abelson JF, Kwan KY, O'Roak BJ et al (2005) Sequence variants in SLITRK1 are associated with Tourette's syndrome. Science 310: 317–320
37. Wang G, van der Walt JM, Mayhew G et al (2008) Variation in the miRNA-433 binding site of FGF20 confers risk for Parkinson disease by overexpression of alpha-synuclein. Am J Hum Genet 82:283–289
38. Chin LJ, Ratner E, Leng S et al (2008) A SNP in a let-7 microRNA complementary site in the KRAS 3′ untranslated region increases non-small cell lung cancer risk. Cancer Res 68:8535–8540
39. Nelson HH, Christensen BC, Plaza SL et al (2010) KRAS mutation, KRAS-LCS6 polymorphism, and non-small cell lung cancer. Lung Cancer 69:51–53
40. Brendle A, Lei H, Brandt A et al (2008) Polymorphisms in predicted microRNA-binding sites in integrin genes and breast cancer: ITGB4 as prognostic marker. Carcinogenesis 29:1394–1399
41. Mishra PJ, Humeniuk R, Mishra PJ et al (2007) A miR-24 microRNA binding-site polymorphism in dihydrofolate reductase gene leads to methotrexate resistance. Proc Natl Acad Sci U S A 104:13513–13518
42. Scotto L, Narayan G, Nandula SV et al (2008) Integrative genomics analysis of chromosome 5p gain in cervical cancer reveals target over-expressed genes, including Drosha. Mol Cancer 7:58
43. Rotunno M, Zhao Y, Bergen AW et al (2010) Inherited polymorphisms in the RNA-mediated interference machinery affect microRNA expression and lung cancer survival. Br J Cancer 103:1870–1874
44. Zhang X, Yang H, Lee JJ et al (2010) MicroRNA-related genetic variations as predictors for risk of second primary tumor and/or recurrence in patients with early-stage head and neck cancer. Carcinogenesis 31:2118–2123
45. Melo SA, Moutinho C, Ropero S et al (2010) A genetic defect in exportin-5 traps precursor microRNAs in the nucleus of cancer cells. Cancer Cell 18:303–315
46. Melo SA, Ropero S, Moutinho C et al (2009) A TARBP2 mutation in human cancer impairs microRNA processing and DICER1 function. Nat Genet 41:365–370
47. Yang H, Dinney CP, Ye Y et al (2008) Evaluation of genetic variants in microRNA-related genes and risk of bladder cancer. Cancer Res 68:2530–2537
48. Horikawa Y, Wood CG, Yang H et al (2008) Single nucleotide polymorphisms of microRNA machinery genes modify the risk of renal cell carcinoma. Clin Cancer Res 14: 7956–7962
49. Liang D, Meyer L, Chang DW et al (2010) Genetic variants in microRNA biosynthesis pathways and binding sites modify ovarian cancer risk, survival, and treatment response. Cancer Res 70:9765–9776
50. Sato F, Tsuchiya S, Meltzer SJ et al (2011) MicroRNAs and epigenetics. FEBS J 278: 1598–1609
51. Pigazzi M, Manara E, Baron E et al (2009) miR-34b targets cyclic AMP-responsive element binding protein in acute myeloid leukemia. Cancer Res 69:2471–2478

52. Roman-Gomez J, Agirre X, Jimenez-Velasco A et al (2009) Epigenetic regulation of microRNAs in acute lymphoblastic leukemia. J Clin Oncol 27:1316–1322
53. Tang JT, Wang JL, Du W et al (2011) MicroRNA 345, a methylation-sensitive microRNA is involved in cell proliferation and invasion in human colorectal cancer. Carcinogenesis 32:1207–1215
54. Brueckner B, Stresemann C, Kuner R et al (2007) The human let-7a-3 locus contains an epigenetically regulated microRNA gene with oncogenic function. Cancer Res 67:1419–1423
55. Lu L, Katsaros D, de la Longrais IA et al (2007) Hypermethylation of let-7a-3 in epithelial ovarian cancer is associated with low insulin-like growth factor-II expression and favorable prognosis. Cancer Res 67:10117–10122
56. Langevin SM, Stone RA, Bunker CH et al (2011) MicroRNA-137 promoter methylation is associated with poorer overall survival in patients with squamous cell carcinoma of the head and neck. Cancer 117:1454–1462
57. Lujambio A, Calin GA, Villanueva A et al (2008) A microRNA DNA methylation signature for human cancer metastasis. Proc Natl Acad Sci U S A 105:13556–13561
58. Wang P, Chen L, Zhang J et al (2013) Methylation-mediated silencing of the miR-124 genes facilitates pancreatic cancer progression and metastasis by targeting Rac1. Oncogene. doi:10.1038/onc.2012.598
59. Gebauer K, Peters I, Dubrowinskaja N et al (2013) Hsa-mir-124-3 CpG island methylation is associated with advanced tumours and disease recurrence of patients with clear cell renal cell carcinoma. Br J Cancer 108:131–138
60. Asuthkar S, Velpula KK, Chetty C et al (2012) Epigenetic regulation of miRNA-211 by MMP-9 governs glioma cell apoptosis, chemosensitivity and radiosensitivity. Oncotarget 3:1439–1454
61. Geng J, Luo H, Pu Y et al (2012) Methylation mediated silencing of miR-23b expression and its role in glioma stem cells. Neurosci Lett 528:185–189
62. Hulf T, Sibbritt T, Wiklund ED et al (2012) Epigenetic-induced repression of microRNA-205 is associated with MED1 activation and a poorer prognosis in localized prostate cancer. Oncogene. doi:10.1038/onc.2012.300
63. Li Y, Kong D, Ahmad A et al (2012) Epigenetic deregulation of miR-29a and miR-1256 by isoflavone contributes to the inhibition of prostate cancer cell growth and invasion. Epigenetics 7:940–949
64. Augoff K, McCue B, Plow EF et al (2012) miR-31 and its host gene lncRNA LOC554202 are regulated by promoter hypermethylation in triple-negative breast cancer. Mol Cancer 11:5
65. Minor J, Wang X, Zhang F et al (2012) Methylation of microRNA-9 is a specific and sensitive biomarker for oral and oropharyngeal squamous cell carcinomas. Oral Oncol 48:73–78
66. Incoronato M, Urso L, Portela A et al (2011) Epigenetic regulation of miR-212 expression in lung cancer. PLoS One 6:e27722
67. Munker R, Calin GA (2011) MicroRNA profiling in cancer. Clin Sci (Lond) 121:141–158
68. Lebanony D, Benjamin H, Gilad S et al (2009) Diagnostic assay based on hsa-miR-205 expression distinguishes squamous from nonsquamous non-small-cell lung carcinoma. J Clin Oncol 27:2030–2037
69. Del Vescovo V, Cantaloni C, Cucino A et al (2011) miR-205 Expression levels in nonsmall cell lung cancer do not always distinguish adenocarcinomas from squamous cell carcinomas. Am J Surg Pathol 35:268–275
70. Navarro A, Gaya A, Martinez A et al (2008) MicroRNA expression profiling in classic Hodgkin lymphoma. Blood 111:2825–2832
71. Ma L, Teruya-Feldstein J, Weinberg RA (2007) Tumour invasion and metastasis initiated by microRNA-10b in breast cancer. Nature 449:682–688
72. Tavazoie SF, Alarcon C, Oskarsson T et al (2008) Endogenous human microRNAs that suppress breast cancer metastasis. Nature 451:147–152
73. Eitan R, Kushnir M, Lithwick-Yanai G et al (2009) Tumor microRNA expression patterns associated with resistance to platinum based chemotherapy and survival in ovarian cancer patients. Gynecol Oncol 114:253–259
74. Furer V, Greenberg JD, Attur M et al (2010) The role of microRNA in rheumatoid arthritis and other autoimmune diseases. Clin Immunol 136:1–15
75. Stanczyk J, Ospelt C, Karouzakis E et al (2011) Altered expression of microRNA-203 in rheumatoid arthritis synovial fibroblasts and its role in fibroblast activation. Arthritis Rheum 63:373–381
76. De Smaele E, Ferretti E, Gulino A (2010) MicroRNAs as biomarkers for CNS cancer and other disorders. Brain Res 1338:100–111
77. De Pietri Tonelli D, Pulvers JN, Haffner C et al (2008) miRNAs are essential for survival and differentiation of newborn neurons but not for expansion of neural progenitors during early neurogenesis in the mouse embryonic neocortex. Development 135:3911–3921
78. Mukai J, Dhilla A, Drew LJ et al (2008) Palmitoylation-dependent neurodevelopmental deficits in a mouse model of 22q11 microdeletion. Nat Neurosci 11:1302–1310
79. Tufekci KU, Genc S, Genc K (2011) The endotoxin-induced neuroinflammation

model of Parkinson's disease. Parkinsons Dis 2011:487450
80. Jeyaseelan K, Lim KY, Armugam A (2008) MicroRNA expression in the blood and brain of rats subjected to transient focal ischemia by middle cerebral artery occlusion. Stroke 39: 959–966
81. Liu DZ, Tian Y, Ander BP et al (2010) Brain and blood microRNA expression profiling of ischemic stroke, intracerebral hemorrhage, and kainate seizures. J Cereb Blood Flow Metab 30:92–101
82. Ouyang YB, Lu Y, Yue S et al (2012) miR-181 regulates GRP78 and influences outcome from cerebral ischemia in vitro and in vivo. Neurobiol Dis 45:555–563
83. Zeng L, Liu J, Wang Y et al (2011) MicroRNA-210 as a novel blood biomarker in acute cerebral ischemia. Front Biosci (Elite Ed) 3:1265–1272
84. Forero DA, van der Ven K, Callaerts P et al (2010) miRNA genes and the brain: implications for psychiatric disorders. Hum Mutat 31:1195–1204
85. Chen H, Wang N, Burmeister M et al (2009) MicroRNA expression changes in lymphoblastoid cell lines in response to lithium treatment. Int J Neuropsychopharmacol 12: 975–981
86. Dorn GW II (2011) MicroRNAs in cardiac disease. Transl Res 157:226–235
87. van Rooij E, Sutherland LB, Liu N et al (2006) A signature pattern of stress-responsive microRNAs that can evoke cardiac hypertrophy and heart failure. Proc Natl Acad Sci U S A 103:18255–18260
88. Hullinger TG, Montgomery RL, Seto AG et al (2012) Inhibition of miR-15 protects against cardiac ischemic injury. Circ Res 110:71–81
89. Lawrie CH, Gal S, Dunlop HM et al (2008) Detection of elevated levels of tumour-associated microRNAs in serum of patients with diffuse large B-cell lymphoma. Br J Haematol 141:672–675
90. Chen X, Ba Y, Ma L et al (2008) Characterization of microRNAs in serum: a novel class of biomarkers for diagnosis of cancer and other diseases. Cell Res 18:997–1006
91. Yuan A, Farber EL, Rapoport AL et al (2009) Transfer of microRNAs by embryonic stem cell microvesicles. PLoS One 4:e4722
92. Kosaka N, Iguchi H, Yoshioka Y et al (2010) Secretory mechanisms and intercellular transfer of microRNAs in living cells. J Biol Chem 285:17442–17452
93. Zernecke A, Bidzhekov K, Noels H et al (2009) Delivery of microRNA-126 by apoptotic bodies induces CXCL12-dependent vascular protection. Sci Signal 2:ra81
94. Vickers KC, Palmisano BT, Shoucri BM et al (2011) MicroRNAs are transported in plasma and delivered to recipient cells by high-density lipoproteins. Nat Cell Biol 13:423–433
95. Arroyo JD, Chevillet JR, Kroh EM et al (2011) Argonaute2 complexes carry a population of circulating microRNAs independent of vesicles in human plasma. Proc Natl Acad Sci U S A 108:5003–5008
96. Hunter MP, Ismail N, Zhang X et al (2008) Detection of microRNA expression in human peripheral blood microvesicles. PLoS One 3:e3694
97. Hu Z, Chen X, Zhao Y et al (2010) Serum microRNA signatures identified in a genome-wide serum microRNA expression profiling predict survival of non-small-cell lung cancer. J Clin Oncol 28:1721–1726
98. Rabinowits G, Gercel-Taylor C, Day JM et al (2009) Exosomal microRNA: a diagnostic marker for lung cancer. Clin Lung Cancer 10:42–46
99. Asaga S, Kuo C, Nguyen T et al (2011) Direct serum assay for microRNA-21 concentrations in early and advanced breast cancer. Clin Chem 57:84–91
100. Heneghan HM, Miller N, Lowery AJ et al (2010) Circulating microRNAs as novel minimally invasive biomarkers for breast cancer. Ann Surg 251:499–505
101. Roth C, Rack B, Muller V et al (2010) Circulating microRNAs as blood-based markers for patients with primary and metastatic breast cancer. Breast Cancer Res 12:R90
102. Zhu W, Qin W, Atasoy U et al (2009) Circulating microRNAs in breast cancer and healthy subjects. BMC Res Notes 2:89
103. Heneghan HM, Miller N, Kelly R et al (2010) Systemic miRNA-195 differentiates breast cancer from other malignancies and is a potential biomarker for detecting noninvasive and early stage disease. Oncologist 15: 673–682
104. Resnick KE, Alder H, Hagan JP et al (2009) The detection of differentially expressed microRNAs from the serum of ovarian cancer patients using a novel real-time PCR platform. Gynecol Oncol 112:55–59
105. Taylor DD, Gercel-Taylor C (2008) MicroRNA signatures of tumor-derived exosomes as diagnostic biomarkers of ovarian cancer. Gynecol Oncol 110:13–21
106. Nugent M, Miller N, Kerin MJ (2011) MicroRNAs in colorectal cancer: function, dysregulation and potential as novel biomarkers. Eur J Surg Oncol 37:649–654
107. Fichtlscherer S, De Rosa S, Fox H et al (2010) Circulating microRNAs in patients with coronary artery disease. Circ Res 107: 677–684

108. Makino K, Jinnin M, Kajihara I et al (2012) Circulating miR-142-3p levels in patients with systemic sclerosis. Clin Exp Dermatol 37:34–39
109. Wang G, Tam LS, Li EK et al (2011) Serum and urinary free microRNA level in patients with systemic lupus erythematosus. Lupus 20:493–500
110. Etheridge A, Lee I, Hood L et al (2011) Extracellular microRNA: a new source of biomarkers. Mutat Res 717:85–90
111. Lusi EA, Passamano M, Guarascio P et al (2009) Innovative electrochemical approach for an early detection of microRNAs. Anal Chem 81:2819–2822

Chapter 4

Introduction to Bioinformatics

Tolga Can

Abstract

Bioinformatics is an interdisciplinary field mainly involving molecular biology and genetics, computer science, mathematics, and statistics. Data intensive, large-scale biological problems are addressed from a computational point of view. The most common problems are modeling biological processes at the molecular level and making inferences from collected data. A bioinformatics solution usually involves the following steps:

- Collect statistics from biological data.
- Build a computational model.
- Solve a computational modeling problem.
- Test and evaluate a computational algorithm.

This chapter gives a brief introduction to bioinformatics by first providing an introduction to biological terminology and then discussing some classical bioinformatics problems organized by the types of data sources. Sequence analysis is the analysis of DNA and protein sequences for clues regarding function and includes subproblems such as identification of homologs, multiple sequence alignment, searching sequence patterns, and evolutionary analyses. Protein structures are three-dimensional data and the associated problems are structure prediction (secondary and tertiary), analysis of protein structures for clues regarding function, and structural alignment. Gene expression data is usually represented as matrices and analysis of microarray data mostly involves statistics analysis, classification, and clustering approaches. Biological networks such as gene regulatory networks, metabolic pathways, and protein–protein interaction networks are usually modeled as graphs and graph theoretic approaches are used to solve associated problems such as construction and analysis of large-scale networks.

Key words Bioinformatics, Sequence analysis, Structure analysis, Microarray data analysis, Biological networks

1 A Brief Introduction to Biological Terminology

The most abundant data in Bioinformatics consists of DNA sequences. DNA is composed of four bases, A, G, C, and T. A DNA sequence can be a coding or a noncoding sequence. There can be several thousand bases in a gene and several million bases in

Malik Yousef and Jens Allmer (eds.), *miRNomics: MicroRNA Biology and Computational Analysis*, Methods in Molecular Biology, vol. 1107, DOI 10.1007/978-1-62703-748-8_4, © Springer Science+Business Media New York 2014

a typical bacterial genome [1]. There are about 3.2 billion bases in the human genome. Protein sequences are sequences of amino acids and there are 20 different types of basic amino acids in nature making a protein sequence a character string composed of a 20-letter alphabet. An average length protein consists of about 300 amino acids [1]. There are millions of known protein sequences. The genotype, i.e., the DNA and the corresponding protein sequences, determines the phenotype (the appearance of an organism and how it performs its functions). The structure of a protein plays an important role in determining its function. Prediction of protein structure from primary protein sequence is one of the biggest unsolved challenges in biophysics and bioinformatics today. There are ab initio techniques that are based on energy-minimization, and there are knowledge-based techniques such as homology modeling and threading.

Gene expression is the process of using the information in the DNA to synthesize a messenger-RNA (transcription). The transcribed messenger RNA (mRNA) is then used to synthesize the corresponding protein via translation. This whole work-flow is also known as the central dogma of molecular biology. Regulation of gene expression, in other words, determining which genes are expressed in a specific tissue or condition and which genes are not, is also a challenge in molecular biology. The area of transcriptomics deals with this problem and latest advances in biotechnology allow researchers to assay gene expression on a large scale in a high-throughput manner. Microarrays (or gene expression arrays, or gene chips) provide massive amount of data about gene activity under a specific condition or in a certain biological sample. Interaction assays also provide information about which proteins interact with each other and allow researchers to gain a systems level view of the cell. Interaction data accumulated in various databases such as MINT [2], Intact [3], and DIP [4] continuously grow and methods that are able to analyze large-scale graphs which can represent this type of information are discussed in Subheading 5.

Analysis of molecular biology data has many challenges such as redundancy and multiplicity, noise, and incompleteness. There are thousands of data sources for molecular biology data. As of January 2012, the Nucleic Acids Research (NAR) online database collection available at http://www.oxfordjournals.org/nar/database/a/ lists 1,380 carefully selected databases covering aspects of molecular and cell biology. The most popular databases are NCBI Entrez [5], EBI Ensemble [6], UCSC Genome Browser [7], and KEGG [8] databases. Although there are initiatives for data standardization such as the Gene Ontology [9] consortium, most of the databases have their own data format and one has to learn specifics of a database to be able to use the data effectively.

2 Sequence Analysis

In this section, the main sequence analysis problems in Bioinformatics are described, such as pairwise sequence alignment, multiple sequence alignment, pattern search, and construction of phylogenetic trees.

2.1 Pairwise Sequence Alignment

Pairwise sequence alignment is the problem of identifying similarities and differences between a pair of DNA or protein sequences [1]. The common assumption is that the sequences have diverged from a common ancestor and the alignment result shows us the conserved (i.e., unchanged) regions and regions that are diverged by evolutionary events such as mutations, insertions, inversions, duplications, and deletions. Figure 1 shows an alignment of two DNA sequences.

The level of similarity between two sequences is measured using a scoring mechanism applied on the obtained alignment. A simple scoring mechanism involves three parameters: match score, mismatch penalty, and gap penalty. Biologically more accurate scoring mechanisms give different match scores to different types of matched amino acids (e.g., scoring matrices such as BLOSUM62 and PAM250) and penalize the opening and extension of a gaps differently (i.e., affine gap penalties). Given a scoring mechanism the goal of sequence alignment is to align two given input sequences such that the obtained score is optimal with respect to the scoring function. The well-known dynamic programming solutions to the sequence alignment problem by Needleman–Wunsch [10] for global alignment and Smith–Waterman [11] for local alignment find the optimal alignment between two sequences in polynomial time ($O(mn)$, where m and n are the lengths of the input sequences). The two variations of the sequence alignment problem, the global alignment problem and the local alignment problem, have very similar dynamic programming solutions. In global alignment every nucleotide or amino

```
AGGCTATCACCTGACCTCCAGGCCGATGCCC
TAGCTATCACGACCGCGGTCGATTTGCCCGAC

-AGGCTATCACCTGACCTCCAGGCCGA--TGCCC---
TAG-CTATCAC--GACCGC--GGTCGATTTGCCCGAC
```

Fig. 1 Alignment of two DNA sequences of same length. The unaligned sequences are shown at the *top* and the aligned sequences are shown at the *bottom*. The *dashes* show the deletions, i.e., gaps, in the respective sequence. Likewise, the nucleotide in the other sequence matched against a gap can be considered insertion. As the two events cannot be differentiated easily, they are usually referred to as indels

acid in each sequence has to be aligned against an amino acid or a gap, whereas in local alignment some parts of one or both sequences may be ignored and a local sub-sequence alignment can be obtained. The dynamic programming solution to the sequence alignment problem makes use of the observation that the alignment score is additive and does not have non-local terms. Therefore, the alignment problem can be divided into subproblems and the scores of the subproblems can be added to obtain the score of the overall alignment. Using this observation one can construct the following recurrence relations to optimally align a pair of DNA or protein sequences for global alignment (Eq. 1).

$$\begin{aligned} F(i,0) &= \sum s(A(k),-) \quad 0 \le k \le i \\ F(0,j) &= \sum s(-,B(k)) \quad 0 \le k \le j \\ F(i,j) &= \max \begin{bmatrix} F(i,j-1)+s(-,B(j)), \\ F(i-1,j)+s(A(i),-), \\ F(i-1,j-1)+s(A(i),B(j)) \end{bmatrix} \end{aligned} \tag{1}$$

Equation 1: the recurrence relations for the dynamic programming solution to global sequence alignment.

In the recurrence relations given above, A and B denote the input sequences and $F(i,j)$ denotes the score of the optimum alignment of the prefixes of A and B including the first i and j characters, respectively. The scoring function $s(x,y)$ calculates the individual score for aligning particular amino acids or nucleotides, x and y, or alignment against a gap, i.e., $s(x,-)$. The recurrence relations above are for the scoring mechanism which assumes a linear gap penalty. The recurrence relations can be modified easily for the affine gap model [1].

Using the recurrence relations given above, one can fill a table named partial scores table by using $F(0,0)=0$ and the first two equations above two initialize the first column and row of the table in case of global alignment. The subsequent entries in the table can be easily filled in by using the third equation and at the end of algorithm execution, the entry $F(m,n)$ indicates the optimum score of globally aligning A and B. In order to construct the actual alignment which shows which nucleotide or amino acid is matched against which, one can trace back the partial scores table and follow a path from $F(m,n)$ to $F(0,0)$ by tracing the cells that determine the value at a particular cell $F(i,j)$. In other words, by noting which of the three terms inside the max function is used to find the maximum value, we can construct the alignment easily. Figure 2 shows a completed partial scores table and the corresponding alignment. The alignment path is indicated as arrows on the partial scores table. The overall optimum score of the alignment is 2. A linear gap penalty of −8 is used in this example. The match scores and mismatch penalties are taken from a scoring matrix which gives different scores and penalties to different types of amino acids.

	-	H	E	A	G	A	W	G	H	E	E
-	0	-8	-16	-24	-32	-40	-48	-56	-64	-72	-80
P	-8	-2	-8	-16	-24	-33	-42	-49	-57	-65	-73
A	-16	-10	-3	-4	-12	-19	-28	-36	-44	-52	-60
W	-24	-18	-11	-6	-7	-15	-4	-12	-21	-29	-37
H	-32	-14	-18	-13	-8	-9	-12	-6	-2	-11	-19
E	-40	-22	-8	-16	-16	-9	-12	-14	-6	4	-5
A	-48	-30	-16	-3	-11	-11	-12	-12	-14	-4	2
E	-56	-38	-24	-11	-6	-12	-14	-15	-12	-8	2

Optimal alignment: HEAGAWGHE-E
-P--AW-HEAE

Fig. 2 The completed partial scores table for the global alignment of two sequences. The alignment path is indicated with *arrows*. The table is filled by using Eq. 1

The local sequence alignment problem can be solved by a small modification to the dynamic programming recurrence relations as shown in Eq. 2 below.

$$
\begin{aligned}
F(i,0) &= 0 \\
F(0,j) &= 0 \\
F(i,j) &= \max \begin{bmatrix} 0 \\ F(i,j-1)+s(-,B(j)), \\ F(i-1,j)+s(A(i),-), \\ F(i-1,j-1)+s(A(i),B(j)) \end{bmatrix}
\end{aligned}
\qquad (2)
$$

Equation 2: the recurrence relations for the dynamic programming solution to local sequence alignment.

The modifications ensure that there are no negative values in the partial scores table; hence, once can start an alignment anywhere within the sequences and end the alignment anywhere. Figure 3 shows the partial scores table for a local alignment of two protein sequences. Linear gap penalty with a gap penalty of −1 is used. Match score is 4 and mismatch score is −2 in this example. The optimum local alignment score is 14 as highlighted by a circle and the alignment path is indicated by arrows.

The running time and space complexities of both algorithms are $O(mn)$. Although polynomial, the quadratic time complexity may be a big bottleneck when aligning whole genomes and when performing database searches. Therefore, algorithms which use several indexing mechanisms and heuristics have been developed among which BLAST [12] and FASTA [13] are two prominent ones which provide faster ways to perform larger-scale alignments.

	-	E	Q	L	L	K	A	L	E	F	K	L
-	0	0	0	0	0	0	0	0	0	0	0	0
K	0	0	0	0	0	4	3	2	1	0	4	3
V	0	0	0	0	0	3	2	1	0	0	3	2
L	0	0	0	4	4	3	2	6	5	4	3	7
E	0	4	3	3	3	2	1	5	10	9	8	7
F	0	3	2	2	2	1	0	4	9	14	13	12
G	0	2	1	1	1	0	0	3	8	13	12	11
Y	0	1	0	0	0	0	0	2	7	12	11	10

Optimal alignment: KA-LEF
K-VLEF

Fig. 3 The completed partial scores table for the local alignment of two sequences. The alignment path is indicated with *arrows*. The table is filled by using Eq. 2

These algorithms do not guarantee optimality; however, they are widely used in practice due to the interactive running-time performances they provide. Another reason for their popularity is that they provide statistical significance measures such as an *E*-value to indicate a level of significance for the alignments.

2.2 Multiple Sequence Alignment

Quite often biologists need to align multiple related sequences simultaneously. The problem of aligning three or more DNA or protein sequences is called the multiple sequence alignment (MSA) problem. MSA tools are one of the most essential tools in molecular biology. Biologists use MSA tools for finding highly conserved subregions or embedded patterns within a set of biological sequences, for estimating evolutionary distance between sequences, and for predicting protein secondary/tertiary structure. An example alignment of eight short protein sequences is shown in Fig. 4.

In the example alignment (Fig. 4), conserved regions and patterns are marked with different shades of gray. Extending the dynamic programming solution for pairwise sequence alignment to multiple sequence alignment, results in a computationally expensive algorithm. For three sequences of length n, the running time complexity is $7n^3$ which is $O(n^3)$. For k sequences, in order to run the dynamic programming solution a k-dimensional matrix needs to be built which results in a running time complexity of $(2^k - 1)(n^k)$ which is $O(2^k n^k)$. Therefore, the dynamic programming approach for alignment between k sequences is impractical due to exponential running time and can only be used for few or very short sets of sequences. Because of this drawback several heuristic solutions have been developed to solve the multiple sequence alignment problem suboptimally in a reasonable amount of time.

Fig. 4 Conserved residues, regions, patterns in a multiple alignment of eight protein sequences are *highlighted*

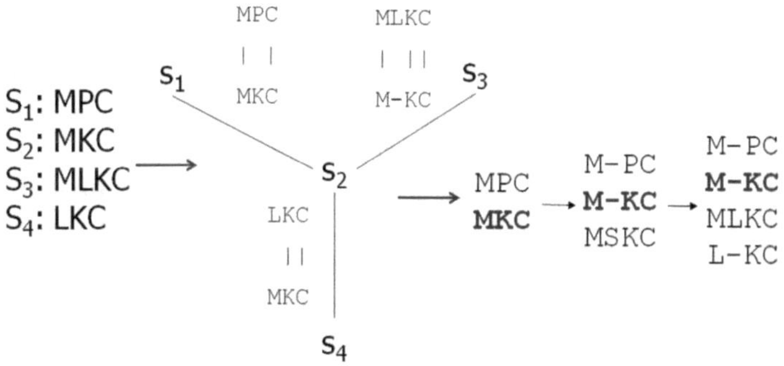

Fig. 5 Star alignment of four sequences. The center sequence is S_2. Each of the other sequences is aligned pairwise to S_2 and these alignments are combined one by one to get the multiple sequence alignment

Most of these heuristics use pairwise alignments between the input sequences to build a multiple sequence alignment progressively. These heuristics are named "progressive alignment approaches." The simplest of progressive alignment methods is the *star alignment* [14]. In star alignment, one of the input sequences is selected as the center and the pairwise alignments of the center sequence to the rest of the sequences are used to construct a multiple sequence alignment progressively. The center sequence is chosen as the most similar sequence to all of the other sequences. One may perform an all-to-all pairwise alignment of input sequences for this purpose and choose the sequence which maximizes the total score of alignments to that sequence. After the center sequence is chosen, the pairwise alignments to the center sequence can be written one after another by using the center sequence as the reference in the alignment. A star alignment example is given in Fig. 5.

In this example, S_2 is selected as the center sequence and the pairwise alignments shown on the left are used to build the multiple sequence alignment shown on the lower right. If the average length of k input sequences is n, the cost of finding the center

All-to-all pairwise alignments

	S_1	S_2	S_3	S_4
S_1	0	11	3	6
S_2		0	7	2
S_3			0	6
S_4				0

Distance Matrix

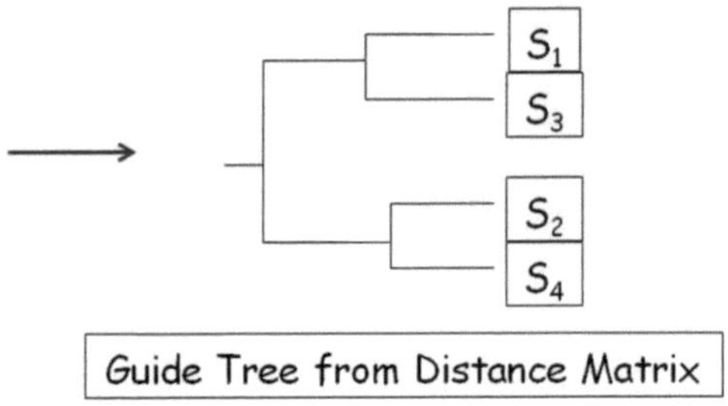

Fig. 6 Multiple alignment of four sequences using a guide tree. The guide tree shows that we should align S_1 to S_3 and S_2 to S_4 and then align the resulting pairwise alignments to get the multiple sequence alignment

sequence is $O(k^2n^2)$, which constitutes the overall cost of the star alignment. The biggest disadvantage of the progressive alignment approach is that decisions made early in the iterations are fixed and propagated to the final alignment. If one makes an incorrect alignment decision due to not foreseeing the rest of the sequences, this error is going to appear in the final alignment (once a gap always a gap). Therefore, the order of pairwise alignments in the progressive alignment approach is very important and affects the final quality of the alignment. Aligning more similar sequences in the early iterations is more likely to produce more accurate alignments. Better progressive alignment techniques, such as ClustalW [15], use a *guide tree*, which is a schedule of pairwise alignments to build the multiple sequence alignment. ClustalW is one of the most popular progressive alignment methods. The guide tree is built using the neighbor joining method [16] which is a method for building phylogenetic trees. Figure 6 shows the main components of ClustalW for an example run of four sequences.

The distance matrix on the left of Fig. 6 shows the pairwise distances between the input sequences. This distance matrix is used to build the guide tree shown on the right. The guide tree provides the schedule of progressive alignments and the multiple sequence alignment of the sequences are constructed by iterative pairwise alignments. Some of the drawbacks of the progressive approach are that they depend on pairwise sequence alignments. They produce inaccurate alignments if the sequences are very distantly related, and care must be made in choosing scoring matrices and penalties. In addition to progressive alignment approaches, there are also *iterative* approaches which make random changes in the final alignment successively as long as the alignment gets better [17].

The result of a multiple alignment of sequences is usually represented as a sequence profile. A simple sequence profile may indicate for each alignment position the composition of amino acids or nucleotides at that position. There are also more elaborate statistical methods to represent profiles, such as Hidden Markov Models (HMMs).

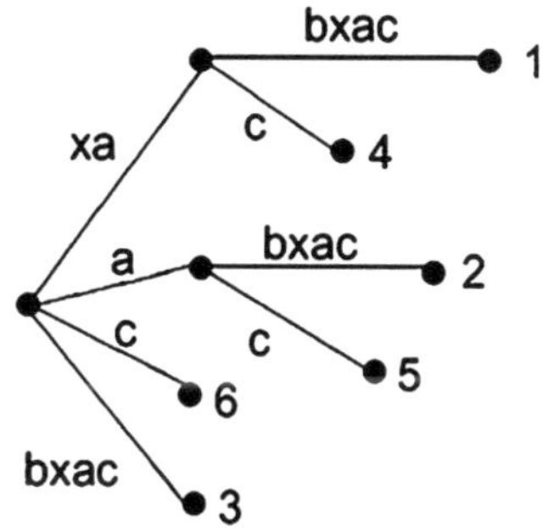

Fig. 7 The suffix tree for the string *xabxac*. The root of the tree is on the *left*. Each suffix can be traced from the root to the corresponding leaf node by a unique path. The leaf node number indicates the corresponding suffix

2.3 Efficient Pattern Matching Using Suffix Trees

Some problems in sequence analysis involve searching a small sequence (a pattern) in a large sequence database, finding the longest common subsequence of two sequences, and finding an oligonucleotide sequence specific to a gene sequence in a set of thousands of genes. Finding a pattern P of length m in a sequence S of length n can be solved simply with a scan of the string S in $O(mn)$ time. However, when S is very long and we want to perform many pattern searches, it would be desirable to have a search algorithm that could take $O(m)$ time. To facilitate this running time we have to preprocess S. The preprocessing step is especially useful in scenarios where the sequence is relatively constant over time (e.g., a genome), and when search is needed for many different patterns. In 1973, Weiner [18] introduced a data structure named suffix trees that allows searching of patterns in large sequences in $O(m)$ time. He also proposed an algorithm to construct the suffix tree in $O(n)$ time. The construction of the suffix tree is an offline onetime cost which can be ignored when multiple searches are performed. Below we give a quadratic-time construction algorithm which is easier to describe and implement. But first we formally define the suffix tree. Let S be a sequence of length n over a fixed alphabet Σ. In biological applications the alphabet usually consists of the four nucleotides for DNA sequences or the 20 amino acids for protein sequences. A suffix tree for S is a tree with n leaves (representing n suffixes) and the following properties:

1. Every internal node other than the root has at least two children.
2. Every edge is labeled with a nonempty substring of S.
3. The edges leaving a given node have labels starting with different letters.
4. The concatenation of the labels of the path from the root to the leaf i spells out the ith suffix of S. We denote the ith suffix by S_i. S_i is the substring of S from the ith character to the last character of S.

Figure 7 shows an example suffix tree which is constructed for the sequence *xabxac*.

Note that if a suffix is a prefix of another suffix we cannot have a tree with the properties defined above. Therefore, we introduce a terminal characters at the end of the sequences, such as \$, which does not exist in the input alphabet. The following quadratic time algorithm can be used to construct a suffix tree for a given input sequence S of length n.

1. Start with a root and a leaf numbered 1, connected by an edge labeled $S\$$.
2. Enter suffixes $S_2\$$, $S_3\$$;...; $S_n\$$ into the tree as follows:
   (a) To insert $S_i = S[i \ldots n]\$$, follow the path from the root matching characters of S_i until the first mismatch at character $S_i[j]$ (which is bound to happen).
   (b) If the matching cannot continue from a node, denote that node by w.
   (c) If the mismatch occurs at the middle of an edge e with a label $S[u \ldots v]$, split e and create a new node w. Replace e by two edges $S[u \ldots u+k-1]$ and $S[u+k \ldots v]$ if the mismatch occurs at character $S[k]$.
   (d) Create a new leaf numbered i, and connect w to it by an edge labeled with $Si[j \ldots |S_i|]$.

Given a sequence S and a pattern P, we search for all occurrences of P in S using the observation that each occurrence of P has to be a prefix of some suffix in S. Each such prefix corresponds to a path starting at the root. So, we try to match P on a path, starting from the root. There are three possible cases.

1. The pattern does not match along the path $\rightarrow$ P does not occur in T.
2. The pattern match ends in a node u of the tree (set $x = u$) $\rightarrow$ All leaves below x represent occurrences of P.
3. The match ends inside an edge (v,w) of the tree (set $x = w$) $\rightarrow$ All leaves below x represent occurrences of P.

It is important to note that the suffix tree method described above performs exact pattern searches. In certain biological problems, where one is looking for similarities rather than exact occurrences, approximate string matching may be sought. In such cases, approximate extensions of suffix trees or other probabilistic methods should be used [19].

2.4 Construction of Phylogenetic Trees

Phylogeny is a term coined by Haeckel in 1866 [20] which means the line of descent or evolutionary development of any plant or animal species or the origin and evolution of a division, group or race of animals or plants. Phylogenetic analyses can be used for understanding evolutionary history like the origin of species, assist in epidemiology of infectious diseases or genetic defects, aid in

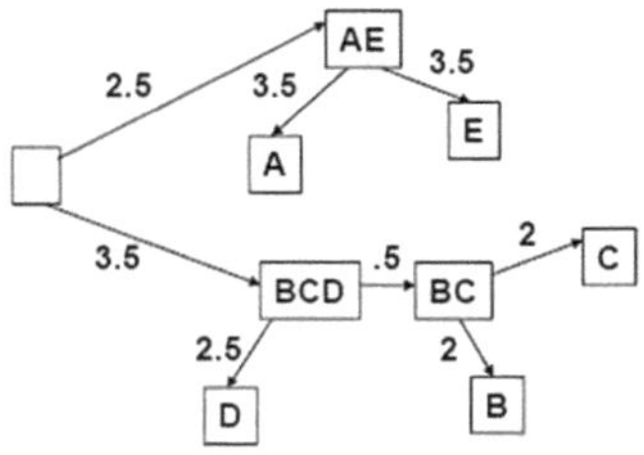

	A	B	C	D	E
A	0	10	12	8	7
B		0	4	4	14
C			0	6	16
D				0	12
E					0

	A	BC	D	E
A	0	11	8	7
BC		0	5	15
D			0	12
E				0

→

	A	BCD	E
A	0	10	7
BCD		0	14
E			0

→

	AE	BCD
AE	0	12
BCD		0

Fig. 8 The UPGMA method to construct a phylogenetic tree for five sequences. The method works on the distance matrix on the *upper right* and constructs the tree on the *upper left*

prediction of function of novel genes, used in biodiversity studies, and can be used for understanding microbial ecologies. The data for building phylogenetic trees may be characteristics such as traits and biomolecular features, or they can be numerical distance estimates such as the edit distance between genomic sequences. The result of a phylogenetic analysis is a phylogenetic tree which shows the relationships among a group of sequences or species in a hierarchical manner. In this regard, phylogenetic trees can be considered as a hierarchical clustering. Below, we describe two distance-based methods, UPGMA [21] and neighbor joining [16], which construct a phylogenetic tree of input sequences given an all-to-all distance (or similarity) matrix between them. Both UPGMA and neighbor joining methods are agglomerative clustering methods which start with all the input sequences in separate clusters and group them into larger clusters in subsequent steps. UPGMA stands for unweighted pair-group method with arithmetic mean. At each iteration, the closest sequences (or groups of sequences) are identified and merged to form a combined larger cluster which makes it a greedy algorithm. In order to proceed, one has to define the new distance values between clusters. In UPGMA, the cluster distance is defined as the average distance of all pairwise distances between two cluster elements, one from each cluster. Figure 8 shows the resulting phylogenetic tree after application of UPGMA on five input sequences. The corresponding distance matrix is shown on the right.

In the first iteration, *B* and *C* are combined to obtain the cluster BC. The total distance of 4 between *B* and *C* is divided equally between the branches that connect *B* and *C* at the intermediate node BC. *B* and *C* are then removed from the original distance matrix and replaced by BC and the distances are updated accordingly (shown on the lower left). BC is then combined with *D* and

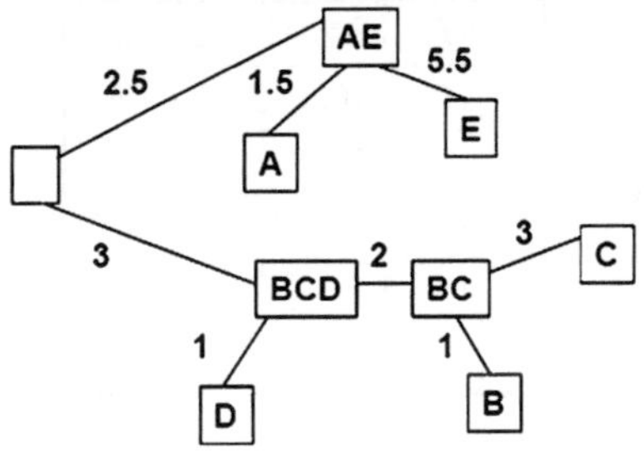

	A	B	C	D	E
A	0	10	12	8	7
B		0	4	4	14
C			0	6	16
D				0	12
E					0

Fig. 9 The phylogenetic tree constructed by the neighbor joining method on the five sequences. The distance matrix is shown on the *right* and the final NJ tree is shown on the *left*

A is combined with *E* in the following step. In the last step BCD is combined with AE. The intermediate distance matrices are shown in Fig. 8. UPGMA produces ultrametric trees, in other words the distance of every leaf to the root is equal in the tree. If the original distance matrix is ultrametric, then the UPGMA method can produce the correct tree; otherwise, it produces an approximate tree. A more accurate algorithm is the neighbor joining algorithm. The neighbor joining algorithm was developed in 1987 by Saitou and Nei [16]. At each iteration, clusters that are close to each other and far from the rest are joined. Neighbor joining always finds the correct tree if the distances are additive. However, in biological practice, the distances may not be additive; hence, this method produces approximate trees since finding the optimum tree, in which the difference between path distances and actual distances is minimum, is an NP-hard problem [22]. The neighbor joining algorithm is given below for n sequences provided as input:

1. For each node i, define u_i as summation of all the distance from i to all other nodes divided by $n-2$.
2. Iterate until two nodes are left.
3. Choose a pair of nodes (i,j) with smallest $D_{ij} - u_i - u_j$.
4. Merge the chosen node to a new node ij and update distance matrix.
   (a) $D_{k,ij} = (D_{i,k} + D_{j,k} - D_{i,j})/2$.
   (b) $D_{i,ij} = (D_{i,j} + u_i - u_j)/2$.
   (c) $D_{j,ij} = D_{i,j} - D_{i,ij}$.
5. Delete nodes i and j.
6. For the final group (i,j), use $D_{i,j}$ as the edge weight.

Figure 9 shows the neighbor joining tree for the same set of sequences as in the UPGMA example (Fig. 8). Note the differences between the two resulting trees.

3 Structure Analysis

Proteins perform their functions within the cell by their structural and physiochemical properties. Therefore, analysis of protein structure is very important to understand their function. Drugs can be designed more accurately if the three-dimensional structures of proteins are known. There are different levels of protein structure: primary structure, secondary structure, tertiary structure, and quaternary structure [1]. Primary structure is the linear sequence of amino acids. Secondary structure is the formation of local shapes such as helices or extended beta strands. Tertiary structure is the global three-dimensional shape of a single polypeptide chain. Quaternary structure is the formation of a molecular complex consisting of multiple polypeptide chains and or prosthetic groups. One of the biggest challenges of structural bioinformatics is the prediction of tertiary structure from primary structure. This problem is named as the protein folding problem, and all the methods that are known today are approximate methods which do no guarantee to find the actual true structure. Therefore, biologists mostly trust the experimentally determined structures when working with protein structures. However, the number of experimentally determined structures is less than 1 % of the known protein sequences. Determination of structure using X-ray crystallography and nucleic magnetic resonance is very time-consuming and expensive. Moreover, crystallization leads to distortion of the protein structure and thus the resulting structure is an approximation of the exact structure of the native protein. This section describes the main approaches used for protein structure prediction and then details the problem of structural alignment.

3.1 Prediction of Protein Structures from Primary Sequence

There are three main approaches for predicting protein structure from protein sequence: ab initio methods, homology modeling methods, and threading methods. In the ab initio methods structure is predicted using pure chemistry and physics knowledge without the use of other information. These techniques exhaustively search the fold space and therefore can only predict structures of small globular proteins (length < 50 amino acids). Homology modeling, on the other hand, makes use of the experimentally determined structures and uses sequence similarity to predict the structure of the target protein. The main steps of homology modeling are as follows:

1. Identify a set of template proteins (with known structures) related to the target protein. This is based on sequence similarity (BLAST, FASTA) with sequence identity of 30 % or more.
2. Align the target sequence with the template proteins. This is based on multiple sequence alignment (ClustalW). Identify conserved regions.

3. Build a model of the protein backbone, taking the backbone of the template structures (conserved regions) as a model.
4. Model the loops. In regions with gaps, use a loop modeling procedure to substitute segments of appropriate length.
5. Add side chains to the model backbone.
6. Evaluate and optimize entire structure.

In order to be able to apply homology modeling approaches, the target protein sequence has to exhibit at least 30 % sequence identity to a protein sequence with known experimental structure. To be able to predict novel sequences with no similarity to sequences with known structures, one can use threading based approaches [1]. In threading, the target sequence is aligned with a library of structures called the template library. The motivation behind threading is that if the template library covers the structural space well, with no sequence similarity present, one can evaluate the fitness of the target sequence on known structures (i.e., folds) and predict its overall structure. The threading problem involves the computation of the optimal alignment score between a given sequence and a template structure, i.e., a fold. If we can solve this problem, then given a sequence, we can try each fold and find the best structure that fits this sequence. Because there are only a few thousands folds, we can find the correct fold for the given sequence. However, it was shown that threading is an NP-hard problem [23]. Therefore, researchers have proposed heuristics which do not guarantee the optimal alignment between the sequence and the fold. The components of a full threading method include the following:

- Template library: Use structures from DB classification categories (PDB).
- Scoring function.
- Single and pairwise energy terms.
- Alignment.
- Consideration of pairwise terms leads to NP-hardness.
- Use heuristics.
- Confidence assessment.
- *Z*-score, *P*-value similar to sequence alignment statistics.
- Improving the final structure.
- Improvement by local threading and multi-structure threading.

3.2 The Structural Alignment Problem

Pairwise structural alignment of protein structures is the problem of finding similar subregions of two given protein structures. The key problem in structural alignment is to find an optimal correspondence between the arrangements of atoms in two molecular structures in order to align them in 3D. Optimality of the alignment is often determined using a root mean square measure of the distances between corresponding atoms in the two molecules although other measures have been proposed. However, the difficulty is that it is not known a priori which atom in the first molecule corresponds to which atom in the second molecule. Furthermore, the two molecules may not even have the same number of atoms. The alignment can be done at various levels such as the atom level, amino acid level, or secondary structure element (SSE) level. During comparison, one can take into account only the geometry or architecture of coordinates or relative positions of amino acids or the sequential order of residues along the backbone may be considered. Some approaches use the physiochemical properties of amino acids but most approaches just treat protein structures as geometrical shapes during comparison. Pairwise structure alignment problem is an NP-hard problem unlike pairwise sequence alignment [23]. Therefore, many different techniques have been proposed for the structural alignment problem. The most widely used measures in assessing the quality of the alignment is the length of the alignment and the RMSD between the two aligned regions. RMSD (root mean squared distance) shows how similar the identified regions are in terms of their shapes and the length of the alignment shows whether the method is capable of identifying more significant larger similarities between structures. An RMSD value of less than 3.0 Å is considered a good structural match. One of the earliest algorithms for structural alignment is an iterative dynamic programming algorithm called STRUCTAL [24]. STRUCTAL was proposed by Subbiah in 1993 and has the following steps:

1. Start with arbitrary alignment of the amino acids in the two protein structures A and B.
2. Superimpose the two structures with respect to the initial alignment in order to minimize RMSD.
3. Compute a structural alignment (SA) matrix where entry (i,j) is the score for the structural similarity between the ith amino acid of A and the jth amino acid of B. Structural similarity is some constant divided by the distance between the ith amino acid of A and the jth amino acid of B.
4. Use Dynamic Programming to compute the next alignment with gap penalty 0.
5. Iterate **steps 2–4** until the overall score converges.
6. Repeat with a number of initial alignments.

Another algorithm for structural alignment is Dali developed by Holm and Sander in 1993 [25]. Dali uses intra-atomic distance matrices to represent three-dimensional proteins structures as two-dimensional matrices. It then compares the intra-atomic distance matrices of the two input protein structures in order to find a structural alignment. There have been many other algorithms like

VAST [26], CE [27], LOCK [28], and TOPS [29] developed for structural alignment, which differ in how they represent the protein structure, extract structural features, and match matching structural features. Some algorithms find short but very similar regions, whereas some algorithms are able to detect larger regions with less similarity. The statistical theory of structural alignments is similar to that of BLAST as many methods compare the likelihood of a match as compared to a random match. However, there is less agreement regarding the score matrix; hence, the z-scores of CE, DALI, and VAST may not be compatible.

4 Microarray Data Analysis

Cells containing the same DNA are still different because of differential gene expression. Only about 40 % of human genes are expressed at any one time. A gene is expressed by transcribing DNA into single-stranded mRNA. Then, the mRNA is translated into a protein. This is known as the central dogma of molecular biology [1]. The microarray technology allows for measuring the level of mRNA expression in a high-throughput manner. Messenger RNA expression represents the dynamic aspects of the cell. In order to measure the mRNA levels, mRNA is isolated and labeled using a fluorescent material. When the mRNA is hybridized to the target, the level of hybridization corresponds to light emission which is measured with a LASER. Therefore, higher concentration means more hybridization which indicates more mRNA.

Microarrays can be used to measure mRNA levels in different tissues, different developmental stages, different disease states, and in response to different treatments. One of the main sources of microarray data is NCBI's Gene Expression Omnibus (GEO) [30]. As of May 2012, there are about 750,000 microarray samples performed in about 30,000 different experiments in the GEO. A typical microarray sample's raw data is about 10–30 Mb.

The main characteristic of microarray data are that they are extremely high dimensional where the number of genes is usually on the order of tens of thousands and the number of experiments is on the order of tens. Microarray data is considered to be noisy due to imperfect hybridization and off-target hybridization. Normalization and thresholding are important for interpreting microarray data (*see* Chapters 16 and 17). Some, mRNA levels may not be read correctly and may be missing from the final output due to a gene failing to hybridize the microarray spot. These characteristics of microarray data make data mining on this data a challenging task and having too many genes leads to many false positive identifications. For exploration purposes, a large set of all relevant genes is desired, whereas for diagnostics or identification of therapeutic targets, a small set of genes is needed.

In a microarray experiment, biologists usually seek for differentially expressed genes between two conditions such as a diseased tissue versus a normal tissue. There are well-developed statistical techniques to identify differentially expressed genes accurately, such as the significance analysis of microarrays [31]. There are also some other data mining related problems in microarray data analysis. Some clustering problems can be identified by trying to find genes with similar function, or trying to subdivide experiments or genes into meaningful classes. Classification problems also exist by trying to correctly classify an unknown experiment or gene into a known class. Also by classifying the patient's microarray sample into a known cancer subtype accurately one can make better treatment decisions for a cancer patient based on gene expression profile.

The rest of this section describes the distance measures used in clustering methods and describes a well-known clustering algorithm: the K-means algorithm [32].

The goal of a clustering algorithm is to group similar genes or samples. In order to identify similarity, one needs a distance or similarity function. The most widely accepted distance measures in microarray analysis are the Euclidean distance and the Pearson's correlation coefficient. For two n dimensional vectors, the Euclidean distance is defined as:

$$d(X,Y) = \sqrt{\sum_{i=1}^{n}(X_i - Y_i)^2} \tag{3}$$

Equation 3: the Euclidean distance between two n dimensional points X and Y.

The upregulation and downregulation behavior of a gene is relative to its basal expression level which may vary among genes. Therefore, the Euclidean distance measure may result in high distances between similarly behaving genes if their basal expression level is different. To overcome this problem, Pearson's correlation coefficient is used. Pearson's correlation coefficient for two n dimensional vectors is defined as:

$$PCC(X,Y) = \frac{\sum_{i=1}^{n}(X_i - \bar{X})(Y_i - \bar{Y})}{\sqrt{\sum_{i=1}^{n}(X_i - \bar{X})^2}\sqrt{\sum_{i=1}^{n}(Y_i - \bar{Y})^2}} \tag{4}$$

Equation 4: the Pearson's correlation coefficient between two genes X and Y with n samples

where $\bar{X}$ and $\bar{Y}$ are the means of the dimensions of the vectors.

Given a distance measure and a microarray dataset, K-means algorithm can be used to group genes or samples to a predetermined number of groups, i.e., *k*. K-means algorithm proceeds as follows:

1. Randomly assign *k* points to *k* clusters.
2. Iterate.
   (a) Assign each point to its nearest cluster (use centroid of clusters to compute distance).
   (b) After all points are assigned to clusters, compute new centroids of the clusters and reassign all the points to the cluster of the closest centroid.

In order to find the natural cluster number in the dataset, different values of *k* may be tried.

There are many other clustering techniques such are hierarchical clustering (e.g., UPGMA [21] and neighbor joining [16]) and self organizing maps (SOM) (*see* Chapter 16 and [33]). Different clustering algorithms produce different clusterings of the data and one can choose to use a clustering method based on the application requirements such as the cluster sizes.

5 Construction and Analysis of Biological Networks

Recent discoveries show that the complex biological functions of higher organisms are due to combinatorial interactions between their proteins. There are signaling pathways, metabolic pathways, protein complexes, gene regulatory networks that take part in the complex organization of the cell. There are many research projects to find the complete repertoire of interactions between proteins in an organism. There are experimental high throughput methods, like yeast-two-hybrid (Y2H) and affinity purification with mass spectrometry (APMS). There are also machine learning methods to predict interactions from indirect data sources such as genomic context, sequence similarity, and microarray expression profiles.

A collection of interactions defines a network and a pathway is a subset of it. We can define a pathway as a biological network that relates to a known physiological process or a complete function. Figure 10 shows an example of a biological network which are usually modeled as graphs.

Results of high-throughput experiments to identify interacting proteins are usually collected in databases such as DIP [4], MINT [2], and IntAct [3]. The results of low-throughput experiments can be found in literature by using text mining techniques. Hundreds of thousands of unstructured free text articles should be processed automatically to extract protein–protein interaction information. The main challenges are that there is no standard naming of genes, proteins, or processes and methods to understand natural language

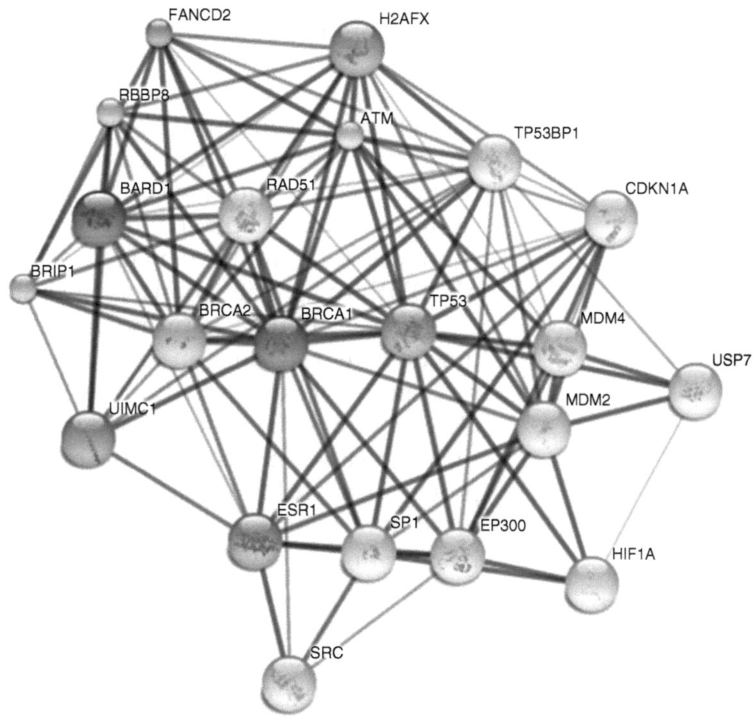

Fig. 10 Functional associations around the BRCA1 gene in human. The snapshot is created from the interactive network visualization tool from the STRING Database at string.embl.de

need to be developed to facilitate the process. The accuracy and coverage of current techniques is limited.

In order to construct large-scale biological networks, predicted and experimentally determined interactions can be combined to generate a large-scale graph of interactions. Large-scale protein–protein interaction networks can be analyzed for predicting members of a partially known protein complex or pathway, for inferring individual genes' functions on the basis of linked neighbors, for finding strongly connected components by detecting clusters to reveal unknown complexes, and for finding the best interaction path between a source and a target gene. Several clustering algorithms such as MCL (Markov clustering) [34], RNSC (restricted neighborhood search clustering) [35], SPC (super paramagnetic clustering) [36], and MCODE (molecular complex detection) [37] were developed to find clusters in biological networks. The MCL algorithm simulates a flow on the input graph and calculates successive powers of the adjacency matrix to reveal clusters in the graph. It has only one parameter which influences the number of clusters that are generated. The RNSC algorithm starts with an initial random clustering and tries to minimize a cost function by iteratively moving vertices between neighboring clusters using several parameters. SPC is a hierarchical algorithm inspired from an analogy with the physical properties of a ferromagnetic model subject to fluctuation at nonzero temperature. MCODE gives a weight to each vertex by its local neighborhood density (using a modified version of clustering coefficient using k-cores). It starts from the top weighted vertex and includes neighborhood vertices with similar weights to the cluster. MCODE has a post-processing step to remove or add new vertices. Iteratively, it continues with the next highest weight vertex in the network. MCODE may provide overlapping clusters.

In a cell, all these networks are intertwined and perform their functions in a dynamic manner. As new components, regulatory mechanisms are discovered; we are able to construct a more complete picture of the genome-wide systems biology of the cell. One such regulatory mechanism is the regulation of the translation of a protein by microRNAs. microRNAs and their targets form another network of regulatory interactions and this network can be combined with gene regulatory networks, which contain transcription factors and their targets, to get a more complete picture of regulatory mechanisms. Further integration of the regulatory mechanisms with protein–protein interaction networks may provide insights into the dynamics of signaling pathways and other processes in the cell.

References

1. Zvelebil M, Baum J (2007) Understanding bioinformatics. Garland Science, New York, NY. ISBN 978-0815340249
2. Chatr-aryamontri A, Ceol A, Palazzi LM et al (2007) MINT: the Molecular INTeraction database. Nucleic Acids Res 35(Suppl 1):D572–D574
3. Kerrien S, Aranda B, Breuza L et al (2012) The IntAct molecular interaction database in 2012. Nucleic Acids Res 40(D1): D841–D846
4. Xenarios I, Rice DW, Salwinski L et al (2000) DIP: the database of interacting proteins. Nucleic Acids Res 28:289–291
5. Maglott D, Ostell J, Pruitt KD, Tatusova T (2010) Entrez Gene: gene-centered information at NCBI. Nucleic Acids Res 33(D1): D54–D58
6. Flicek P, Amode MR, Barrell D et al (2012) Ensemble 2012. Nucleic Acids Res 40(D1): D84–D90
7. Kent WJ, Sugnet CW, Furey TS et al (2002) The human genome browser at UCSC. Genome Res 12(6):996–1006
8. Kanehisa M, Goto S, Sato Y et al (2012) KEGG for integration and interpretation of large-scale molecular datasets. Nucleic Acids Res 40(D1):D109–D114
9. The Gene Ontology Consortium (2000) Gene ontology: tool for the unification of biology. Nat Genet 25(1):25–29
10. Needleman SB, Wunsch CD (1970) A general method applicable to the search for similarities in the amino acid sequence of two proteins. J Mol Biol 48(3):443–453
11. Smith TF, Waterman MS (1981) Identification of common molecular subsequences. J Mol Biol 147:195–197
12. Altschul S, Gish W, Miller W et al (1990) Basic local alignment search tool. J Mol Biol 215(3):403–410
13. Lipman DJ, Pearson WR (1985) Rapid and sensitive protein similarity searches. Science 227(4693):1435–1441
14. Bafna V, Lawler EL, Pevzner PA (1993) Approximation algorithms for multiple sequence alignment. Theor Comput Sci 182(1–2):233–244
15. Chenna R, Sugawara H, Koike T et al (2003) Multiple sequence alignment with the Clustal series of programs. Nucleic Acids Res 31(13): 3497–3500
16. Saitou N, Nei M (1987) The neighbor-joining method: a new method for reconstructing phylogenetic trees. Mol Biol Evol 4(4): 406–425

17. Chakrabarti S, Lanczycki CJ, Panchenko AR et al (2006) State of the art: refinement of multiple sequence alignments. BMC Bioinformatics 7:499
18. Weiner, P. (1973) Linear pattern matching algorithm, *14th Annual IEEE Symposium on Switching and Automata Theory*, 15–17 October, 1973, USA, pp 1–11
19. Cobbs AL (1995) Fast approximate matching using suffix trees. Combinatorial pattern matching, vol 937, Lecture notes in computer science. Springer, New York, NY, pp 41–54
20. Haeckel E (1868) The history of creation, vol 1, 3rd edn. Trench & Co., London, Translated by E. Ray Lankester, Kegan Paul
21. Sokal R, Michener C (1958) A statistical method for evaluating systematic relationships. Univ Kans Sci Bull 38:1409–1438
22. Day WHE (1986) Computational complexity of inferring phylogenies from dissimilarity matrices. Bull Math Biol 49:461–467
23. Lathrop RH (1994) The protein threading problem with sequence amino acid interaction preferences is NP-complete. Protein Eng 7(9): 1059–1068
24. Subbiah S, Laurents DV, Levitt M (1993) Structural similarity of DNA-binding domains of bacteriophage repressors and the globin core. Curr Biol 3:141–148
25. Holm L, Sander C (1993) Protein structure comparison by alignment of distance matrices. J Mol Biol 233(1):123–138
26. Gibrat JF, Madej T, Bryant SH (1996) Surprising similarities in structure comparison. Curr Opin Struct Biol 6(3):377–385
27. Shindyalov IN, Bourne PE (1998) Protein structure alignment by incremental combinatorial extension of the optimum path. Protein Eng 11(9):739–747
28. Singh AP, Brutlag DL (1997) Hierarchical protein structure superposition using both secondary structure and atomic representations. In Proc. Fifth Int. Conf. on Intell. Sys. for Mol. Biol. AAAI Press, Menlo Park, CA, pp 284–293
29. Viksna J, Gilbert D (2001) Pattern matching and pattern discovery algorithms for protein topologies. Algorithms in bioinformatics: first international workshop, WABI 2001 proceedings, vol 2149, Lecture notes in computer science. Springer, New York, NY, pp 98–111
30. Edgar R, Domrachev M, Lash AE (2002) Gene expression omnibus: NCBI gene expression and hybridization array data repository. Nucleic Acids Res 30(1):207–210
31. Tusher VG, Tibshirani R, Chu G (2001) Significance analysis of microarrays applied to the ionizing radiation response. Proc Natl Acad Sci 98(9):5116–5121
32. Lloyd SP (1982) Least squares quantization in PCM. IEEE Trans Inform Theor 28(2): 129–137
33. Kohonen T (1982) Self-organized formation of topologically correct feature maps. Biol Cybern 43(1):59–69
34. van Dongen, S. (2000) Graph clustering by flow simulation. Ph.D. thesis, University of Utrecht, May 2000
35. King AD, Pržulj N, Jurisica I (2004) Protein complex prediction via cost-based clustering. Bioinformatics 20(17):3013–3020
36. Blatt M, Wiseman S, Domany E (1996) Superparamagnetic clustering of data. Phys Rev Lett 76:3251–3254
37. Bader GD, Hogue CW (2003) An automated method for finding molecular complexes in large protein interaction networks. BMC Bioinformatics 4:2

Chapter 5

MicroRNA and Noncoding RNA-Related Data Sources

Patrizio Arrigo

Abstract

Noncoding RNAs (ncRNAs) are ribonucleic acids capable of controlling different genetic and metabolic functions. These molecules have been recently organized into different classes, and among them microRNAs (miRNAs) are extensively studied. MicroRNAs are short oligomers mainly involved in posttranscriptional gene silencing. The specific research field, focused on structural and functional characterization of microRNAs, is commonly called *mirnomics*. The exploitation of the interest in microRNAs has stimulated the organization of several databases that are often integrated with analytical tools in order to predict microRNA targets, or to find those miRNAs capable to inhibit the expression of a specific protein. This work attempts to provide an overview of accessible information about microRNAs and other noncoding RNAs that has been gathered in curated databases.

Key words miRNA, ncRNAs, Bioinformatics, Databases

1 Introduction

Experiments of gene silencing, involving microRNAs, such as those of Fire and Mello [1] or McCaffrey [2] have opened the opportunity to study the role of these noncoding RNAs on the gene expression process. The role of these molecules in cellular homeostasis is continuously acquiring significance. These kind of ribonucleic acids are able to act as signal transducers through the inhibition of protein expression at the posttranscriptional level. The rapid advancement of molecular biology has allowed to demonstrate the existence of many different noncoding RNAs (ncRNAs). These molecules have been classified into different families, and the main classes are the following:

1. Transfer RNAs (tRNAs) which are involved in translational processes.
2. Small interfering RNAs (siRNAs) which are involved in RNA silencing.

Malik Yousef and Jens Allmer (eds.), *miRNomics: MicroRNA Biology and Computational Analysis*, Methods in Molecular Biology, vol. 1107, DOI 10.1007/978-1-62703-748-8_5,

3. MicroRNA (miRNAs) which are involved in translational silencing.
4. Small nuclear RNAs (snRNAs) which are located in the nucleus and are involved in RNA splicing.
5. Small nucleolar RNAs (snoRNAs) which are in the nucleolus and are involved in RNA modification.
6. Small Cajal-body specific RNAs (scaRNAs) which are involved in RNA modification.
7. Small guide RNAs (gRNAs) which are involved in RNA editing.
8. Piwi interacting RNAs (piRNAs) which are involved in gene silencing.

This classification underlines how ncRNAs constitute a pivotal element in modulating genetic information both at the transcriptional and the epigenetic level. In addition, they act on protein expression, influencing mRNA processing and editing. MicroRNAs are, among the previously listed classes (Fig. 1), the most extensively studied ncRNAs. A large corpus of experimental findings demonstrates the role of microRNAs in modulation of physiological

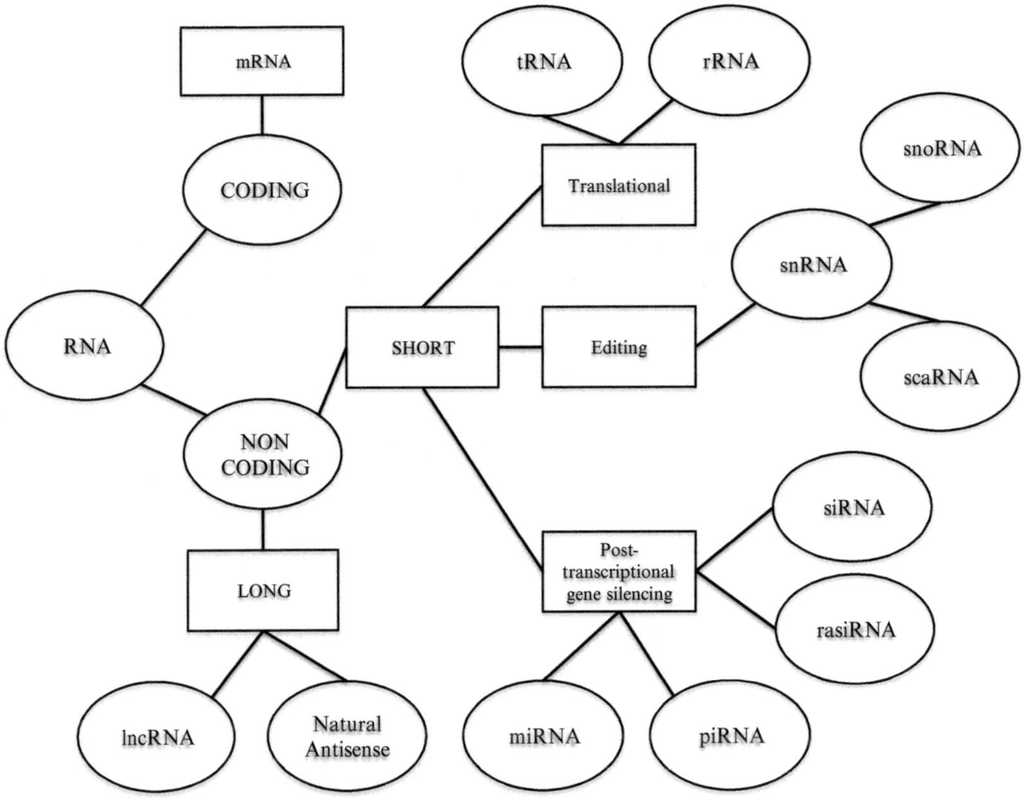

Fig. 1 RNA classification represented in a tree-like structure

processes. An altered microRNA profile has been determined in many pathological conditions [3]. One of the major difficulties that researchers must face during the analysis of microRNA function, is the multiplicity of targets of a single microRNA. To highlight this point it seems important to briefly sketch the biogenesis and anatomy of a microRNA although it has been detailed in previous chapters in this volume. MicroRNA biogenesis is commonly split into two phases: (a) nuclear phase and (b) cytoplasmic phase. In the nuclear phase, a transcriptional process produces a long ribonucleotide sequence, called primary microRNA (pri-miRNA), that is enzymatically processed [4]. The nuclear enzymatic cleavage by Drosha produces a short sequence, the precursor (pre-miRNA), which folds back on itself and creates a stem-loop structure. The pre-miRNA is thereafter shuttled into the cytoplasm by Exportin-5. The cytoplasmic phase is mainly centered on activation (maturation), targeting and degradation of microRNAs. The loading and processing of microRNA precursors, by a multi-enzyme complex called RISC, is a critical step in posttranscriptional gene silencing. Information about the maturation process is embedded in the compositional and structural features of the precursor (pre-miRNA). Compositional properties seem to influence the loading by RISC and the activation kinetic and selection of microRNA isoforms. The argonaute family includes several members and they are responsible to deliver a mature miRNA to its target mRNA. In the pre-miRNA single base mismatches, the distance from the two ends, and the terminal asymmetry seem to influence precursor stability. It is outside the scope of this paper to extensively describe the biophysical and functional properties of noncoding RNAs and some of the issues are discussed in other chapters of this volume. The miRNA–mRNA recognition is mainly dependent on the complementarity between a special nucleotide motif, embedded in the mature microRNA, called "seed," that shows a very high evolutionary conservation [5]. The "seed" motif is generally used to group microRNAs into different families. The "seed" however is not the only responsible player in miRNA–mRNA interactions. Other particular positions in the precursor sequence can have a role in target recognition and hybridization efficiency. As previously mentioned, the microRNA–target relation is a one-to-many correspondence which makes the overall silencing process a many-to-many relationship since a target can be silenced by many miRNAs independently or cooperatively. The multiplicity of putative targets is a strong constraint for their experimental validation. MicroRNA target prioritization is one crucial task in *mirnomics.* The miRNA–mRNA interaction in animals is based on imperfect matching and also on the existence of multiple potential targets. A very important factor that can influence the microRNA functionality is genetic variability. The discovery of circulating miRNAs [6] suggests that a single microRNA could be able to influence the function of other cells not only in the neighborhood of the source cell. It is out of the scope of this chapter to deeply analyze these factors; nevertheless, it is important to underline the complexity of the functional analysis of microRNAs.

In the following sections I will present a closer look at RNA data sources and will consider some of the topics introduced above in the appropriate sections.

2 Noncoding RNA Data Resources

The number of Molecular biology databases increased immensely in the last decade as demonstrated by the special database issues of Nucleic Acids Research (41(D1), 2013). In these publications there are a limited number of RNA-specific databases, but genomic information about noncoding RNA, in particular microRNAs, can be retrieved from generalist data repositories. The great interest generated by the function of noncoding RNAs, however, has stimulated the development of specialized databases. The availability of reliable specialized data repositories is a fundamental support for target recognition of miRNAs. Short ncRNAs (e.g., microRNAs or siRNAs) are extensively studied and, as consequence, more relevant databases are focused on these kinds of ncRNAs. In this paper, I would like to describe not only microRNA specific databases but also those that can give additional information useful to investigate the role of microRNAs in posttranscriptional control. Selected databases can be roughly organized into four classes: structural (Subheading 2.1), general RNA database (Subheading 2.2), specialized ncRNAs repositories (Subheading 2.3), ncRNA specific proteins (Subheading 2.4).

2.1 Structural Databases

Current information about the structure of microRNAs or other ncRNAs is rather poor. I take into account the more common structural databases (Table 1). Structural protein database (PDB), for a long time, has standardized the protocol for experimental data submission. The structural genomic consortium [7] has extended the level of integration. Despite these advances, the experimental structural information about RNAs and in particular microRNAs, as well as other regulatory ncRNAs, is currently inadequate [8]. The structural characterization of noncoding RNA is important to understand elementary mechanisms that are involved in posttranscriptional gene control. A quite exhaustive survey of structural nucleic acid data repository is provided in a recent review by Washietl and Hofacker [9].

The organization of structural information about ncRNAs is still a challenge for computer scientists. I would like to focus the attention on the nucleic acid database (NDB). It is a specialized repository containing 3D structure of nucleic acids (DNA and

Table 1
List of relevant structural databases for RNAs

Structural databases					
Name	**URL**	**Data type**	**Frequency of update**	**Analytical tools available**	**Reference**
Nucleic acid database	ndbserver.rutgers.edu	x-ray NMR	High	DownSW WebS	[11]
RNA base-pair structure	bps.rutgers.edu/bps	x-ray NMR	Low	No (see NDB)	[12]
NCBI structure	www.ncbi.nlm.nih.gov/structure	x-ray NMR	High	WebS	[13]
SCOR database	Scor.berkeley.edu	x-ray NMR	Low	No	[14]
RNA strand	www.rnasoft.ca/strand	2D Fold	Low	Yes	[15]

The table contains information about the type of data (column III) and their updating frequency (**High** = periodic maintenance, **Medium** = not regularly maintained; **Low** = no regular updates). This information depends on the date of last released version. In the case of BPS database the Low updating level is related to the availability of new knowledge in the NDB repository. The column of Analytical tools available indicates if the site offers some Web services (*WebS*) or it contains also downloadable software or data (*DownSW*)

RNA) either alone or bound to proteins or with ligands. Each structure is recorded taking the Cambridge structural database guidelines into account. These rules are integrated with nucleic acid specific conformational information [10]. The user can combine different options to perform a search. The same NDB [11] Web site contains also the *Base Pairing Database* (BPS) [12]. This repository includes information about base-pairing among nucleic acids. This information is very useful when for example investigating the secondary structure of RNAs. BPS contains about 400 RNA structures and it is interesting to note that it allows the estimation of some properties of triplets or higher order nucleotide features. The NCBI structure database [13], unlike the preceding ones, contains also synthetic sequences. NDB, NCBI structure, and PDB are cross-referenced. A structural classification of RNAs has been attempted by the *SCOR* database [14]. This repository has a layout similar to the SCOP database but it is specific for RNAs. Unfortunately, it seems that this database is currently not being maintained. The *STRAND* database [15] is a collection of different secondary RNA conformations. This repository has the great advantage to integrate different data sources. The query form is quite easy to use and it allows the activation of different search options: (a) RNA type, (b) data source, and (c) dimension of local conformation such as hairpin or bulge. The selected databases offer a general framework of useful structural resources for RNA analysis. It is important also to underline that it is necessary to improve the capability to use structural information to optimize the design of synthetic oligonucleotides that mimic the function of those biological ones contained in the databases. The enlargement of structural knowledge is strongly dependent on availability of experimental methods that allow obtaining long RNA sequences to resolve their structure. The majority, of deposited sequences, does not allow obtaining 3D conformational information. In particular, the scarcity of resolved structures of RNAs complicates the prediction of the spatial structure of a pre-miRNA and other nCRNAs.

2.2 General RNA and RNA-Related Databases

This group of databases includes general purpose databases and I further include also those repositories that can be useful to support the investigation of posttranscriptional gene silencing in this section. I have selected, among available RNA-related databases, a set of repositories that can offer a reasonable survey about current accessible knowledge. The selected databases are listed in Table 2. This small catalog comprises genomic information about noncoding RNAs, mRNA decay signals (Poly-A) among other information. All these repositories can be useful to investigate the function of microRNAs, siRNAs, or other ncRNAs.

The *ncRNA* database is a comprehensive noncoding RNA database [16] and it is built from GENBANK, FANTOM3, and H-InvDB (http://Jbirc.jbic.or.jp/hinv/ahg-db/index.sp) as the primary sources. The GtRNAdb resource contains information about genes that code for tRNAs [17]. The transfer RNAs are the oldest type of noncoding RNAs that have been identified. I have inserted this database for historical reasons, but *GtRNAdb* is not directly related with ncRNAs such as microRNAs or siRNAs. It is known, however, that tRNAs contain hairpins similar to pre-miRNAs so that the database may be of value for computational studies (*see* other chapters). The Poly-A database (*PolyA-db*) is a collection of poly-adenylation sites in human, mouse, and other eukaryotes [18]. The majority of experiments have identified potential miRNAs binding sites in 3′UTR regions. Taking this into account, the knowledge about poly-A site displacement can give suitable information about the regulatory context in which a microRNA operates [19]. The Rfam database is maintained by the same organization that has also implemented the microRNA registry (Sanger Institute) [20]. This repository contains different RNA families not only microRNAs and snoRNAs. The information about microRNAs is embedded in the microRNA registry. Rfam is now an element of the Wiki project RNA. The *RNAdb* is another collection of noncoding RNAs [21], which allows the selection of different search options through a simple user-interface. *The RNA Modification Database* is a comprehensive collection of posttranscriptionally modified nucleotides [22]. Information, about this kind of changes is fundamental to optimize the design of siRNAs,

Table 2
List of Repositories containing general information about ncRNAs

General-purpose RNA-related databases					
Name	**URL**	**Data type**	**Update frequency**	**Analytical tools**	**Ref**
Noncoding RNA database	http://biobases.ibch .poznan.pl/ncRNA	Primary Sequence Annotation	Low	No	[16]
Genomic tRNA database	http://Gtrnadb.ucsc.edu	Primary Sequence Annotation	Low	WebS	[17]
Poly-A database	http://exon.umdnj.edu /polya_db	Primary Sequence Annotation	Low	DownSW	[18]
Rfam	http://Rfam.sanger.ac.uk	Primary Sequence Annotation	High	WebS	[20]
RNAdb	http://Research.imb.uq .edu.au/rnadb	Primary Sequence Annotation	Low	DownSW	[21]
RNA Modification Database	http://Rna-mdb.cas .albany.wdu/RNAmods	MS data RPHPLC	Low	WebS	[22]
Half life mRNA database	No Web service	Microarray data	Low	No	[23]
DeepBase	http://deepbase.sysu .edu.cn	Primary sequence (NGS) Annotation	Medium	WebS DownSW	[24]
ncRNAImprint	http://rnaqueen.sysu .edu.cn	Primary sequence Annotation	Low	None	[25]
UTRdb	http://utrdb.ba.itb.cnr.it	Primary sequence Annotation	Medium	WebS DownSW	[27]
fRNAdb (Functional RNA db)	http://www.ncrna.org	Primary Sequence Annotation	Medium	DownSW	[28]
Long noncoding RNA	http://ncrnadb.com	Primary sequence	Medium	No	[29]
TRANSTERM	http://mRNA.otago .ac.uk/Transterm	Primary Sequence Annotation	Medium	WebS	[30]
HuSiDa	http://www.human -siRNA-database.net	Primary Sequence Annotation	Low	None	[31]

The meaning of the terms, in columns four and five, is the same as in Table 1

antagomirs, or other synthetic functional oligomers. This database permits to identify the potential modification for all the possible nucleotides in the RNA sequences, not only canonical bases. For each modification, a short annotation is given. The Web site further allows to perform some analysis on custom sequences. An important feature is the possibility to cross-reference this database with chemical data by the chemical abstract registry code. The Half-life mRNA database [23] is currently not available through the net, but it contains information about physiological mRNA decay which could provide relevant contextual information when planning microRNA experiments. *DeepBase* is a comprehensive ncRNA database that stores information from next generation sequencing (NGS) measurements [24]. This database contains information about special type of ncRNAs such as promoter associated RNAs (pasRNAs) or repeat associated RNAs (rasiRNA). Among other Web resources, several screening tools are available.

A very peculiar library is *ncRNAImprint* [25]. Genomic imprinting is an epigenetic process that involves DNA methylation and histone modification. The genes are expressed by a single allele without modification of the genetic sequence. The genomic imprint markers are established in the germ line and are maintained in all somatic cells. The ncRNAImprint database contains information about ncRNAs that have been demonstrated to be associate with genomic imprinting. The knowledge about the role of ncRNAs in genomic imprinting can support the investigation of epigenetic mechanisms of several diseases. The 3′UTR is considered the preferred microRNA's targeting region and is therefore the object of many studies although targets could be in other regions as well [26]. The *UTRdb resource* organizes the annotated sequences of 3′UTRs for several different organism [27] and contains information about experimentally validated microRNAs. An interesting repository is the fRNAdb [28] which collects information about transcripts that do not code for proteins, which may help researchers to investigate the biogenesis of ncRNAs. A special database, taking their peculiarities into account, is dedicated to long noncoding RNAs. The lcRNAdb [29] collects only information about this class of regulatory RNAs. *Transterm* [30] is a database similar to UTRdb. Transterm collects the available cis-regulatory elements that can be displaced in the 5′ and 3′UTR regions of a messenger RNA. The Web service allows the prediction of these regulatory motifs in a custom sequence. *HuSiDa* [31] is a specialized repository that stores information about functional small interfering RNAs (siRNAs). MicroRNAs and siRNAs show a similar functionality and also a similar activation pathway. In particular, the endosiRNAs are small endogenous oligonucleotides capable of modulating gene expression. The HuSiDa resource provides potentially useful information about this class of molecules taking different biological features into account.

2.3 MicroRNA Specific Databases

MicroRNAs are the focus of this chapter, but I would like to underline that it could be relevant to integrate microRNA specific information from other repositories as well. This section describes some microRNA specific databases (Table 3). The miRNA registry

Table 3
A sample of microRNA-related repositories

Selected sample of microRNA databases					
Name	**URL**	**Data type**	**Update frequency**	**Analytical tools**	**Reference**
MicroRNAs registry (miRBase)	http://www.mirbase.org/	Primary Sequence Annotation	High	WebS DownSW	[32]
MirDB	http://mirdb.org/miRDB/	Primary sequence Annotation	High	WebS DownSW	[33]
miRGen	http://diana.pcbi.upenn.edu/cgi-bin/miRGen/v3/Targets.cgi	Primary sequence Annotation	High	WebS DownSW	[34]
TarBase	http://diana.cslab.ece.ntua.gr/tarbase/	Primary sequence Annotation	Medium	WebS	[35]
PMRD	http://bioinformatics.cau.edu.cn/PMRD	Primary Sequence Annotation	Medium	WebS DownSW	[36]
miRNAMap	http://mirnamap.mbc.nctu.edu.tw/	Primary sequence Annotation	Medium	WebS DownSW	[37]
UCbase and miRfunc	http://microrna.osu.edu/.UCbase4	Primary Sequence Annotation	Low	WebS	[38]
PmiRKB	http://bis.zju.edu.cn/pmirkb/	Primary sequence Annotation	Low	WebS	[39]
miReg	http://www.iioab-mireg.webs.com/	Primary sequence Annotation	Medium	WebS	[40]
Patrocles	http://www.patrocles.org/	Primary sequence SNPs	Medium	WebS	[41]

(continued)

Table 3
(continued)

Selected sample of microRNA databases					
Name	**URL**	**Data type**	**Update frequency**	**Analytical tools**	**Reference**
TransmiR	http://cmbi.bjmu.edu.cn/transmir	Primary Sequence Annotation	High	WebS	[42]
dPORE-miRNA	http://cbrc.kaust.edu.sa/dpore	Primary Sequence Annotation	Low	WebS	[43]
miRGator	http://genome.ewha.ac.kr/miRGator/	Annotation	Medium	WebS	[44]
miR2Disease	http://www.miR2Disease.org	Annotation	Medium	WebS DownSW	[45]
dbSMR	http://miracle.igib.res.in/dbSMR	Annotation	Low	WebS	[46]
PolymiRTS database	http://compbio.utmem.edu/miRSNP	Annotation	Medium	WebS DownSW	[47]
PhenomiR	http://mips.helmholtz-muenchen.de/phenomir	Annotation	Low	WebS DownS	[48]
S-MED	http://www.oncomir.umn.edu/	Annotation	Low	WebS	[51]
mESAdb	http://konulab.fen.bilkent.edu.tr/mirna/	Annotation	Low	WebS	[52]

This table shows a survey of the heterogeneity of microRNA databases. The meaning of the terms, in columns four and five, is the same as in Table 1

has historically provided rules for miRNA nomenclature [32], but has since been superseded by miRBase. The provided nomenclature allows the assignment of an efficient identifier to a microRNA. The microRNA registry is the pivotal database for microRNA analysis. The first release of the database was published in 2004 and it has been continuously updated. It offers the possibility to select sequences on the basis of taxonomy. This fundamental microRNA database collects both precursor and mature miRNA forms. The microRNA registry allows the selection of sequences using different options such as organism, microRNA clusters, or custom sequences.

MirDB [33] is a library, based on the microRNA registry information, that allows the screening of potential miRNA targets or, on the contrary, the prediction of those miRNAs that are able to target a specific gene. This database is accessible, but the last release dates back to 2009 and it may not be well maintained at this point. *MiRGen* is an integrated database [34] that contains information about microRNAs, their targets, and genomic information. It is comparable with the microRNA registry. Today, MiRGen is a module of the *TarBase* system. The TarBase system [35] supplies information about microRNAs, validated targets, and some data about experimental protocols used to validate the targets. MiRGen, in the TarBase system, enables the selection of a microRNA or otherwise a transcription factor that could be targeted by a specific microRNA which I believe is a relevant feature. Upon selecting a specific microRNA, TarBase supplies a list of putative transcription factors that can be bound by the microRNA. The Plant microRNA database (PMRD) is a library focused on plant specific microRNAs [36]. Information about plant microRNAs is also available in the microRNA registry, but PMRD organizes the annotation of microRNAs. The database includes the possibility to obtain information about microRNA expression in several plants. PMRD has, like other databases, the possibility to analyze custom sequences. *MiRNAMap* is an integrated database [37] that allows searching both microRNAs and their targets. The data is organized in respect to biological taxonomy. *MiRNAMap* introduces the possibility to retrieve, for precursors or mature miRNAs, different kinds of knowledge such as chromosomal location or genes with multiple targets. Conserved genomic sequences can be targeted by microRNAs and a collection of conserved sequences is stored in *UCbase & mirFun* [38]. Information about the degree of conservation could help to select potential targets. *PmiRKB* is a comprehensive database of plant microRNAs [39]. It appears to be more basic than PMRD; the main advantage of this database, however, is the knowledge about microRNA associated polymorphisms. *MiREG* [40] is a repository that stores information about regulatory elements that can influence microRNA expression. This database is manually curated and contains data about transcription factors, drugs, xenobiotics, and physical stress factors. The information is extracted from literature and this origin is a relevant constraint for its development. The *Patrocles* library [41] is a collection of DNA polymorphic sequences involved in microRNA targeting. Genetic variants influence the targeting efficiency. The identification of microRNA polymorphisms provides valuable information when prospective microRNAs are used as disease biomarkers. A genetic variant can affect a microRNA, its target mRNAs, or both. The knowledge about polymorphisms permits the prioritizing of the targets taking genetic information into account.

MicroRNAs can influence not only protein expression by interaction with mRNA, but also epigenetic processes. An alteration of the microRNA mediated transcriptional process is often associated with pathological conditions (*see* Chapters 1, 2, and 19). As is true for protein-coding genes, the transcription of miRNAs is regulated by transcription factors (TFs).

The *TransmiR* database [42] permits to identify the TFs that specifically control pri-miRNA production. This information is important for the investigation of the nuclear initial phase of microRNA biogenesis. It is also useful for associating aberrant pri-miRNAs and disease progression. The information contained is extracted from literature. The user can perform a search using the acronym of a transcription factor or the microRNA code. The *dPORE-miRNA* repository [43] has been developed in order to integrate information about transcription of microRNA genes and their polymorphisms. Its user interface is quite user friendly and allows to perform a search using different options such as microRNA name, disease, or SNPs. An integrated system allows the user to obtain a general information framework about microRNAs and their function. *MirGator* is a system developed to provide an outlook about microRNAs [44]. MirGator has three main search options: (a) browsing the data in the repository, (b) selecting association studies, and (c) analyzing the crosslink between genes and microRNAs. The availability of experimental expression data is one of the relevant features of this system. It is important to underline that the last three databases also contain information about pathological conditions associated with a specific microRNA. This kind of information is valuable for the investigation of the role of microRNAs from a molecular medicine perspective. The *miR2dDisease* repository has been developed in order to support translational researches [45]. This database is manually curated and it can be mined using options similar to previously described database (microRNA code, disease name, target protein name, etc.). Like *MirGator*, this database is focused on human pathologies, but *miR2Disease* does not currently include expression data. The *dbSMR* is a collection of polymorphisms that affect a microRNA's capability to recognize and interact with its specific target site [46]. It is possible to perform a search by using standard options such as microRNA code or target gene name.

PolymiRTS is another microRNA database that is mainly devoted to the investigation of the effect of genetic variability on microRNA functionality [47]. This database allows the finding of chromosomal location, expression of quantitative trait loci (eQTL) data, validated targets, and other information suitable for planning experimental studies. *PhenomiR* is another microRNA data collection that has been developed in with a genetic perspective in mind [48]. It contains experimental information about microRNA expression, both in regard to normal and pathological conditions. This database is manually curated and it takes available literature into account. *PhenomiR* integrates different data sources such as OMIM [49] or gene ontology [50]. An interesting feature is the information about microRNA fold change in a specific tissue or cell line. Knowledge about the expression level of a microRNA is important during the target validation phase.

The sarcoma microRNA expression database (S-MED) is an oncological microRNA database [51]. This repository organizes microRNA expression data related to different types of sarcomas. It allows, like *PhenoiR*, the selection of microRNAs with relevant positive or negative expression change. The *mESAdb* is an integrated database also suitable to screen the role of a microRNA and its targets, for a specific biological function [52]. The system combines the microRNA registry with other data sources such as metabolic KEGG [53] and epidemiological [54] resources. The mESAdb database has the peculiarity to permit a search based on compositional features of a specific microRNA. It is possible to mine the database using dinucleotides or larger motifs. The search options are as follows: (a) association between motif and expression level, (b) comparison of different expression experiments, and (c) association between specific motifs and biological function.

2.4 Data Sources for Proteins Involved in MicroRNA Processing

The last group of ncRNA-related databases includes those libraries that store information about proteins involved in the microRNA pathway (Table 4) since I think it is important to have information about macromolecules that are involved in microRNA processing. The first selected database is starBase. This repository collects information about microRNAs, and about proteins such as AGO or other binding proteins [55] involved in the miRNA pathway. StarBase allows the screening of CDS sequences in order to identify potential target sites suitable for a validation by Degradome sequencing. This technique, also known as parallel analysis of RNA ends (PARE), is applied to screen potential microRNA cleavage sites. The presence of these domains provides additional information suitable for microRNA target identification. The investigation of protein–ncRNA interactions can also be supported by the *NPInter* database [56]. The developers have introduced a

Table 4
This table indicates some repositories that contain information about proteins that can interact with RNAs

Databases of ncRNAS-related proteins (miRNA)					
Name	**URL**	**Data type**	**Updating frequency**	**Analytical tools**	**Reference**
starBase	http://starbase.sysu.edu.cn/	Annotation	Medium	WebS DownSW	[55]
NPInter	http://bioinfo.ibp.ac.cn/NPInter	Annotation	Low	WebS	[56]
CLPZ	http://www.clipz.unibas.ch	Microarray Annotation	High	WebS	[57]

These databases are mainly addressed to give information about ncRNAs. The meaning of the terms, in columns four and five, is the same as in Table 1

classification of ncRNAs taking their interactions into account. They have split ncRNAs into the following categories:

- ncRNAs physically interacting with a protein (spatial connectivity).
- ncRNAs regulating the protein expression by interaction with mRNA.
- ncRNAs indirectly regulating gene expression by interaction with DNA (transcription).
- ncRNAs expression is controlled by a protein.
- ncRNAs that synergistically act with a protein in order to affect the protein function.
- ncRNAs functionality is influenced by a protein.
- ncRNAs that directly interact, at the genomic level, with a protein coding gene.
- special heterogeneous interactions between ncRNA and proteins.

This database allows, for several organisms, to select a class of ncRNAs such as miRNAs or snoRNAs and obtain, if available, some knowledge about their potential interaction with proteins. This kind of information could have great relevance, even though the amount of currently available information is still poor. The *CLIPZ* database allows to retrieve experimental information about RNA binding proteins [57]. The data were obtained by cross-linking immunoprecipitation. This technique allows to screen the site of RNA-binding proteins [58]. Information about these functional sites allows to design RNAse protected probes.

3 Conclusions

The aim of this paper is to offer a rather exhaustive survey on available repositories that contain valuable information for studies involving miRNAs. I have omitted some data sources in cases where it was impossible to access them. I have focused this paper on those repositories that are easily reachable through the Internet or ones that contain very special information. I would like to make some general considerations about microRNAs and other ncRNAs. First of all I would like to highlight the information redundancy. In many cases the libraries use the same data sources, but they differentiate on the basis of mining tools. The second consideration is the tendency to develop integrative bioinformatic systems for ncRNA analysis. It is important to underline that ncRNAs have acquired great importance in molecular medicine. As a consequence of this, knowledge about the involvement of ncRNAs, in particular of siRNAs or miRNAs, in pathological processes is helpful to develop diagnostic and therapy systems. It is also noticeable that small ncRNAs constitute a possible way to investigate the function of a gene in its particular functional pathway. This survey also underlines the role of genetic information in order to optimize target prioritization. The lack of conformational data about ncRNAs is a limiting factor when attempting structural bioinformatics analysis for ncRNA–protein interaction prediction. As a final consideration, I would like to point out the need of integrative bioinformatics mining tools that are able to perform a parallel knowledge discovery using the majority of accessible data repositories. I also would like to underline that the rapid evolving development of specialized databases can modify the present landscape of resources for ncRNAs mining and analysis.

References

1. Timmons L, Tabara H, Mello CC et al (2003) Inducible systemic RNA silencing in Caenorhabditis elegans. Mol Biol Cell 14:2972–2983
2. McCaffrey AP, Meuse L, Pham T-TT et al (2002) RNA interference in adult mice. Nature 418:38–39
3. Urbich C, Kuehbacher A, Dimmeler S (2008) Role of microRNAs in vascular diseases, inflammation, and angiogenesis. Cardiovasc Res 79:581–588
4. Bartel DP (2004) MicroRNAs: genomics, biogenesis, mechanism, and function. Cell 116:281–297
5. Wheeler BM, Heimberg AM, Moy VN et al (2009) The deep evolution of metazoan microRNAs. Evol Dev 11:50–68
6. Etheridge A, Lee I, Hood L et al (2011) Extracellular microRNA: a new source of biomarkers. Mut Res 717:85–90
7. Terwilliger TC (2011) The success of structural genomics. J Struct Funct Genom 12:43–44
8. Laederach A (2007) Informatics challenges in structured RNA. Brief Bioinform 8:294–303
9. Washietl S, Hofacker IL (2010) Nucleic acid sequence and structure databases. Methods Mol Biol (Clifton, NJ) 609:3–15
10. Dickerson RE (1989) Definitions and nomenclature of nucleic acid structure components. Nucleic Acids Res 17:1797–1803
11. Berman HM, Olson WK, Beveridge DL et al (1992) The nucleic acid database. A comprehensive relational database of three-dimensional structures of nucleic acids. Biophys J 63:751–759
12. Xin Y, Olson WK (2009) BPS: a database of RNA base-pair structures. Nucleic Acids Res 37:D83–D88

13. Wang Y, Addess KJ, Chen J et al (2007) MMDB: annotating protein sequences with Entrez's 3D-structure database. Nucleic Acids Res 35:D298–D300
14. Tamura M, Hendrix DK, Klosterman PS et al (2004) SCOR: structural classification of RNA, version 2.0. Nucleic Acids Res 32: D182–D184
15. Andronescu M, Bereg V, Hoos HH et al (2008) RNA STRAND: the RNA secondary structure and statistical analysis database. BMC Bioinformatics 9:340
16. Szymanski M, Erdmann VA, Barciszewski J (2007) Noncoding RNAs database (ncRNAdb). Nucleic Acids Res 35:D162–D164
17. Chan PP, Lowe TM (2009) GtRNAdb: a database of transfer RNA genes detected in genomic sequence. Nucleic Acids Res 37:D93–D97
18. Lee JY, Yeh I, Park JY et al (2007) PolyA_DB 2: mRNA polyadenylation sites in vertebrate genes. Nucleic Acids Res 35:D165–D168
19. Eulalio A, Huntzinger E, Nishihara T et al (2009) Deadenylation is a widespread effect of miRNA regulation. RNA (New York, NY) 15:21–32
20. Gardner PP, Daub J, Tate J et al (2011) Rfam: wikipedia, clans and the "decimal" release. Nucleic Acids Res 39:D141–D145
21. Pang KC, Stephen S, Engström PG et al (2005) RNAdb – a comprehensive mammalian noncoding RNA database. Nucleic Acids Res 33:D125–D130
22. Limbach PA, Crain PF, McCloskey JA (1994) Summary: the modified nucleosides of RNA. Nucleic Acids Res 22:2183–2196
23. Sharova LV, Sharov AA, Nedorezov T et al (2009) Database for mRNA half-life of 19 977 genes obtained by DNA microarray analysis of pluripotent and differentiating mouse embryonic stem cells. DNA Res 16:45–58
24. Yang J-H, Shao P, Zhou H et al (2010) deepBase: a database for deeply annotating and mining deep sequencing data. Nucleic Acids Res 38:D123–D130
25. Zhang Y, Guan D-G, Yang J-H et al (2010) ncRNAimprint: a comprehensive database of mammalian imprinted noncoding RNAs. RNA (New York, NY) 16:1889–1901
26. Ørom UA, Nielsen FC, Lund AH (2008) MicroRNA-10a binds the 5'UTR of ribosomal protein mRNAs and enhances their translation. Molecular cell 30:460–471
27. Grillo G, Turi A, Licciulli F et al (2010) UTRdb and UTRsite (RELEASE 2010): a collection of sequences and regulatory motifs of the untranslated regions of eukaryotic mRNAs. Nucleic Acids Res 38:D75–D80
28. Kin T, Yamada K, Terai G et al (2007) fRNAdb: a platform for mining/annotating functional RNA candidates from non-coding RNA sequences. Nucleic Acids Res 35:D145–D148
29. Amaral PP, Clark MB, Gascoigne DK et al (2011) lncRNAdb: a reference database for long noncoding RNAs. Nucleic Acids Res 39:D146–D151
30. Jacobs GH, Chen A, Stevens SG et al (2009) Transterm: a database to aid the analysis of regulatory sequences in mRNAs. Nucleic Acids Res 37:D72–D76
31. Truss M, Swat M, Kielbasa SM et al (2005) HuSiDa–the human siRNA database: an open-access database for published functional siRNA sequences and technical details of efficient transfer into recipient cells. Nucleic Acids Res 33:D108–D111
32. Griffiths-Jones S (2010) miRBase: microRNA sequences and annotation. Current Protoc Bioinform Chapter 12: Unit 12.9.1–10
33. Wang X (2008) miRDB: a microRNA target prediction and functional annotation database with a wiki interface. RNA (New York, NY) 14:1012–1017
34. Megraw M, Sethupathy P, Corda B et al (2007) miRGen: a database for the study of animal microRNA genomic organization and function. Nucleic Acids Res 35:D149–D155
35. Sethupathy P, Corda B, Hatzigeorgiou AG (2006) TarBase: a comprehensive database of experimentally supported animal microRNA targets. RNA 12:192–197
36. Zhang Z, Yu J, Li D et al (2010) PMRD: plant microRNA database. Nucleic Acids Res 38:D806–D813
37. Hsu PWC, Huang H-D, Hsu S-D et al (2006) miRNAMap: genomic maps of microRNA genes and their target genes in mammalian genomes. Nucleic Acids Res 34:D135–D139
38. Taccioli C, Fabbri E, Visone R et al (2009) UCbase & miRfunc: a database of ultraconserved sequences and microRNA function. Nucleic Acids Res 37:D41–D48
39. Meng Y, Gou L, Chen D et al (2011) PmiRKB: a plant microRNA knowledge base. Nucleic Acids Res 39:D181–D187
40. Barh D, Bhat D, Viero C (2010) miReg: a resource for microRNA regulation. J Integr Bioinform 7(1):144
41. Hiard S, Charlier C, Coppieters W et al (2010) Patrocles: a database of polymorphic miRNA-mediated gene regulation in vertebrates. Nucleic Acids Res 38:D640–D651
42. Wang J, Lu M, Qiu C et al (2010) TransmiR: a transcription factor-microRNA regulation database. Nucleic Acids Res 38:D119–D122
43. Schmeier S, Schaefer U, MacPherson CR et al (2011) dPORE-miRNA: polymorphic regulation of microRNA genes. PloS one 6:e16657
44. Nam S, Kim B, Shin S et al (2008) miRGator: an integrated system for functional annota-

tion of microRNAs. Nucleic Acids Res 36:D159–D164
45. Jiang Q, Wang Y, Hao Y et al (2009) miR2Disease: a manually curated database for microRNA deregulation in human disease. Nucleic Acids Res 37:D98–D104
46. Hariharan M, Scaria V, Brahmachari SK (2009) dbSMR: a novel resource of genome-wide SNPs affecting microRNA mediated regulation. BMC Bioinformatics 10:108
47. Ziebarth JD, Bhattacharya A, Chen A et al (2012) PolymiRTS database 2.0: linking polymorphisms in microRNA target sites with human diseases and complex traits. Nucleic Acids Res 40:D216–D221
48. Ruepp A, Kowarsch A, Schmidl D et al (2010) PhenomiR: a knowledgebase for microRNA expression in diseases and biological processes. Genome biology 11:R6
49. Hamosh A, Scott AF, Amberger JS et al (2005) Online mendelian inheritance in man (OMIM), a knowledgebase of human genes and genetic disorders. Nucleic Acids Res 33:D514–D517
50. The Gene Ontology Consortium (2008) The gene ontology project in 2008. Nucleic Acids Res 36:D440–D444
51. Sarver AL, Phalak R, Thayanithy V et al (2010) S-MED: sarcoma microRNA expression database. Laboratory investigation 90:753–761
52. Kaya KD, Karakülah G, Yakicier CM et al (2011) mESAdb: microRNA expression and sequence analysis database. Nucleic Acids Res 39:D170–D180
53. Kanehisa M, Goto S, Sato Y et al (2012) KEGG for integration and interpretation of large-scale molecular data sets. Nucleic Acids Res 40:D109–D114
54. Yu W, Gwinn M, Clyne M et al (2008) A navigator for human genome epidemiology. Nature genetics 40:124–125
55. Yang J-H, Li J-H, Shao P et al (2011) starBase: a database for exploring microRNA-mRNA interaction maps from Argonaute CLIP-Seq and Degradome-Seq data. Nucleic Acids Res 39:D202–D209
56. Wu T, Wang J, Liu C et al (2006) NPInter: the noncoding RNAs and protein related biomacromolecules interaction database. Nucleic Acids Res 34:D150–D152
57. Khorshid M, Rodak C, Zavolan M (2011) CLIPZ: a database and analysis environment for experimentally determined binding sites of RNA-binding proteins. Nucleic Acids Res 39:D245–D252
58. Kishore S, Jaskiewicz L, Burger L et al (2011) A quantitative analysis of CLIP methods for identifying binding sites of RNA-binding proteins. Nature methods 8:559–564

Chapter 6

High-Throughput Approaches for MicroRNA Expression Analysis

Bala Gür Dedeoğlu

Abstract

Profiling microRNA (miRNA) expression is of widespread interest due to their critical roles in diverse biological processes, including development, cell proliferation, differentiation, and apoptosis. Profiling can be achieved via three major methods: amplification-based (real-time quantitative PCR, qRT-PCR), hybridization-based (microarrays), and sequencing-based (next-generation sequencing (NGS)) technologies. The gold standard is qRT-PCR and serves as a platform for single reverse PCR amplification experiments and for a large number of miRNAs in parallel, both by multiplexing and plate based arrays. Currently, qRT-PCR is used for the validation of miRNA profiling results from other platforms. Hybridization based miRNA profiling by microarrays has become a widely used method especially for biomarker and therapeutic target identification. The data obtained from microarrays also enables functional prediction of miRNAs by correlating miRNA expression patterns to corresponding mRNA and protein profiles. Additionally, miRNA profiling strategies based on deep sequencing allow both the identification of novel miRNAs and relative quantification of miRNAs. Each miRNA profiling strategy has specific strengths and challenges that have to be considered depending on the nature of the research context.

In this chapter the high-throughput approaches that can be applied to microRNA profiling are discussed starting from small-scale qRT-PCR technology to a wider one, NGS.

Key words miRNA profiling, qRT-PCR, Microarray, Next-generation sequencing (NGS)

1 Introduction

MicroRNAs (miRNAs) are short (~22 nucleotides long), noncoding regulatory RNA molecules that regulate gene expression post-transcriptionally. After the discovery of the first miRNA in the roundworm *Caenorhabditis elegans*, these short regulatory RNAs have been found to have crucial regulatory roles in gene expression in plants and animals [1, 2]. About 3 % of human genes encode for miRNAs, and up to 60 % of human genes are putative targets of miRNAs [3].

Profiling the expression of miRNAs has revealed the roles of miRNAs in diverse biomolecular processes like development [4], tissue differentiation [5], cell proliferation, and apoptosis [6, 7].

Malik Yousef and Jens Allmer (eds.), *miRNomics: MicroRNA Biology and Computational Analysis*, Methods in Molecular Biology, vol. 1107, DOI 10.1007/978-1-62703-748-8_6, © Springer Science+Business Media New York 2014

MicroRNAs are reportedly involved in 94 human diseases [8] including autoimmune and psychiatric disorders, neurological diseases, and cancer [9–13]. They were found to have roles in cancer as biomarkers for identifying the tissue differentiation state of cancers of unknown tissue origin [14] and for classifying human cancers [15, 16]. Additionally since miRNAs were found to be circulating in blood plasma and serum in detectable amounts they can also be used as noninvasive biomarkers for some diseases [17].

Despite some challenges, three major approaches for the profiling of expression status of miRNAs are currently well established and studied: real-time quantitative PCR (qRT-PCR) [18, 19], hybridization based methods (DNA microarrays) [20, 21] and high-throughput sequencing (next-generation sequencing (NGS), deep sequencing) [22, 23].

In this chapter the high-throughput approaches that can be applied to microRNA profiling are discussed starting from a small-scale qRT-PCR technology to a wider one, NGS.

2 Array Systems for qRT-PCR

Real-time PCR, also known as quantitative PCR (qRT-PCR), is considered to be the gold standard for gene expression measurement. This method enables a quantitative, real-time analysis of microRNA expression with sensitivity, specificity, and robustness. Additionally, accuracy, ease of use, and short duration of the protocol makes qRT-PCR a preferred method. Several companies have developed qRT-PCR-based assays with high-throughput profiling capabilities for the detection of miRNA expression by taking into account the aforementioned considerations [18, 24, 25] and also have increased throughput with the introduction of microfluidic array platforms which allow the analysis of thousands of miRNAs at the same time [26].

Profiling using qRT-PCR-based methodology is more rapid compared to other platforms and makes a wide range of samples, from cultured cells to formalin-fixed, paraffin-embedded (FFPE) tissues accessible with the method while requiring limited input. Since the serum or plasma from blood contains only small amounts of miRNAs, the detection of these miRNAs requires a highly sensitive and accurate method, and thus qRT-PCR is a preferred method to be used for the identification of circulating miRNA biomarkers [27].

One of the companies that released high-throughput profiling for human, mouse, and rat miRNAs is Life Technologies. They released TaqMan® OpenArray® Human MicroRNA Panel for high-throughput profiling of human microRNAs and the TaqMan OpenArray Rodent MicroRNA Panel for high-throughput profiling of mouse and rat microRNAs. The panels have six logs of dynamic range with a capacity of three samples on each array.

These panels yield the highest concordance with TaqMan MicroRNA Assays, which is a commonly used validation tool for hybridization-based methods. Utilizing nanoliter fluidics, TaqMan OpenArray MicroRNA Panels offer researchers a fast, easy, and affordable system to validate data [28, 29].

Qiagen is one of the pioneers who has implemented many improvements in the field of miRNA research. With their next-generation miScript PCR System for miRNA profiling and quantification they offer alternative possibilities for miRNA research. The miScript II RT Kit, with its unique dual-buffer system, quantifies mature miRNA exclusively or simultaneously quantifies miRNA, pre-miRNA, and mRNA from the same cDNA sample. Qiagen also offers profiling pathway-focused panels of miRNAs or entire miRNomes using miScript miRNA PCR Arrays [30]. Exiqon also released similar panels, *microRNA Focus PCR Panels*, which are collections of qPCR assays for microRNAs that are believed to have a significant role in specific diseases, development, cells, and body fluids [31].

If the sensitivity is the major concern of the experiments in miRNA profiling, locked nucleic acid (LNA™) technology, which is the base of Exiqon's microRNA qPCR assay design, delivers high sensitivity. LNAs are a class of RNA analogues in which the 2′ oxygen and the 4′ carbon positions in the ribose ring are connected or "locked" to create increased thermal stability relative to DNA or RNA when they complement with DNA or RNA [32]. The increase in thermal stability allows shortening of PCR primers and consequently two microRNA-specific PCR primers per miRNA can be designed, rather than only one miRNA-specific primer.

2.1 Normalization Strategies for MicroRNA Quantitative Real-Time (qPCR) Arrays

MicroRNAs represent only a small fraction of the total RNA and this fraction may largely vary across samples. As changes in the miRNA expression can be clinically or biologically significant, normalization of data is one of the most important and challenging issues that have to be considered when profiling (for method details *see* Chapter 8). Data normalization is crucially important for obtaining accurate results. The goal of normalization is to adjust the data to remove technical bias across samples that is not related to the biological condition, thereby identifying the relevant biological differences. Various approaches have been used in the literature, and the discrepancies between miRNA-profiling studies may be due to the application of different normalization approaches. The challenge in normalization is finding out the best normalizer (reference) for the study of interest. There is no common housekeeping gene for every study, but each study has its own specific reference gene. Normalizing to an accurate reference gene can remove differences due to sampling, input, and quality of RNA and can identify true changes in gene expression. A good normalizer has to have the expression with target in the cells of interest and it is the best if the quantification is in the same range as the target of interest [33–35]. One of the two common approaches, cell number, cannot really be used as a normalization factor when dealing with tissue samples and the other approach, normalizing to 18/28S ribosomal RNA (rRNA), can present a challenge for miRNA-enriched samples where rRNA is absent.

Currently, many groups use predefined endogenous controls, reference miRNAs [36], and small nuclear or small nucleolar RNAs as normalizers for miRNA expression analysis [37]. These may not serve as great reference genes because small nuclear RNAs like U6 do not share the same properties as miRNAs in terms of their transcription, processing, and tissue-specific expression patterns. Therefore, especially in experiments analyzing potential defects in miRNA processing and regulation, other small RNAs could be misleading when used as reference genes and it may be best to normalize genes with reference genes belonging to the same RNA class [33].

One of the strategies for normalization is to use a stably expressed gene as a reference control according to the existing data, such as miR-16 [38], small nuclear/nucleolar RNAs RNU6 [39], RNU44, and RNU48 [33]. Since these reference genes cannot ensure constant expression under all experimental conditions the best way to approach analysis of miRNA expression data is through global mean normalization of a set of reference genes. This method takes a minimum of three stable housekeeping genes and takes the geometric mean to provide a reliable normalization factor that can control for outliers and differences in abundance among genes [33, 40].

A second approach is the usage of spiked-in synthetic control miRNAs, such as cel-miR-39, cel-miR-54, and cel-miR-238 [41]. They are introduced into the RNA sample at a range of known input amounts and a stable reference control is obtained. This approach has the advantages of providing quality control, leading to absolute quantification, correcting for many aspects of technical variation, and providing normalization over a range of signal intensities as the spiked-in miRNAs represent a range of input amounts. Kang et al. suggested combining the spiked-in control approach with endogenous normalizer approaches as an ideal normalization strategy [27].

Global mean normalization method is the third strategy for miRNA expression data. This method uses the average expression level of all miRNAs detected in a sample as a normalization factor. The assumption for this strategy is that the mean expression level of all miRNAs in a sample is constant when using the same total RNA input. The sample may be both from a control or a patient. Mestdagh et al. [42] demonstrated that the global mean method is better than using endogenous small miRNAs, such as nuclear/nucleolar RNAs in getting rid of technical variation while keeping biological variation. This method is suitable to normalize genome-wide miRNA profiling without the need for selecting a specific reference control. However, it is not suitable for studying only a few miRNAs. The qBaseplus and the GenEx analysis software are two available programs for global mean normalization analysis [27].

3 Microarray Technology

MicroRNA microarray hybridization is a widely used technique for miRNA quantification since it allows measuring a large number of miRNAs simultaneously. Currently, several platforms of microRNA microarrays are commercially available. Affymetrix, Agilent, Exiqon, Life Technologies, and Illumina are the mostly used platforms for miRNA expression detection.

Affymetrix has recently released the new GeneChip® miRNA 3.0 array to keep pace with the discovery of new and novel miRNAs. This new array offers 100 % coverage to miRBase version 17 with 153 organisms, 1,733 human mature miRNAs, 2,216 snoRNA and scaRNAs, and 1,658 pre-miRNAs. Mouse and rat mature miRNAs and pre-miRNAs are also included on the chip [43]. The new GeneChip® miRNA 3.0 array requires an input of only 130–500 ng of isolated total RNA.

In Agilent Human microRNA Microarrays system each slide contains eight 60K microarrays or eight 15K microarrays printed using Agilent's 60-mer SurePrint technology. SurePrint technology allows as many as eight arrays to reside on a single slide, allowing experiments to range from a genome-wide view to more focused studies. Human microRNA Microarrays have 1,205 human and 144 human viral miRNAs on a single chip regularly updated from miRBase 16.0 [44].

Exiqon is one of the pioneers that introduced improvements in the field of miRNA research. Exiqon's microRNA array, miRCURY LNA™ microarray, is highly specific and sensitive even for AT-rich microRNAs. In addition, they offer great reproducibility with 99 % correlation between arrays and a dynamic range greater than five orders of magnitude. The seventh generation of the miRCURY LNA™ microRNA array contains 3,100 capture probes and covers all human, mouse, and rat microRNAs annotated in miRBase 18.0. It also contains viral microRNAs related to the included species [31].

A recent innovation in miRNA profiling, based on hybridization, is the Nanostring nCounter, in which a multiplexed probe library is created using two sequence specific capture probes that are tailored to each miRNA of interest. An important advantage of this method is the ability to discriminate between similar variants with high accuracy [32, 45].

As mentioned above, currently, several platforms of microRNA microarray chips are commercially available and each of them has its advantages and disadvantages over each other. Several studies compared the repeatability and comparability of microRNA microarray platforms to find out which hybridization based platform to choose for miRNA profiling [25, 46, 47]. Sato et al. compared five different platforms (Agilent, Ambion, Exiqon, Invitrogen, and Toray) of microRNA microarrays according to their repeatability and comparability and additionally, they compared the results of microarray data with that of qRT-PCR (Taqman). At the end they reported high intra-platform repeatability and comparability in microRNA microarray and qRT-PCR. However, the commercially available microRNA microarrays failed to show good inter platform concordance probably due to major differences in stringency of detection call criteria between different platforms [25]. On the other hand, Sah et al. compared four platforms (Ambion, Agilent, Exiqon, and Illumina) and revealed the strengths and weaknesses for each platform. They concluded Ambion and Agilent platforms as sensitive while Illumina and Exiqon are more specific [46].

It is seems essential to investigate the correlation and reproducibility among miRNA technologies, since they are new and due to significant differences in the probe design, experimental protocols, and data analysis. To increase the reliability of miRNA profiling platforms, the technical and analytical variability have to be identified and the studies have to focus on minimizing these variabilities [48].

4 Next-Generation Sequencing

NGS is a powerful approach for discovering new miRNAs and profiling their expression status in biological samples. NGS technologies pioneered the sequencing of the complete set of miRNAs present in an RNA sample which led to the discovery of novel miRNAs, identification of sequence isoforms, and prediction of potential mRNA targets. NGS can be used to determine RNA expression levels more accurately compared to the hybridization-based technologies since it is possible to determine the absolute quantity of every molecule in a cell population and directly compare results between experiments. Expression levels determined by this method were found to be correlated with qRT-PCR experiments [49, 50].

454 Life Sciences was the first company that developed sequencing technology by synthesis in 2005 [51]. Since then, several NGS platforms (Illumina Genome Analyzer, Illumina, Inc., San Diego, CA, USA; SOLID, Life Technologies Corporation, Carlsbad, CA, USA) have been developed and they were used in different fields of biological research including detection of miRNA expression and identification of novel miRNAs.

Table 1
Comparison of leading NGS platforms

	Platform		
	Illumina/Solexa GA	**Roche/454 GS FLX Titanium**	**ABI/SOLID**
Sequencing type	Sequencing by synthesis	Pyrosequencing	Ligation-based sequencing
Amplification	Bridge PCR	Emulsion PCR	Emulsion PCR
Read length (bases)	50–100	350[a]	35–75
Basepairs/run	2–3 Gb	400 Mb	3–6 Gb
Run time (days)	4	0.35 (7 h)	7
Comments	Most widely used platform in the field. Low multiplexing capability of samples	Longer reads improve mapping in repetitive regions; fast run times. High reagent cost; high error rates in homo-polymer repeats	Good data quality. Long run times

[a]Average read length

Table 1 compares the main characteristics of three popular NGS platforms. Since NGS technology is so dynamic it has to be kept in mind that each value is changing constantly with the release of newer models.

The NGS of miRNAs is composed of two main parts, experimental part and data analysis by bioinformatic tools. Each part is as important as the other since the production of high quality data for accurate data analysis can just be provided by a careful experiment design and handling. In general, an RNA population is converted to a library of cDNA fragments, which have adapters attached to one or both ends. Then each molecule is sequenced in a high-throughput manner [52]. The detailed procedure is given in for the detailed procedure see section 4.1 Small RNA Cloning for Deep Sequencing. On the other hand, analysis of sequencing data is complex, requires in-depth bioinformatics and complicated sequence algorithms. While more user-friendly programs are being introduced for this growing field, sequencing can provide an overwhelming amount of read data that can be difficult to decode and translate into miRNA profiles. Although there are increasing number of web servers and independent programs available for miRNA profiling and novel miRNA discovery from NGS data, a standard approach has not yet emerged. The tools miRanalyzer [53, 54] and mirTools [55] are two Web servers for NGS data analysis, while miRDeep [56] and miRExpress [57] are two examples for standalone programs.

4.1 Small RNA Cloning for Deep Sequencing

The essential step for the identification of novel miRNAs or comparative studies is the generation of a small RNA library using NGS technologies. Different cloning approaches were discussed in the literature but each method can be applied to different NGS platforms by minor modifications like using platform specific adapters.

In one of the methodologies small RNA fractions are first polyadenylated using RNA polymerase. An RNA oligo linker is then added to the 5′ ends of the polyadenylated small RNAs and reverse transcription is performed using an oligo with a linked sequence in its 5′ end and polyT in its 3′ end. The reverse transcribed small RNA cDNAs are then amplified with a pair of 5′ and 3′ linker primers [58]. These cDNAs can be sequenced by 454 sequencing platform by incorporating 454 adapters at the ends of small RNA cDNAs by a second round of PCR.

Recently Malone et al. have reported a protocol that details the process of small RNA cloning for sequencing on the Illumina/Solexa sequencing platform, which can be easily modified for use on other next-generation platforms (e.g., SOLiD, 454). The procedure reported is designed to clone canonical small RNA molecules with 5′-monophosphate and 3′-hydroxyl termini [59].

The following "cloning small RNAs for sequencing" procedure is modified from three different protocols and can be applied to different sequencing platforms; Illumina/Solexa, 454, and SOLID [58–60].

4.1.1 RNA Preparation

Total RNA isolated by TRIzol can be used to proceed. The only concern is how much RNA will be isolated and whether it will be sufficient for downstream applications. If the sequencing platform is Illumina/Solexa, 1–5 μg of total RNA is enough while it requires 20–50 μg of total RNA for 454 sequencing. If polyadenylation of RNAs will be performed for cloning, small RNA extraction kits can be used for RNA isolation step (e.g., mirVana miRNA isolation kit, Ambion).

4.1.2 Separation of Small RNAs

The separation of small RNAs is performed by polyacrylamide urea gel electrophoresis (PAA). The thickness of the gel depends on the amount of total RNA being run. As a guide, 1.0 mm for <20 μg and 1.5 mm for >20 μg total RNA can be used. After separation by PAA it is appropriate to stain the gel with SYBR Gold® (only in cases where the amount of small RNAs is sufficient to allow direct visualization). The gel is then visualized under a phosphoimager and gel slices containing the RNAs of desired size are cut out with a clean scalpel.

For the extraction of small RNAs from the gel, fourfold to sixfold volume of 0.4 M NaCl is added onto the gel slices and incubated by shaking overnight at room temperature.

4.1.3 First Ligation

After precipitation of the small RNAs, the 3′-end ligation is performed (also called first ligation). The 3′ adaptor is called Modban [AMP-5′p-5′p/CTGTAGGCACCATCAATdi-deoxyC-3′ (IDT, Integrated DNA Technologies)]. To perform a proper ligation reaction truncated version of T4 RNA ligase 2 and ATP-free T4 RNA ligase buffer have to be used. The first ligation product is obtained by PAA gel electrophoresis.

4.1.4 Second Ligation

The first ligation product is used as a substrate for the 5′ end ligation reaction also known as second ligation. For Illumina/Solexa platform Solexa linker and for 454 sequencing "Nelson's Linker" is used as an adapter.

At the end of the second ligation, the 5′- and 3′-ligated RNA product are either purified with 15 % PAA/urea gel electrophoresis or without any gel extraction. The product is confirmed with PCR following reverse transcription.

Solexa Linker: 5′-rArCrArCrUrCrUrUrUrCrCrCrUrArCrArCrGr ArCrGrCrUrCrUrUr-CrCrGrArUrC-3′

Nelson's linker: 5′ATCGTrArGrGrCrArCrCrUrGrArArA 3′

4.1.5 Reverse Transcription

Reverse transcription (RT) reaction is primed by BanOne, which is the complement of Modban, the 3′ linker. After RT reaction PCR is performed by using 5′ and 3′ linker specific primers to obtain small RNA libraries.

Size selection and gel extraction of small RNA libraries are resolved by UV light on a 2 % low-melting agarose gel stained with ethidium bromide. The bands at the expected sizes are cut out of the gel and stored as small RNA libraries for sequencing reactions.

BanOne: 5′-ATTGATGGTGCCTACAG-3′ (3′ primer for reverse transcription).

5 Comparison of Three High-Throughput Technologies

Although three high-throughput systems (qRT-PCR, microarray and NGS) are in use for miRNA profiling, it has been suggested that NGS will likely replace the use of microarrays in the future. The introduction of NGS into miRNA research can be attributed to its ability to read short fragments in a high-throughput manner [61]. Even though the prospective of NGS sequencing technology has been established the technology has not yet fully matured especially when compared to the microarray and qRT-PCR, which have been in use for more than 20 years. In addition, NGS application in quantitative gene expression is challenging due to cost, labor, and time consumption concerns and also professional bioinformatic support is needed for data analysis [62]. As summarized in Table 2, NGS not only profiles known miRNAs but it is able to identify unknown miRNAs which are beyond the capabilities of qRT-PCR and microarrays. Additionally NGS has high accuracy in distinguishing miRNAs that are very similar in sequence, as well as isomiRs but substantial computational support is needed for data analysis as a drawback. Microarrays have typically lower specificity than qRT-PCR or RNA sequencing and it is difficult to use them for absolute quantification. The qRT-PCR approach, on the other

Table 2
Comparison of miRNA profiling platforms

	Method		
	qRT-PCR	**Microarray**	**Next-generation sequencing**
Throughput	Medium to high	High	Ultra high
Principle	PCR amplification	Hybridization	Sequencing
Time	<6 h	2 days	1–2 week
Sample input	10 ng–500 ng	100 ng–1 μg	500 ng–10 μg
Dynamic range	10^6	10^3–10^4	10^4–10^7
Application	Relative and absolute quantification; validation of other miRNA profiling approaches	Profiling known miRNAs across samples	De novo identification of small RNAs, simultaneous relative quantification of different small RNA species

hand, is the best technique for absolute quantification among the aforementioned and it is sensitive and specific but it only provides medium throughput with respect to the number of samples processed per day.

Several studies have addressed the strengths and limitations of these three principal platforms for miRNA profiling. In the study of Willenbrock et al. the relative and absolute RNA quantification was performed using Exiqon's LNA-based microarrays and Illumina's GA-II sequencing platform. The authors assessed the sensitivity and reproducibility of these two platforms [63]. Consequently, microarray expression analysis was found to be both highly specific and very sensitive for quantification of microRNAs compared to NGS with respect to absolute RNA expression quantification, which is in agreement with a study by Ach and colleagues [64]. On the other hand, both technologies produced highly reproducible expression data and performed well in relative gene expression studies.

In another study Wang et al. [62] systematically analyzed three representative microRNA profiling platforms: Locked Nucleic Acid (LNA) microarray, beads array, and TaqMan quantitative real-time PCR low density array (TLDA) and at the end they could not identify any of the technology to be superior to the others due to their low inter platform consistency. Thus, they suggest selecting a platform according to available facilities, budget, interests, and loyalties. They concluded the study by emphasizing the importance of developing specific normalization methods for miRNA profiling in order to improve the accuracy of validating data and to provide the possibility of data integration across platforms [62].

6 Conclusion

It is not always the platform, but the sample type or RNA preparation technique that affects the quality of miRNA profiling. These technical issues have to be considered while choosing the right platform for profiling miRNAs. As mentioned in the study of Jensen et al., while the qRT-PCR based platforms are sensitive enough to reproducibly detect miRNAs at the abundance levels found in human plasma, the array-based platforms are not [65]. But when the miRNA levels are high although qRT-PCR based platforms performed well in terms of specificity, reproducibility, and recovery, at low miRNA levels, as in plasma, not every qRT-PCR array reproduces the same specificity and quality [65]. As very well summarized in the review of Pritchard et al. [32], choosing a right miRNA profiling platform depends on the type and the purpose of the research and each step has to be kept in mind to obtain accurate results.

References

1. Lee RC, Ambros V (2001) An extensive class of small RNAs in Caenorhabditis elegans. Science 294(5543):862–864
2. Lagos-Quintana M, Rauhut R, Lendeckel W et al (2001) Identification of novel genes coding for small expressed RNAs. Science 294(5543):853–858
3. Git A, Dvinge H, Salmon-Divon M et al (2010) Systematic comparison of microarray profiling, real-time PCR, and next-generation sequencing technologies for measuring differential microRNA expression. RNA 16(5):991–1006
4. Stefani G, Slack FJ (2008) Small noncoding RNAs in animal development. Nat Rev Mol Cell Biol 9:219–230
5. Shi L, Reid LH, Jones WD et al (2006) The MicroArray quality control (MAQC) project shows inter- and intraplatform reproducibility of gene expression measurements. Nat Biotechnol 24:1151–1161
6. Bueno MJ, de Castro IP, Malumbres M (2008) Control of cell proliferation pathways by microRNAs. Cell Cycle 7:3143–3148
7. Jovanovic M, Hengartner MO (2006) miRNAs and apoptosis: RNAs to die for. Oncogene 25:6176–6187
8. Jiang Q, Wang Y, Hao Y et al (2009) miR2Disease: a manually curated database for microRNA deregulation in human disease. Nucleic Acids Res 37:D98–D104
9. Tili E, Michaille JJ, Costinean S et al (2008) MicroRNAs, the immune system and rheumatic disease. Nat Clin Pract Rheum 4:534–541
10. Barbato C, Giorge C, Catalanotto C, Cogoni C (2008) Thinking about RNA? MicroRNAs in the brain. Mamm Genome 19:541–551
11. Lai CY, Yu SL, Hsieh MH et al (2011) MicroRNA expression aberration as potential peripheral blood biomarkers for schizophrenia. PLoS One 6:e21635
12. Buckley PG, Alcock L, Bryan K et al (2010) Chromosomal and microRNA expression patterns reveal biologically distinct subgroups of 11q-neuroblastoma. Clin Cancer Res 16:2971–2978
13. Guerau-de-Arellano M, Alder H, Ozer H et al (2011) miRNA profiling for biomarker discovery in multiple sclerosis: from microarray to deep sequencing. J Neuroimmunol 248:32–39
14. Rosenfeld N, Aharonov R, Meiri E et al (2008) MicroRNAs accurately identify cancer tissue origin. Nat Biotech 26:462–469
15. Lu J, Getz G, Miska EA et al (2005) MicroRNA expression profiles classify human cancers. Nature 435:834–838
16. Ferracin M, Pedriali M, Veronese A et al (2011) MicroRNA profiling for the identification of cancers with unknown primary tissue-of-origin. J Pathol 225:43–53
17. Boeri M, Verri C, Conte D et al (2011) MicroRNA signatures in tissues and plasma predict development and prognosis of computed tomography detected lung cancer. Proc Natl Acad Sci U S A 108:3713–3718
18. Chen C, Ridzon DA, Broomer AJ et al (2005) Real-time quantification of microRNAs by stem-loop RT-PCR. Nucleic Acids Res 33(20):e179
19. Shi R, Chiang VL (2005) Facile means for quantifying microRNA expression by real-time PCR. Biotechniques 39(4):519–525
20. Liu CG, Calin GA, Volinia S, Croce CM et al (2008) MicroRNA expression profiling using microarrays. Nat Protoc 3(4):563–578

21. Li W, Ruan K (2009) MicroRNA detection by microarray. Anal Bioanal Chem 394(4): 1117–1124
22. Hafner M, Landgraf P, Ludwig J et al (2008) Identification of microRNAs and other small regulatory RNAs using cDNA library sequencing. Methods 44(1):3–12
23. Motameny S, Wolters S, Nürnberg P et al (2010) Next generation sequencing of miRNAs – strategies, resources and methods. Genes 1:70–84
24. Kauppinen S, Vester B, Wengel J (2006) Locked nucleic acid: high-affinity targeting of complementary RNA for RNomics. Handb Exp Pharmacol 173:405–422
25. Sato F, Tsuchiya S, Terasawa K et al (2009) Intra-platform repeatability and inter-platform comparability of microRNA microarray technology. PLoS One 4:e5540
26. Jang JS, Simon VA, Feddersen RM et al (2011) Quantitative miRNA expression analysis using fluidigm microfluidics dynamic arrays. BMC Genom 12:144
27. Kang K, Peng X, Luo J et al (2012) Identification of circulating miRNA biomarkers based on global quantitative real-time PCR profiling. J Anim Sci Biotechnol 3(1):4
28. http://tools.invitrogen.com/content/sfs/manuals/cms_092509.pdf
29. Hurley J, Roberts D, Bond A et al (2012) Stem-loop RT-qPCR for microRNA expression profiling. Methods Mol Biol 822:33–52
30. http://www.qiagen.com/
31. http://www.exiqon.com
32. Pritchard CC, Cheng HH, Tewari M (2012) MicroRNA profiling: approaches and considerations. Nat Rev Genet 13(5):358–369
33. Vandesompele J, De Preter K, Pattyn F et al (2002) Accurate normalization of real-time quantitative RT-PCR data by geometric averaging of multiple internal control genes. Genome Biol 3:RESEARCH0034
34. Bustin SA, Benes V, Garson JA et al (2009) The MIQE guide- lines: minimum information for publication of quantitative real-time PCR experiments. Clin Chem 55:611–622
35. Bustin SA (2010) Why the need for qPCR publication guidelines? The case for MIQE. Methods 50:217–226
36. Peltier HJ, Latham GJ (2008) Normalization of microRNA expression levels in quantitative RT-PCR assays: identification of suitable reference RNA targets in normal and cancerous human solid tissues. RNA 14:844–852
37. Benes V, Castoldi M (2010) Expression profiling of microRNA using real-time quantitative PCR, how to use it and what is available. Methods 50:244–249
38. Wei J, Gao W, Zhu CJ et al (2011) Identification of plasma microRNA-21 as a biomarker for early detection and chemosensitivity of non-small cell lung cancer. Chin J Cancer 30: 407–414
39. Ji F, Yang B, Peng X et al (2011) Circulating microRNAs in hepatitis B virus-infected patients. J Viral Hepat 18:242–251
40. Meyer SU, Pfaffl MW, Ulbrich SE (2010) Normalization strategies for microRNA profiling experiments: a 'normal' way to a hidden layer of complexity? Biotechnol Lett 32:1777–1788
41. Brase JC, Johannes M, Schlomm T et al (2011) Circulating miRNAs are correlated with tumor progression in prostate cancer. Int J Cancer 128:608–616
42. Mestdagh P, Van Vlierberghe P, De Weer A et al (2009) A novel and universal method for microRNA RT-qPCR data normalization. Genome Biol 10:64–74
43. http://www.affymetrix.com
44. http://www.genomics.agilent.com
45. Geiss GK, Bumgarner RE, Birditt B et al (2008) Direct multiplexed measurement of gene expression with color-coded probe pairs. Nat Biotechnol 26(3):317–325
46. Sah S, McCall MN, Eveleigh D et al (2010) Performance evaluation of commercial miRNA expression array platforms. BMC Res Notes 3:80
47. Yauk CL, Rowan-Carroll A, Stead JD et al (2010) Cross-platform analysis of global microRNA expression technologies. BMC Genomics 11:330
48. Nelson PT, Wang WX, Wilfred BR et al (2008) Technical variables in high-throughput miRNA expression profiling: much work remains to be done. Biochim Biophys Acta 1779(11):758–765
49. Nagalakshmi U, Wang Z, Waern K et al (2008) The transcriptional landscape of the yeast genome defined by RNA sequencing. Science 320(5881):1344–1349
50. Mortazavi A, Williams BA, McCue K et al (2008) Mapping and quantifying mammalian transcriptomes by RNA-Seq. Nat Methods 5(7):621–628
51. Margulies M, Egholm M, Altman WE et al (2005) Genome sequencing in microfabricated high-density picolitre reactors. Nature 437(7057):376–380
52. Wang Z, Gerstein M, Snyder M (2009) RNA-Seq: a revolutionary tool for transcriptomics. Nat Rev Genet 10(1):57–63
53. Hackenberg M, Sturm M, Langenberger D et al (2009) miRanalyzer: a microRNA detection and analysis tool for next-generation

sequencing experiments. Nucleic Acids Res 37:W68–W76

54. Hackenberg M, Rodríguez-Ezpeleta N, Aransay AM (2011) miRanalyzer: an update on the detection and analysis of microRNAs in high-throughput sequencing experiments. Nucleic Acids Res 39:W132–W138
55. Zhu E, Zhao F, Xu G et al (2010) mirTools: microRNA profiling and discovery based on high-throughput sequencing. Nucleic Acids Res 38:W392–W397
56. Friedländer MR, Chen W, Adamidi C et al (2008) Discovering microRNAs from deep sequencing data using miRDeep. Nat Biotechnol 26(4):407–415
57. Wang WC, Lin FM, Chang WC et al (2009) miRExpress: analyzing high-throughput sequencing data for profiling microRNA expression. BMC Bioinformatics 10:328
58. Ro S, Yan W (2010) Small RNA cloning. Methods Mol Biol 629:273–285
59. Malone C, Brennecke J, Czech B et al (2012) Preparation of small RNA libraries for high-throughput sequencing. Cold Spring Harb Protoc 2012(10):1067–1077, pii: pdb.prot071431
60. http://www.genoseq.ucla.edu/images/a/a9/SmallRNA.pdf
61. Morozova O, Marra MA (2008) Applications of next-generation sequencing technologies in functional genomics. Genomics 92:255–264
62. Wang B, Howel P, Bruheim S et al (2011) Systematic evaluation of three microRNA profiling platforms: microarray, beads array, and quantitative real-time PCR array. PLoS One 6(2):e17167
63. Willenbrock H, Salomon J, Søkilde R et al (2009) Quantitative miRNA expression analysis: comparing microarrays with next-generation sequencing. RNA 15(11): 2028–2034
64. Ach RA, Wang H, Curry B (2008) Measuring microRNAs: comparisons of microarray and quantitative PCR measurements, and of different total RNA prep methods. BMC Biotechnol 8:69
65. Jensen SG, Lamy P, Rasmussen MH et al (2011) Evaluation of two commercial global miRNA expression profiling platforms for detection of less abundant miRNAs. BMC Genomics 12:435

Chapter 7

Introduction to Machine Learning

Yalin Baştanlar and Mustafa Özuysal

Abstract

The machine learning field, which can be briefly defined as enabling computers make successful predictions using past experiences, has exhibited an impressive development recently with the help of the rapid increase in the storage capacity and processing power of computers. Together with many other disciplines, machine learning methods have been widely employed in bioinformatics. The difficulties and cost of biological analyses have led to the development of sophisticated machine learning approaches for this application area. In this chapter, we first review the fundamental concepts of machine learning such as feature assessment, unsupervised versus supervised learning and types of classification. Then, we point out the main issues of designing machine learning experiments and their performance evaluation. Finally, we introduce some supervised learning methods.

Key words Machine learning, Supervised learning, Unsupervised learning, Clustering, Classification, Regression, Model complexity, Model evaluation, Performance metrics, Dimensionality reduction

1 Introduction

1.1 What Is Machine Learning?

In many scientific disciplines, the primary objective is to model the relationship between a set of observable quantities (inputs) and another set of variables that are related to these (outputs). Once such a mathematical model is determined, it is possible to predict the value of the desired variables by measuring the observables. Unfortunately, many real-world phenomena are too complex to model directly as a closed form input–output relationship. Machine learning provides techniques that can automatically build a computational model of these complex relationships by processing the available data and maximizing a problem dependent performance criterion. The automatic process of model building is called "training" and the data used for training purposes is called "training data." The trained model can provide new insights into how input variables are mapped to the output and it can be used to make predictions for novel input values that were not part of the training data.

Malik Yousef and Jens Allmer (eds.), *miRNomics: MicroRNA Biology and Computational Analysis*, Methods in Molecular Biology, vol. 1107, DOI 10.1007/978-1-62703-748-8_7, © Springer Science+Business Media New York 2014

To be able to learn an accurate model, machine learning algorithms often require large amounts of training data. Therefore, an important first step in using machine learning techniques is to collect a large set of representative training examples and store it in a form that is suitable for computational purposes. Recent advances in digital data gathering, storage, and processing capacity have made the application of machine learning possible in many domains such as medical diagnosis, bioinformatics, chemical informatics, social network analysis, stock market analysis, and robotics.

There is usually more than one computational model that can be trained for a given machine learning problem. Unfortunately, there is no fixed rule to select a particular model or an algorithm. The performance of a specific model depends on many factors such as the amount and quality of training data, the complexity and form of the relationship between the input and output variables, and computational constraints such as available training time and memory. Depending on the problem, it is often necessary to try different models and algorithms to find the most suitable ones. Fortunately, there are standard software packages that combine different algorithms into the same framework such as [1–4]. Once the available data is prepared in a suitable format, these packages make it simpler to try the different alternatives.

As an example, consider the problem of labeling a candidate nucleotide sequence as miRNA or not. One simple approach would be to determine a set of short nucleotide sequences that are parts of the known miRNA and non-miRNA sequences and to construct a set of rules based on the existence of these nucleotide "words." For example, one such rule can state that a sequence containing "AGCACU" is more likely to be a miRNA than not. Then one could simply label candidate sequences using these rules. In practice, constructing such a rule based system is very difficult as there are many possible nucleotide words and the mapping is very complex. Instead of manually specifying a complex set of rules, machine learning methods can automatically build a statistical model using these nucleotide words. These models can then be trained using large samples of biological data since the training process is automated. For machine learning, such rules (here a nucleotide hexamer) are determined from features which need to be defined for the input data.

1.2 What Are Features?

The observable quantities that are input to a machine learning algorithm are called "features." The algorithm learns a mapping from these features to the desired output variables by tuning the model parameters using the available training data. Therefore, it is important that the features are relevant to the prediction of the outputs.

For some machine learning problems, there are thousands of features that can be used to predict the output variables, e.g., gene expression in microarray experiments can be considered as features (*see* Chapters 6, 17, and 18). However, using all available features may not be the best approach. Features that are loosely related to the output might adversely affect the learning process by decreasing the effect of the important ones. Features that are strongly coupled with other features do not provide extra information and unnecessarily bias the result. These can further lower training performance by straining computational resources such as time and memory.

The first step in selecting good features is using expert judgment. An expert that knows the problem domain well can select a compact set of relevant features for input to the machine learning algorithm. This is especially important in the data gathering stage since collecting training data can be time consuming and costly. However, extra caution is required not to eliminate potentially important features. It is important to note that feature selection and extraction requires experience and is often an iterative process. As additional insight into the problem is gained, it might be necessary to add or remove features to improve the performance [5]. It is also possible to automate this feature selection and extraction process. Such automated techniques are detailed in Subheading 2.5.

For the miRNA identification problem, features can be the existence or the frequency of a selected set of nucleotide sequence "words" of small length within the candidate sequence. Again it is important to include all the available information that might help with the prediction. So in actuality, more features such as those that describe the number of base pairs, bulges, loops and asymmetric loops in different parts of the candidate sequence may also be included in the analysis [6].

After features, that well model the problem, have been defined, machine learning algorithms need to be chosen and in the field of miRNA detection supervised methods have been applied widely (*see* Chapters 10, 12, and 15–18).

1.3 What Is Unsupervised Versus Supervised Learning?

Machine learning techniques can be broadly classified into two main categories depending on whether the output values are required to be present in the training data.

1.3.1 Unsupervised Learning

Unsupervised learning techniques require only the input feature values in the training data and the learning algorithm discovers hidden structure in the training data based on them. Clustering techniques that try to partition the data into coherent groups fall into this category. In bioinformatics, these techniques are used for problems such as microarray and gene expression analysis. In general, market segment analysis, grouping people according to their social behavior, and categorization of articles according to their topic are popular tasks involving clustering. Typical clustering algorithms are K-means [7], hierarchical clustering [8], and spectral clustering [9].

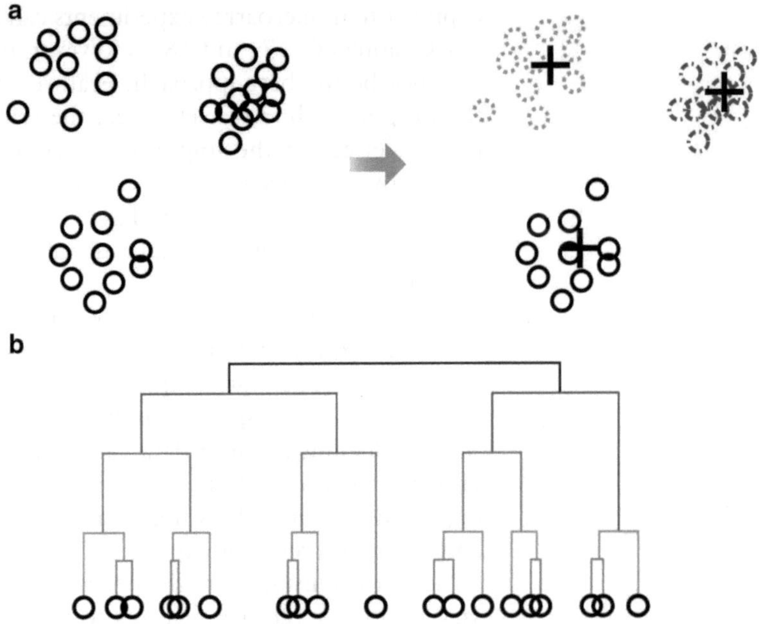

Fig. 1 Unsupervised clustering of data points (marked with *circles*). (**a**) The K-means algorithm groups the data into a given number of clusters (*K*) such that each data point is closer to the mean of its own cluster (depicted by *plus signs*) than any other cluster's centroid. (**b**) The hierarchical clustering method performs multiple rounds of clustering; merging the closest clusters or dividing the clusters of points at each round. The resulting clusters can be analyzed at multiple scales to find meaningful structures in the data distribution

It is not possible to directly measure the performance of clustering because the correct output labels are not known a priori. Instead, the performance depends on whether interesting trends in the data have been captured by the clusters or not. Since the output labels are not needed, it is often easier to collect a large training dataset for unsupervised algorithms.

Figure 1a shows an example result of clustering using the K-means algorithm. Let us briefly explain the steps of the algorithm. Firstly, user needs to define the number of clusters and initializes the centroid of each cluster (usually performed in a random manner). Then, each sample is assigned to the closest cluster centroid (cluster assignment step) and cluster centroids are recomputed using assigned samples (move centroids step). These two steps are iteratively performed until no further changes occur. Hopefully, in the end the clusters are well separated. However, K-means can get stuck in local optimum due to an unlucky initialization. Also, it is not very effective when the number of clusters (K) is not clear.

Hierarchical clustering is more suitable for the cases where the clusters are not well separated, i.e., the number of clusters is not obvious.

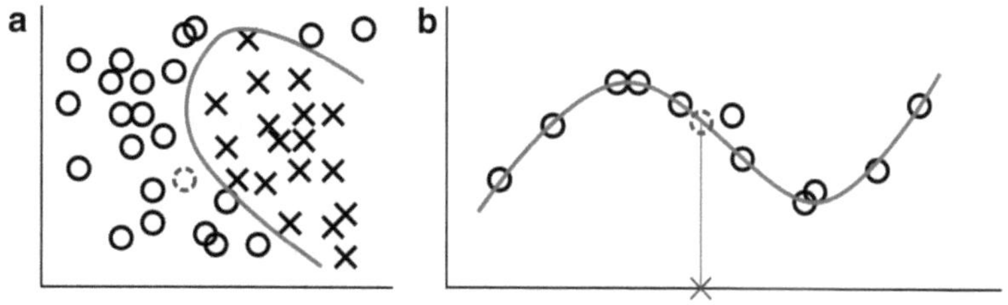

Fig. 2 Supervised machine learning problems. (**a**) In a classification problem, the training data belongs to one of several possible classes (the *solid circles* or *crosses*). A decision boundary (the *curve*) that best separates these data points is learned during training. At testing time, a novel data point (*dashed circle*) is classified as belonging to one of the classes depending on which side of the decision boundary it is on. (**b**) The goal in regression problems is to find a mapping from the inputs to the continuous output variable. A regression function (the *solid curve*) is fit to the training data (the *solid circles*). Afterwards it can be used to transform novel inputs (the *cross*) into output predictions (the *dashed circle*)

It performs multiple rounds of clustering; merging the closest clusters or divides the clusters at each round. Figure 1b shows a so-called dendrogram which can represent the result of hierarchical clustering. Any desired number of clusters can be obtained by "cutting" the dendrogram at the desired level.

1.3.2 Supervised Learning

Supervised learning methods require the value of the output variable for each training sample to be known. As a result, each training sample comes in the form of a pair of input and output values. The algorithm then trains a model that predicts the value of the output variables from the input variables using the defined features in the process. If the output variables are continuous valued then the predictive model is called a "regression function." For example, predicting the air temperature at a certain time of the year is a regression problem. If the output variables take a discrete set of values then the predictive model is called a "classifier." A typical classification problem is automated medical diagnosis for which a patient's data need to be classified as having a certain disease; or whether a given input is a miRNA. Figure 2 illustrates these two kinds of problems.

For supervised learning problems, it is possible to quantify the performance of the learned model by measuring the difference between the known output values and the predicted ones. However, the error for this performance evaluation must not be measured on the training data but on a separate test set. This ensures that the algorithm performance on novel data can be estimated correctly and gives an idea about the generalization of the learned model. The training and test procedures are discussed in more detail in Subheading 2.2.

Since it is much easier to gather unlabeled data, there are also semi-supervised learning methods that combine a small supervised training dataset with a larger unsupervised one. While training a predictive model, these algorithms can exploit both the supervised output values and the data distribution in the unlabeled data. However, these algorithms make additional assumptions to take advantage of the unlabeled data, which may or may not be suitable for the problem at hand [10].

As pointed out above, supervised learning with discrete results is called classification and the number of expected classes affects the choice of machine learning algorithm.

1.4 What Are Multi-class, Binary Classifications, and One-Class Density Estimation?

There are many machine learning problems with the objective of classifying the inputs into one of two categories. Often one category represents data points with a special property and the other category plays the role of a "background" class that includes everything else. Usually, the class representing the category of interest is termed the "positive" class and the background class is called the "negative" one. The miRNA identification problem is such an example with a category that represents miRNA sequences and another representing those that are not.

Classifiers that use machine learning techniques are often designed for such binary classification problems. This greatly simplifies their design and analysis. During training, these classifiers learn a decision boundary (*see* Fig. 2a) in the feature space that separates data points of the two classes as well as possible. Once the training is complete, they can predict the class of a new data point by comparing its location in the feature space with the learned decision boundary.

When there are more than two possible classes, the classification problem is said to be multi-class. Some machine learning techniques such as Decision Trees (DTs) can naturally handle the existence of multiple classes. Others such as Support Vector Machines (SVMs) can only handle binary problems in their original design. There are several ways to extend a binary classifier to handle multiple classes. A general approach is to turn a multi-class problem into multiple binary classification problems each in the form of "one class against all the others." When classifying test data, all binary classifiers are evaluated and the one with the highest confidence score wins (Fig. 3).

1.4.1 One Class Density Estimation

For some classification problems, it is not possible to collect reliable data belonging to one of the classes. Assume that you are working on diagnosing a rare type of cancer using some features obtained from the body cell. To do that, you develop a machine learning algorithm which would give "positive" as a result when the patient has cancer. To perform an effective training for your algorithm you would try to collect as many samples as possible.

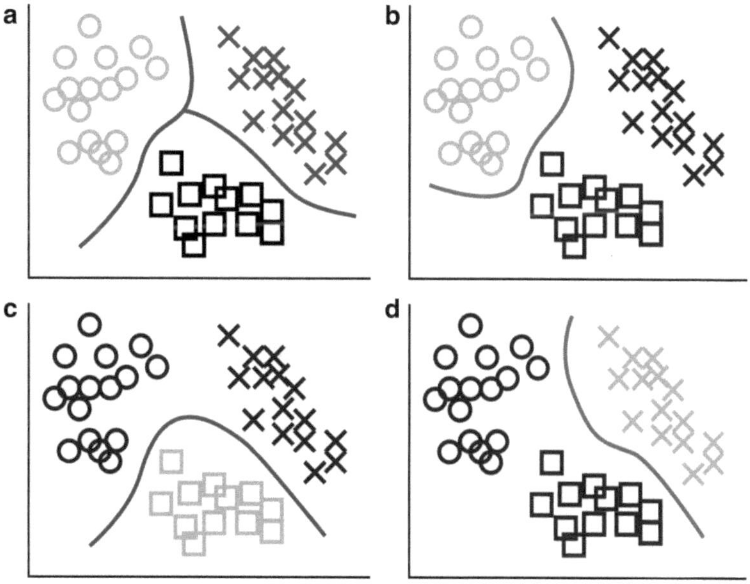

Fig. 3 Multi-class classification. (**a**) Some classification algorithms can handle multiple classes naturally to fit a complex decision boundary that separates all the classes from each other. (**b–d**) Some algorithms are designed to work with only two classes. In this case a multi-class problem can be decomposed into several binary classification problems with separate decision boundaries

However, in such a case, you probably would end up with many more "negative" samples than the "positive" ones. In other words, your dataset does not have a balanced amount of samples from different classes.

Estimation of one-class densities is also referred to as "anomaly detection" since the rare class indicates an anomaly within the huge amount of "normal" samples. Labeling a sample as "normal" is not as easy as labeling an "anomalous" one, because concealed anomalies may exist. A machine learning algorithm in such a case is trained to discover the common properties of the normal class to distinguish the anomalous samples from the rest.

For example, microRNAs can be identified experimentally but it is not currently feasible to clearly state that a given hairpin from a genome is not a miRNA (*see* Chapter 10) so the miRNA classification problem is also essentially a one class density estimation problem.

2 Design of Machine Learning Experiments

2.1 Model Complexity and Generalization

When given a dataset and a machine learning technique, we need to perform experiments to examine if the algorithm is working properly on the data and to gain insight on how to improve its performance.

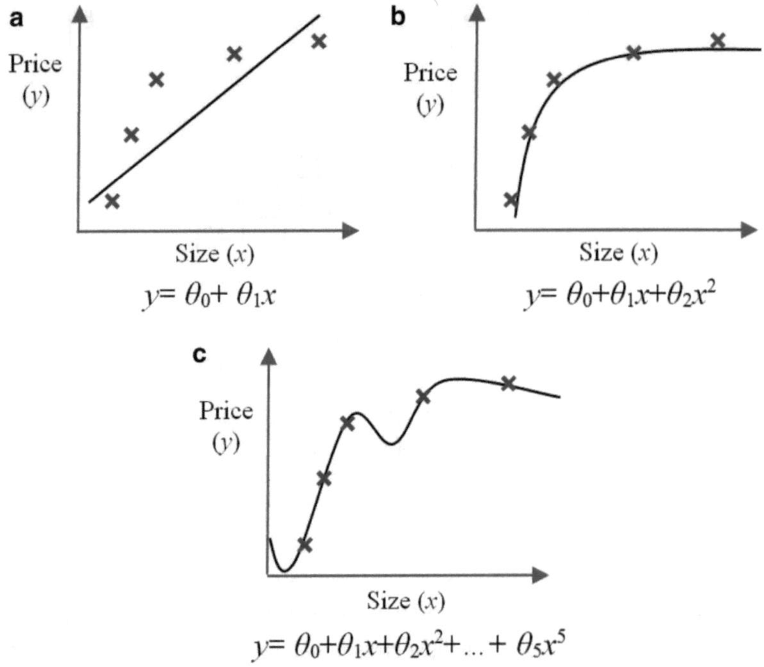

Fig. 4 Three hypotheses with different complexities for the house price prediction problem. (**a**) Underfit, hypothesis is a *line*. (**b**) Proper fit, hypothesis is a second degree *polynomial*. (**c**) Overfit, hypothesis is a fifth degree *polynomial*

We evaluate several hypotheses (models) to choose the best among them and this process is called model selection.

We consider the fact that the aim of a machine learning algorithm is to generate the correct output for sample data points outside the training set. The ability of the model to predict correct output for new samples after trained on the training set is defined as generalization. For best generalization, we should match the complexity of the model with the complexity of the function underlying the data [11].

To give a concrete example, let us consider a regression problem to predict the price of a house when its size is given. For the sake of simplicity, the size is the only feature in this example. In Fig. 4, crosses represent the data that we use to train our model and three models (hypotheses) with different complexities are shown with solid lines/curves.

If the hypothesis is not complex enough to model the samples, we have underfitting. Figure 4a shows the result of fitting a line to the data sampled from a high order polynomial. As we increase the complexity (the number of θ parameters) of our model, the training error decreases and we reach a better fit as shown in Fig. 4b. The error, here, can be defined as the sum of the squared distances between the data samples and the polynomial model (Fig. 5).

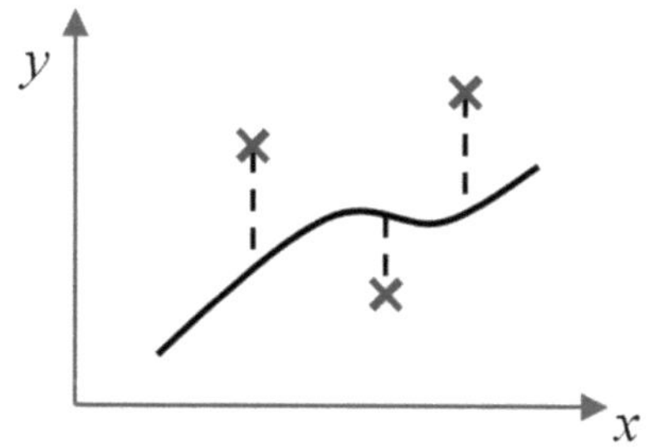

Fig. 5 The representation of the regression error. *Dashed lines* show the distances between the data samples and the polynomial. The error is the sum of the squares of these distances. Note that these are not the shortest distances to the model but the distances in *y* coordinate. This is correct since our estimate is the *y* coordinate (price) for a given *x* coordinate (size)

On the other hand, if the hypothesis is too complex and the data are not enough to constrain it, we may again end up with an improper model. For example, fitting a fifth order polynomial to some data sampled from a lower order polynomial (Fig. 4c). This is called overfitting. The hypothesis may fit the training set very well and we have quite low training error, however it fails to generalize to new samples (predicting prices of other houses).

For a given model complexity, the overfitting problem becomes less severe as the size of the dataset increases [12]. Ideally, when we have enough samples, a higher order polynomial becomes close to a lower order polynomial after training, so it resembles a proper fit. However, in most cases we cannot guarantee the sufficiency of data. Moreover, most of the time, the complexities of the model and data distribution cannot be visually compared like in this toy example. Therefore, we use other methods to evaluate the model as we will see in the following.

2.2 Using the Dataset for Evaluation

As explained previously, the generalization ability of a model can only be evaluated using samples outside the training set. To respect this requirement, we usually divide the training data into two parts. We use one part for training (e.g.: fitting a polynomial), and the remaining part is used to compute the error for that model to test its generalization.

The first part is called the training set and usually represents the bigger portion of the data (say 70 %). The second part is called the validation set. The model that gives the least error on the validation set is assumed to be the best.

In our regression example, to find a proper degree of the polynomial (this is the complexity of our model), we evaluate a number of candidate polynomials of different degrees (d), and find the coefficients (θ) using the training set for each of these polynomials (hypotheses/models). Let us denote the parameters

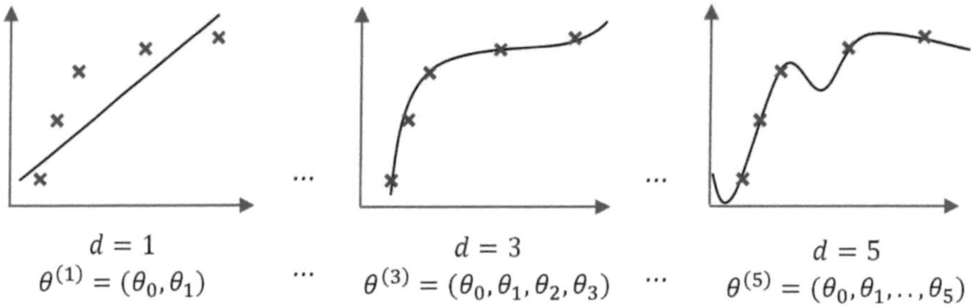

Fig. 6 Candidate models with different degrees for the polynomial regression problem. In this example, we evaluate *polynomials* with degree from *1* to *5*

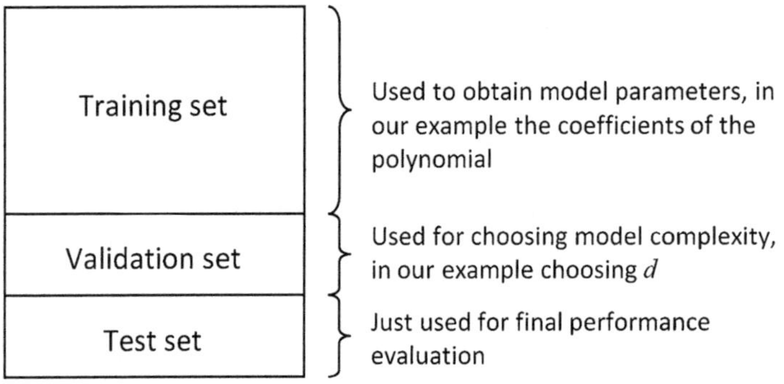

Fig. 7 Approximate distribution of the entire dataset for training, validation, and test sets

for the *i*th order polynomial with $\theta^{(i)}$. Figure 6 shows some of the candidate polynomials with different degrees. The next step is to compute the errors of these models on the validation set, and take the one that has the least validation error as the best polynomial. Let us say the model with $d = 3$ and $\theta^{(3)}$ generated the lowest error. In this case, we choose the third order polynomial as our model.

Although we have chosen the best performing model, we still do not know the real performance of the model since we have used the training set to estimate the parameters and the validation set to decide on the model complexity d (which is essentially a parameter as well). The test should contain new data. Therefore, while reporting the expected performance of our trained model, we use a third set, the test set, containing examples not used in training or validation.

In practice, most of the data is used for training and about one fifth for validation and again one fifth for testing (Fig. 7). Referring to our regression example, if we chose a third order polynomial as our model then, computing the error of $\theta^{(3)}$ on the test set gives us a fair error measure of the selected model.

2.2.1 Bias Versus Variance

In short, a model with high bias is an "underfit" one and a model with high variance is an "overfit" one. To better visualize the

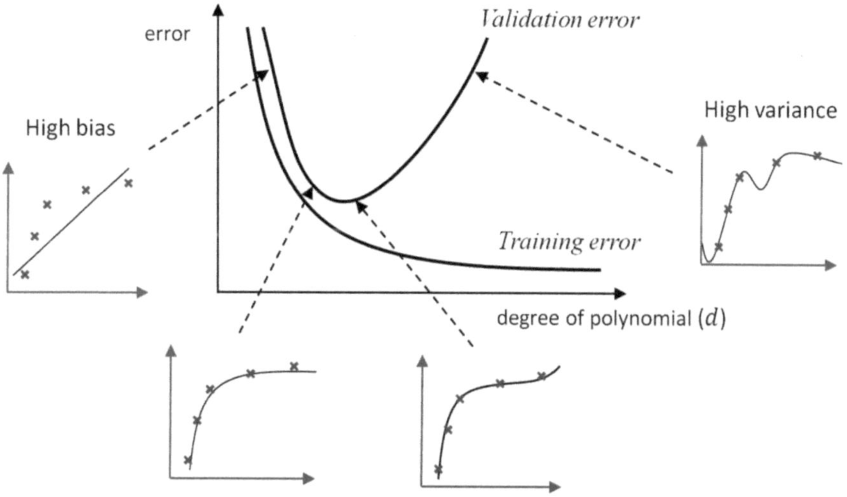

Fig. 8 Training and validation error curves for increasing model complexity. Simple models (small *d* in our example) have the risk of high bias, where the error is high both in training and validation sets. Complex models have the risk of high variance (fluctuations on the polynomial), where the training error is low since the model fits better to the training data, but the validation error is high

relationship between training and validation (generalization) errors, let us examine Fig. 8. Starting from a less complex model, as the complexity of the model increases, the training error decreases since the model fits better to the data. When we consider the error on the validation set, it initially decreases as it possibly fits better to the validation set as well. But then, as we move further to more complex models it increases again. High variance (fluctuations on the polynomial) in the complex models causes a poor fit (overfitting) for the novel data in the validation set. The bottom of the bowl on the validation error curve is the point where the generalization error is the minimum.

2.3 Dimensionality Reduction

The polynomial regression example in the previous section had just one input variable (x). For practical applications of machine learning, on the other hand, we have to deal with spaces of high dimensionality consisting of many input features [12].

This *multivariate* structure of the data generally causes problems for computation or visualization, and therefore this situation is referred to as the *curse of dimensionality* [13].

Dimensionality reduction is one of the major tasks in the analysis of multidimensional data, which is the step that we decrease the number of dimensions/features. The main motivations for performing dimensionality reduction are the following:

- Computation is faster with fewer features. Genomic data can be given as an example. If all the genes in a genome are considered as features, then we would have several thousand features.

- If we find out that one or more features are not discriminative, eliminating them saves time, effort, and increases prediction accuracy.
- Two or three dimensional projections help us (1) to visually represent our data to gain insight, (2) to screen data for obvious outliers and (3) to observe cluster tendencies when using unsupervised learning.

Since we wish to minimize the information loss to be caused by dimensionality reduction, we try to eliminate the least distinctive/informative features. For instance, a feature that is highly correlated with another one can be considered as redundant.

There are two main methods for reducing dimensionality: *feature selection* and *feature extraction*. In feature selection, we find k of the d dimensions (features) that give us the most information and we eliminate the other dimensions. Feature selection methods can be roughly divided into two categories: filters and wrappers. Filters extract feature relevancies via various scoring techniques without using a learning model and select a subset of features using these scores. Filters are computationally simple and fast. Some of the popular filter approaches are mutual information [14], chi-square [15], and information gain [16].

Wrappers, on the other hand, conduct a search for good features using the learning algorithm itself as part of the function [17]. Wrapper techniques provide interaction between feature subset and learning model, but are computationally expensive when compared to filters. The two approaches here are forward selection and backward elimination. Forward selection refers to a search that begins with an empty set of features. At each step, for all features, we add a feature in the feature subset and we train our model on the training set and calculate error on the validation set. Then, we choose the feature which causes the greatest decrease in error and permanently put it in the feature subset. This continues until no further improvement occurs. In backward elimination, we start with the full set of features and we remove one feature at a time. We eliminate the one, removing of which causes the least error increase [11].

In feature extraction, we transform the original d dimensions to a new set of dimensions and select k of these new dimensions. A popular technique to do the latter is principal component analysis (PCA), where we analyze the data and come up with the most informative components (dimensions).

Normally the reduction is performed for much higher dimensionalities but to visualize the process let us consider an example where three dimensional data is reduced to two dimensions. As mentioned above, the main idea is to capture the most informative dimensions. Figure 9a shows a set of data points in three dimensions (features). The data points roughly constitute a plane and

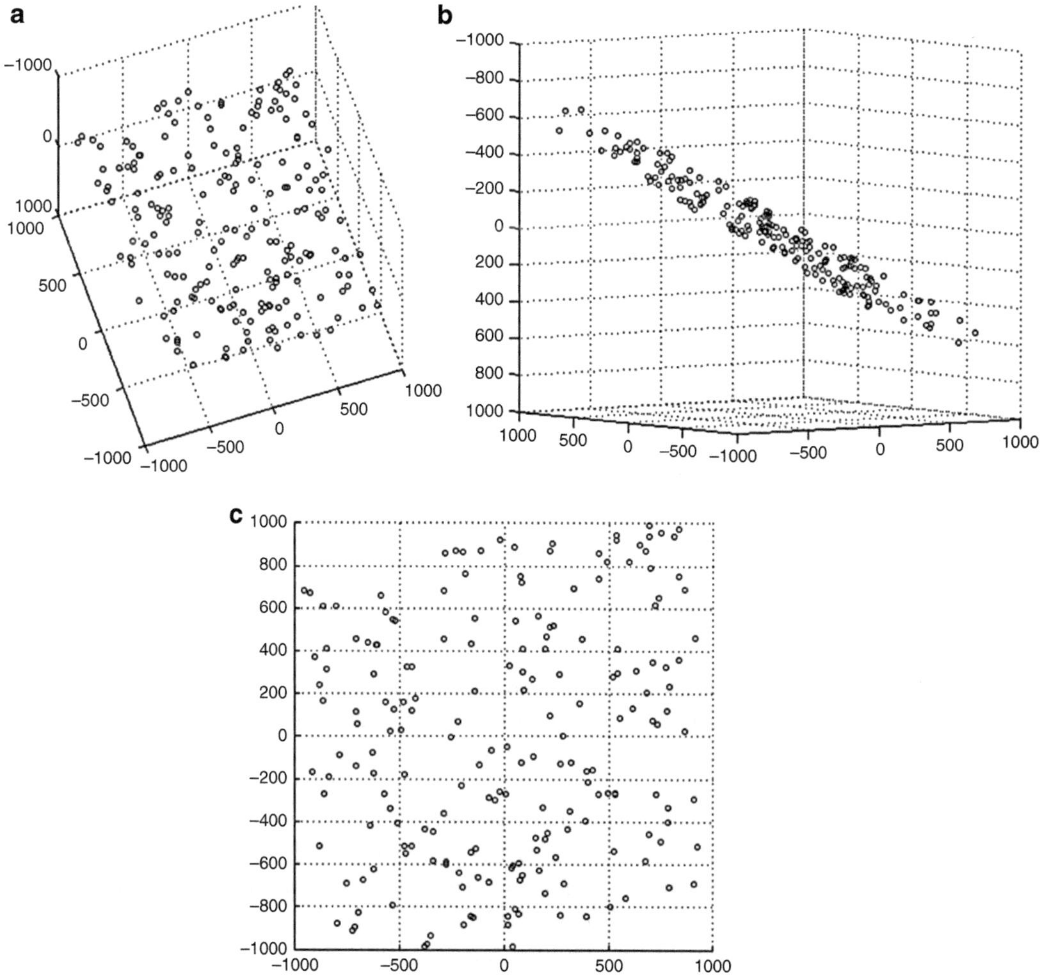

Fig. 9 3D to 2D dimensionality reduction example. (**a**) A set of data points in three dimensions (features). (**b**) The same data points viewed from another angle to emphasize that the points roughly constitute a plane. (**c**) 2D data points after the redundant dimension is removed with principal component analysis (PCA)

this fact is shown in Fig. 9b where the same data points are viewed from another angle. This means that the distinction between the data points can be represented in 2D and a third dimension does not add much information to this distinction because all the data points have approximately the same value on that dimension. The reader should note that selecting two features of the original data does not accomplish the desired reduction because the plane that the data is on, i.e., the redundant dimension is not one of the original *x*, *y* and *z* dimensions (axes) but a combination of these. PCA helps us to transform our data to a new set of three dimensions and in that space we can omit the redundant dimension to obtain a new 2D dataset (Fig. 9c).

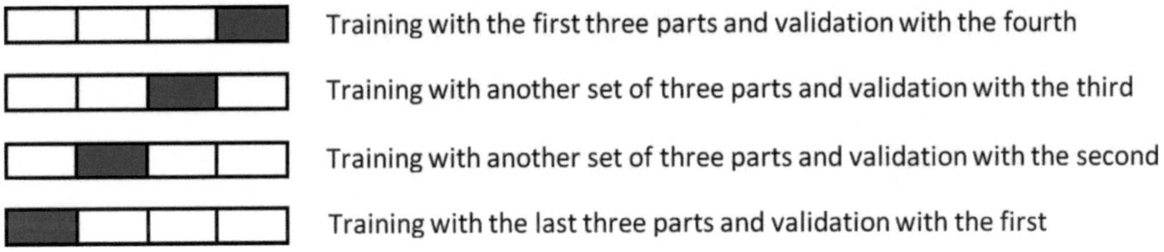

Fig. 10 Illustration of the partition of the dataset for K-fold Cross-Validation with $K=4$

Table 1
The so-called *confusion matrix* shows the four possible situations that can occur according to the truth values of the actual and the predicted class

	Predicted class label	Actual positive	Actual negative
Actual class label		+1	−1
Predicted positive	+1	True positive	False positive
Predicted negative	−1	False negative	True negative

2.4 Randomization and Cross-validation

Randomization is required to ensure that the result of the learning process is independent of the selection of training data [11]. This is a typical problem in real-world experiments. For instance, a part of some measurement data may have been taken when the device was in a certain state (slightly different tuning etc.).

As mentioned earlier, we need to divide our training data to obtain the training and validation sets (after sparing some part as the test set). We would like to ensure the random sampling of these sets from the data we have. If the dataset is large enough, we can randomly divide it into K parts, and then randomly divide each part into two as the training and validation sets. This means we repeat the experiment K times. Unfortunately, datasets are rarely large enough to do this. So randomization is accomplished by repeated use of the same data split differently; this is called *cross-validation.*

2.4.1 K-Fold Cross-validation

Illustrated in Fig. 10 for $K=4$, the dataset is divided randomly into K equal-sized parts. Then, $K-1$ parts are used to train a set of models and the remaining part is used as a validation set to evaluate those models. This procedure is repeated for all K possible choices [12]. As K increases, the percentage of the training set increases and we get more robust estimators, but the validation set becomes smaller. Therefore, a K value that ensures randomization is a good choice and larger values should be avoided.

2.5 Robust Performance Metrics

2.5.1 Precision–Recall

Let us first introduce the so-called *confusion matrix* (Table 1). As shown at the bottom-right portion of the table, there are four possible cases. For a positive example, if the prediction is also positive, it is a true positive; if our prediction is negative for an actually

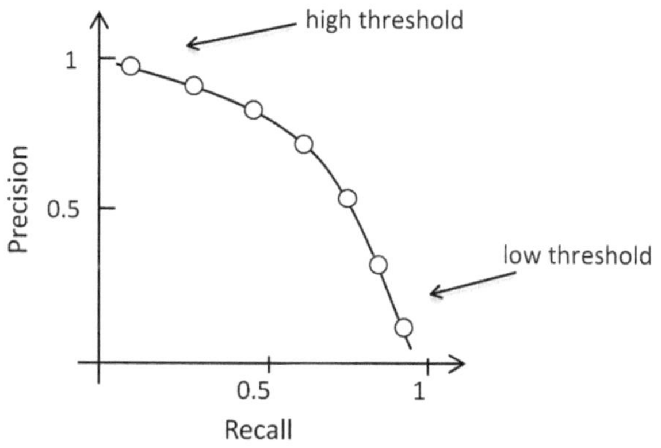

Fig. 11 Precision–Recall curve representation. *Circles* denote results of trials with different thresholds

positive example, it is a false negative. For a negative example, if the prediction is also negative, we have a true negative, and we have a false positive if we predict a negative example as positive.

For the cancer diagnosis example, where "positive" denotes having cancer, a false positive is wrongly making a cancer diagnosis for a healthy patient and a false negative is missing a patient actually having cancer. Precision tells us what fraction actually has cancer of all patients where we predicted "positive." Recall tells us what fraction we correctly detected as having cancer of all patients that actually have cancer.

With the definitions above, we can write

$$Precision = \frac{\#\,True\ Positives}{\#\,Predicted\ Positives} \qquad Recall = \frac{\#\,True\ Positives}{\#\,Actual\ Positives}.$$

The values in the confusion matrix, as well as precision and recall, change as we modify our detection algorithm's threshold, which defines at which probability a sample is labeled as "positive." Different threshold probability values can be chosen for different tasks or for preferred behavior regarding the same task.

Precision–Recall (PR) curves are generally used in the community for performance evaluation. A typical PR curve is shown in Fig. 11, where the circles denote results of trials with different thresholds. Changing the threshold of the algorithm moves us on the curve.

With a low threshold, we tend to predict "positive" for the data samples and our recall gets closer to 1 since we miss few actual positives (patients with cancer), however our precision is quite low since there are many false positives. On the other hand, if we choose a high threshold and we only predict "positive" for the most probable samples, our precision is high since the number of true and predicted positives become close to each other. But this time, recall is low. Ideally we want to keep both precision and recall high, this corresponds to the area under the curve.

Table 2
The F_1 scores of three different algorithms used for a detection problem

	Precision (*P*)	Recall (*R*)	F_1 Score
Algorithm 1	0.5	0.4	0.444
Algorithm 2	0.7	0.1	0.175
Algorithm 3	0.02	1.0	0.0392

The algorithm with precision and recall values close to each other has a higher F_1 score

A measure that was proposed to compare different precision–recall pairs (different thresholds) is the *F* score:

$$F_\beta = \left(1+\beta^2\right)\frac{Precision \times Recall}{\left(\beta^2 \times Precision\right) + Recall}$$

Most commonly, F_1 score is used (where $\beta = 1$). That is the *F* score corresponding to the harmonic mean of precision and recall:

$$F_1 = \frac{2 \times Precision \times Recall}{Precision + Recall}.$$

To concretize its effectiveness, the F_1 score for different algorithms is tabulated in Table 2, where the algorithm having precision and recall values close to each other has a distinctively higher F_1 score.

2.5.2 Specificity and Sensitivity

From another perspective but with the same aim, there are the two measures of *sensitivity* and *specificity*. Sensitivity is the same as recall. Specificity measures the proportion of negatives which are correctly identified, i.e., true negatives divided by the total number of negatives. One can also draw sensitivity vs. specificity curve using different thresholds.

2.5.3 ROC Curve

Receiver Operating Characteristics (ROC) curve is another graphical plot to illustrate the performance comparison of different methods. It is created by plotting, at various thresholds, the fraction of true positives out of the positives (TPR = true positive rate) vs. the fraction of false positives out of the negatives (FPR = false positive rate). TPR is the same as recall and sensitivity; FPR is 1 − specificity.

3 Supervised Machine Learning Methods

3.1 Probabilistic Classification Methods

A popular classification approach is to model the relationship between features and the class of the data points using probabilities. Let us denote the features as $x_i (i = 1, \ldots, M)$ and the feature vector for a data point then becomes:

$$\boldsymbol{x} = [x_1, x_2, \ldots, x_M].$$

We can write, for a data point, the probability of belonging to each of the N classes $(c_1, c_2, \ldots, c_N)$ as

$$P(C = c_1 \mid \boldsymbol{x}), P(C = c_2 \mid \boldsymbol{x}), \ldots, P(C = c_N \mid \boldsymbol{x}).$$

Given a new data point, it is classified as the class with the maximum probability,

$$c^* = \arg\max_{c_j} P(C = c_j \mid \boldsymbol{x}) \quad \mathit{and} \quad j = 1, \ldots, N.$$

The probability for each class can be computed using the Bayes' rule

$$P(C \mid \boldsymbol{x}) = \frac{P(\boldsymbol{x} \mid C) P(C)}{P(\boldsymbol{x})} = \frac{P(\boldsymbol{x} \mid C) P(C)}{\sum_{c_i \in C} P(\boldsymbol{x} \mid C = c_i) P(C = c_i)}.$$

As a toy example for the miRNA identification problem, assume that we have received a collection of nucleotide sequences. Of these 1,000 are miRNAs and 1,500 are not miRNAs that we call negative samples. The nucleotide motif "AGCACU" exists in 900 of the miRNA sequences and only 50 of the negative samples. We will have a single feature x that is equal to 1 if the candidate sequence contains "AGCACU" and 0 otherwise. We can compute the relevant probabilities as follows:

$$P(x = 0 \mid C = \mathit{miRNA}) = \frac{1{,}000 - 900}{1{,}000} = 0.1, \quad P(x = 1 \mid C = \mathit{miRNA}) = \frac{900}{1{,}000} = 0.9,$$

$$P(x = 0 \mid C = \mathit{negative}) = \frac{1{,}500 - 50}{1{,}500} = 0.967, \quad P(x = 1 \mid C = \mathit{negative}) = \frac{50}{1{,}500} = 0.033,$$

$$P(C = \mathit{miRNA}) = \frac{1{,}000}{2{,}500} = 0.40, \quad P(C = \mathit{negative}) = \frac{1{,}500}{2{,}500} = 0.60,$$

$$P(x = 0) = \sum_{c \in \left\{ \substack{\mathit{miRNA}, \\ \mathit{negative}} \right\}} P(x = 0 \mid C = c) P(C = c) = 0.1 \times 0.4 + 0.967 \times 0.6 = 0.62,$$

$$P(x = 1) = \sum_{c \in \left\{ \substack{\mathit{miRNA}, \\ \mathit{negative}} \right\}} P(x = 1 \mid C = c) P(C = c) = 0.9 \times 0.4 + 0.033 \times 0.60 = 0.38.$$

Given these probabilities, it is possible to compute the probability of each class $P(C|x)$ for a novel candidate sequence as follows:

$$P(C = miRNA \mid x = 0) = \frac{P(x = 0 \mid C = miRNA)P(C = miRNA)}{P(x = 0)} = \frac{0.1 \times 0.4}{0.62} = 0.06,$$

$$P(C = negative \mid x = 0) = \frac{P(x = 0 \mid C = negative)P(C = negative)}{P(x = 0)} = \frac{0.967 \times 0.6}{0.62} = 0.94,$$

$$P(C = miRNA \mid x = 1) = \frac{P(x = 1 \mid C = miRNA)P(C = miRNA)}{P(x = 1)} = \frac{0.9 \times 0.4}{0.38} = 0.95.$$

$$P(C = negative \mid x = 1) = \frac{P(x = 1 \mid C = negative)P(C = negative)}{P(x = 1)} = \frac{0.033 \times 0.6}{0.38} = 0.05.$$

With the calculated probabilities, if a given nucleotide sequence contains the word "AGCACU," it will be classified as miRNA because $P(C = \text{miRNA}|x = 1) > P(C = \text{negative}|x = 1)$.

Since we are taking the maximum over the classes, the $P(x)$ term in the denominator does not affect the predicted class and we can simplify the classification rule. It can be directly written as:

$$c^* = \arg\max_{c_j} P(\boldsymbol{x} \mid C = c_j)P(C = c_j) \quad and \quad j = 1, \ldots, N.$$

During training $P(x|C)$ and $P(C)$ are estimated from the training data points for each class.

When the features are continuous, there are many ways to model the probability $P(x|C)$. Parametric models assume a particular form for the probability that is controlled by several parameters as shown in Fig. 12a. A commonly used model is the Gaussian distribution $\mathcal{N}(\boldsymbol{\mu}, \Sigma)$ controlled by its mean $\boldsymbol{\mu}$ and covariance matrix Σ. The control parameters can be estimated from the training data. If a single Gaussian distribution is too simple to model the feature distribution, a mixture of Gaussians can be estimated by the Expectation Maximization (EM) technique [18]. The mixture is formed by a weighted sum of Gaussian distributions as $\sum_{i=1}^{K} \alpha_k \mathcal{N}(\mu_k, \Sigma_k)$, where α_k are the mixing coefficients that weigh the contribution from each Gaussian component. Unlike a single Gaussian, a mixture model can have multiple modes and therefore it is more general.

Another alternative is to model the probability $P(x|C)$ with nonparametric methods that do not have control parameters. A histogram over the feature space can estimate the data density in various regions of the partitioned feature space. As illustrated in Fig. 12b, by weighting the contribution from each sample over a range of histogram bins, the computed histogram can be smoothed to reduce errors in the sparsely populated regions of the feature space.

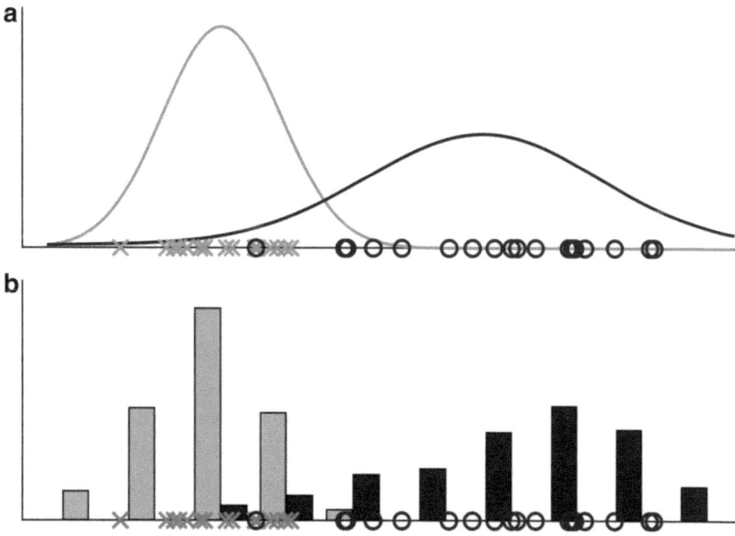

Fig. 12 Modeling $P(x|C)$ from the training data (the *crosses* and the *circles*). (**a**) Parametric models (such as the Gaussian curves depicted above) are functions that are controlled by a set of control parameters. The values of these parameters are estimated from the training data points. (**b**) Nonparametric models are histograms that are computed by separating the input feature space into distinct bins. Each training data point contributes to several bins around itself according to a local weighting function

3.1.1 Naïve Bayes

One common way to simplify the modeling of the joint feature probability $P(x|C)$ is to assume independence between features. In exact form, we can write

$$P(\boldsymbol{x} \mid C) = P(x_1 \mid x_2,\ldots,x_M,C)\cdot P(x_2 \mid x_3,\ldots,x_M,C),\ldots,P(x_M \mid C).$$

If we assume independence between features $x_1,\ldots,x_M$ then the joint probability reduces to

$$P(\boldsymbol{x} \mid C) \cong P(x_1 \mid C)\cdot P(x_2 \mid C),\ldots,P(x_M \mid C) = \prod_{j=1}^{M} P(x_j \mid C).$$

This simplification is called the naïve Bayes assumption and a classifier using such a model is called a naïve Bayes classifier. Although the independence assumption is quite strong and does not hold in general, naïve Bayes classifiers perform remarkably well for a wide range of problems. Moreover, the training can be very efficiently performed since each feature probability can be computed independently.

The independence assumption can improve training accuracy by reducing the number of model parameters that needs to be estimated. Consider a learning problem with F features that are real numbers. If the full joint probability is modeled as a multidimensional Gaussian, then $(F^2+3F)/2$ parameters are required to represent the mean and the covariance matrix. With the naïve Bayes assumption, this reduces to $2F$ parameters since each feature probability can be modeled by a one-dimensional Gaussian distribution. Hence, the model parameters can be more reliably estimated with limited training data.

Still, it is necessary to exercise caution when estimating the feature probabilities with a small number of data points. Since the feature probabilities are multiplied, a single zero probability for a feature can overcome strong evidence from several other features. As a precaution, virtual training samples that are uniformly distributed across the feature space and the classes can be used to make sure that no feature probability is ever exactly zero.

Naïve Bayes classifiers are also used in the miRNA identification task [6]. For a chosen dictionary of M nucleotide words that are likely to be present in miRNAs, a binary feature w_i represents whether word i exists in the nucleotide sequence, i.e., $w_i = 1$ if the word i is part of the sequence, $w_i = 0$ otherwise. For each such feature, the probabilities $P(w_i = 0|C = \text{miRNA})$, $P(w_i = 1|C = \text{miRNA})$, $P(w_i = 0|C = \text{negative})$, and $P(w_i = 1|C = \text{negative})$ are all estimated from a large sample of miRNA positive and negative examples. Given the words in a novel candidate sequence, a feature vector $w = [w_1, w_2, \ldots, w_M]$ can be extracted. The joint probabilities for both the "miRNA" and "negative" classes can be calculated as

$$P(\boldsymbol{w} \mid C = miRNA) = \prod_{i=1}^{M} P(w_i \mid C = miRNA) \text{ and}$$

$$P(\boldsymbol{w} \mid C = negative) = \prod_{i=1}^{M} P(w_i \mid C = negative).$$

The larger of these probabilities determines the label of the candidate nucleotide sequence. Of course, in an actual miRNA system, more complex features determined by experts are included in the statistical model (*see* Chapters 9, 10, and 12 and [6]).

3.2 Linear Discriminant Functions

The probabilistic approach outlined above first models the class conditional probabilities $P(x|C)$ then bases its classification estimates on the class with the maximum probability. In the case of a binary classification problem this amounts to first computing the ratio

$$\gamma = \frac{P(\boldsymbol{x} \mid C = c_1)}{P(\boldsymbol{x} \mid C = c_2)}.$$

And then, choosing c_1 if $\gamma > 1$, and c_2 otherwise. The region of the feature space where $\gamma = 1$ forms the decision boundary.

Depending on the form of $P(\boldsymbol{x}|C)$, the decision boundary can be very complex or just a simple hyper-plane. A hyper-plane in a d dimensional space is a $d-1$ dimensional flat region with the equation $a_1x_1 + a_2x_2 + \cdots + a_dx_d = \boldsymbol{a}^T\boldsymbol{x} = c$. A hyper-plane in two dimensions is a line and a hyper-plane in three dimensions is a plane. If $P(\boldsymbol{x}|C)$ is a Gaussian distribution with the same covariance matrix Σ for both classes then the decision boundary is exactly a hyper-plane [12]. This linear decision boundary can be written as

$$y=\boldsymbol{a}^{\mathrm{T}}\boldsymbol{x}+c, \text{ with } \boldsymbol{a}=\Sigma^{-1}(\boldsymbol{\mu}_1-\boldsymbol{\mu}_2) \text{ and } c=-\tfrac{1}{2}\mu_1^{\mathrm{T}}\Sigma^{-1}\mu_1+\tfrac{1}{2}\mu_2^{\mathrm{T}}\Sigma^{-1}\mu_2.$$

3.2.1 Fisher's Linear Discriminant

In case the form of $P(x|C)$ is not Gaussian, we can still exploit the formulation above to find a linear separation boundary that distinguishes between the classes in an optimal way. Fisher's linear discriminant method achieves this by projecting the input data points onto a hyper-plane such that their data distributions are as far apart from each other as possible. The projection that maximizes the separation between distributions is given as $y=\boldsymbol{a}^{\mathrm{T}}\boldsymbol{x}$ with $\boldsymbol{a}=(\Sigma_1+\Sigma_2)^{-1}(\boldsymbol{\mu}_1-\boldsymbol{\mu}_2)$. Note that the data distributions do not need to have the same covariance matrix. If they do, then the Fisher's linear discriminant is equivalent to the Gaussian formulation above. Once the data is projected into one dimension y, a threshold is computed so that the prediction error is minimized on the training data. Both the Fisher's linear discriminant and the Gaussian formulation above can be easily generalized for multi-class classification problems.

3.3 Support Vector Machines

For two linearly separable classes with unknown distributions, there is no unique decision boundary. As long as the selected hyper-plane separates the training samples of the two classes, it can be chosen as the decision boundary. However, it is prudent to select a decision boundary that does not pass too close to the training samples to account for the limited training data and errors in data collection.

Support Vector Machines (SVMs) rely on maximizing the margin of error to select the best hyper-plane. The margin is determined by a set of hyper-planes parallel to the decision boundary on the positive and negative sides of the discriminant function each at the same distance to the boundary as depicted by Fig. 13. When the margin is maximized, the training data points that are closest to the decision boundary are on the margin hyper-planes. These training data points are called the "support vectors."

Since the margins and the decision boundary are only determined by the support vectors, the SVM classification rule can be written as a function of these points. If we have a two class problem and we label the classes with $\{-1,+1\}$, the ith training data point can be written as $X^i=\{\boldsymbol{x}^i,y^i\}$, where x^i is the feature vector and $y^i\in\{-1,+1\}$ is the supervised class label. The support vector decision boundary corresponds to the hyper-plane equation $\gamma=\boldsymbol{w}^{\mathrm{T}}\boldsymbol{x}+c$ and the weight vector is given by $\boldsymbol{w}=\sum_i \alpha^i y^i \boldsymbol{x}^i$ with $\alpha^i=0$ for the training samples that are not support vectors.

In practice, it is usually not possible to completely separate all training samples by a hyper-plane and some training samples can end up on the wrong side of the decision boundary or within the margin.

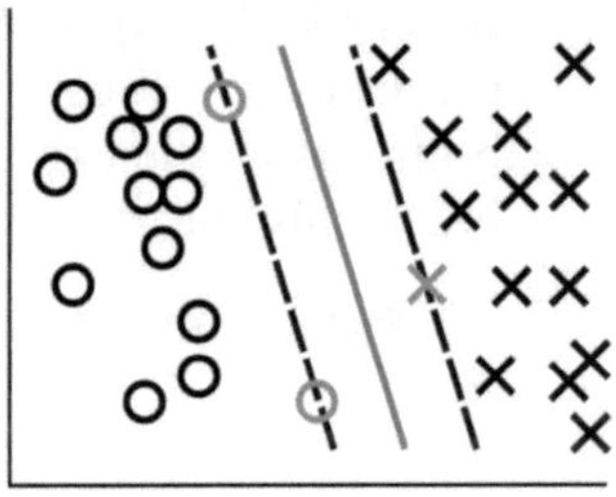

Fig. 13 The decision boundary of an SVM (the *solid line*) is determined by the support vectors (the *lighter circles* and *crosses*) that lie on the margin hyperplanes (the *dashed lines*). The support vectors and the decision boundary are selected to maximize the margin

The SVM formulation is softened to allow such data points. A complexity parameter, usually denoted by *C*, controls how much these points are penalized. A higher penalty means a more complex model with potentially more support vectors. The value of this parameter should be set using a validation dataset as discussed before.

3.3.1 Kernels

Most real-world problems involve data distributions that are not linearly separable. One possible approach around this problem is to perform a nonlinear transformation of the input features into a higher dimensional space as $\boldsymbol{x} \rightarrow \Phi(\boldsymbol{x})$. This transformation can make the class data distributions linearly separable. Then the linear SVM can be trained in the new space since the linear decision boundary in this space corresponds to a nonlinear curve in the original feature space as illustrated by Fig. 14.

The SVM formulation is particularly suitable to this kind of transformation since the feature vectors always appear in the form of dot products of two data points as $K(\boldsymbol{x},\hat{\boldsymbol{x}}) = \boldsymbol{x}^{\mathrm{T}}\hat{\boldsymbol{x}}$. This dot product is called a "kernel" and it is a measure of similarity between the two data points $\boldsymbol{x}$ and $\hat{\boldsymbol{x}}$. So even if the transformation $\Phi(\boldsymbol{x})$ is complex and very high dimensional, after the transformation the dot product $K(\boldsymbol{x},\hat{\boldsymbol{x}}) = \Phi(\boldsymbol{x})^{\mathrm{T}}\Phi(\hat{\boldsymbol{x}})$ may have a simple form. Indeed, the form of $K(\boldsymbol{x},\hat{\boldsymbol{x}})$ can be set directly without ever computing the nonlinear mapping $\Phi(\boldsymbol{x})$, provided that the kernel form satisfies some mathematical constraints [19].

For example, the Gaussian kernel can be written as $K(\boldsymbol{x},\hat{\boldsymbol{x}}) = e^{-\frac{\|\boldsymbol{x}-\hat{\boldsymbol{x}}\|^2}{2\sigma^2}}$. Other kernels that are commonly used are the polynomial, the sigmoid and the radial basis function (RBF) kernels. Since each kernel corresponds to a different nonlinear transformation of the input space, it is not possible to know which one will be the best choice for a particular machine learning problem. Also kernels usually have parameters that define the shape and the complexity of the nonlinear transformation such as $\boldsymbol{\sigma}$ for the Gaussian kernel. Both the type and the parameters of the kernel should be selected using a validation set as described before in Subheading 2.3.

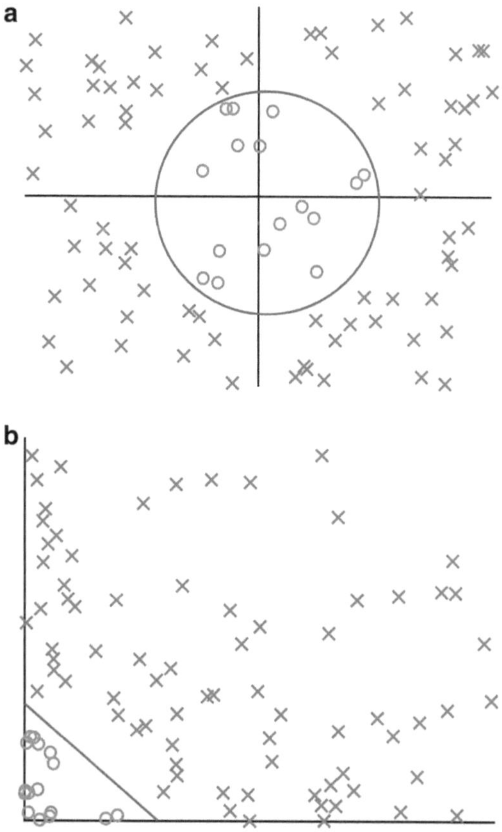

Fig. 14 Nonlinear transformations can linearize the decision boundary between classes. (**a**) In the original feature space, the classes are not linearly separable. (**b**) Each feature value is squared to perform a nonlinear mapping of the data points. In the transformed feature space, the decision boundary is a *simple line*

4 Conclusion

Machine learning techniques provide exciting new ways to exploit the available computational power and data in a variety of scientific domains. They can analyze huge amounts of data in a relatively short time that is not possible by manual labor. This provides opportunities for scientists to develop new experimental procedures and to channel their efforts on the most promising questions of their problem domain.

However, automated solutions are not a replacement for good scientific judgment. Like any other tool, a machine learning technique needs to be utilized in a careful manner to make the most out of its use. It is better to start with the simpler methods to judge problem difficulty and to gain more insight about algorithm behavior. It is also important to try a few different algorithms and compare their performances. As discussed in Subheading 2, experiments to test the generalization ability of a model should be designed properly considering the important aspects such as choosing the training samples and randomization.

References

1. RapidMiner -- Data mining, ETL, OLAP, BI, http://sourceforge.net/projects/rapidminer/
2. scikit-learn: machine learning in Python, http://scikit-learn.org/stable/
3. The SHOGUN machine learning toolbox, http://www.shogun-toolbox.org/
4. Weka 3 - Data mining with open source machine learning software in Java, http://www.cs.waikato.ac.nz/ml/weka/
5. Guyon I, Elisseeff A (2003) An introduction to variable and feature selection. J Mach Learn Res 3:1157–1182
6. Yousef M, Nebozhyn M, Shatkay H et al (2006) Combining multi-species genomic data for microRNA identification using a Naïve Bayes classifier. Bioinformatics 22: 1325–1334
7. MacQueen J (1967) Some methods for classification and analysis of multivariate observations. Proceedings of the fifth Berkeley symposium on mathematical statistics and probability. University of California Press, Los Angeles, CA, pp 281–297
8. Hastie T, Tibshirani R, Friedman JH (2003) The elements of statistical learning. Springer, New York, NY
9. Ng AY, Jordan MI, Weiss Y et al (2002) On spectral clustering: analysis and an algorithm. Adv Neural Inform Process Syst 2:849–856
10. Chapelle O, Schölkopf B, Zien A (eds) (2010) Semi-supervised learning. The MIT Press, Cambridge, MA
11. Alpaydın E (2010) Introduction to machine learning. The MIT Press, Cambridge, MA
12. Bishop C (2006) Pattern recognition and machine learning. Springer, New York, NY
13. Bellman RE (1961) Adaptive control processes: a guided tour. Princeton University Press, Princeton, NJ
14. Liu H, Sun J, Liu L et al (2009) Feature selection with dynamic mutual information. Pattern Recogn 42:1330–1339
15. Chen Y-T, Chen MC (2011) Using chi-square statistics to measure similarities for text categorization. Expert Syst Appl 38:3085–3090
16. Lee C, Lee GG (2006) Information gain and divergence-based feature selection for machine learning-based text categorization. Inform Process Manag 42:155–165
17. Kohavi R, John GH (1997) Wrappers for feature subset selection. Artif Intell 97:273–324
18. Dempster AP, Laird NM, Rubin DB (1977) Maximum likelihood from incomplete data via the EM algorithm. J Roy Stat Soc B 39:1–38
19. Schlkopf B, Smola AJ (2001) Learning with kernels: support vector machines, regularization, optimization, and beyond. The MIT Press, Cambridge, MA

Chapter 8

Introduction to Statistical Methods for MicroRNA Analysis

Gökmen Zararsiz and Erdal Coşgun

Abstract

MicroRNA profiling is an important task to investigate miRNA functions and recent technologies such as microarray, single nucleotide polymorphism (SNP), quantitative real-time PCR (qPCR), and next-generation sequencing (NGS) have played a major role for miRNA analysis. In this chapter, we give an overview on statistical approaches for gene expressions, SNP, qPCR, and NGS data including preliminary analyses (pre-processing, differential expression, classification, clustering, exploration of interactions, and the use of ontologies). Our goal is to outline the key approaches with a brief discussion of problems avenues for their solutions and to give some examples for real-world use. Readers will be able to understand the different data formats (expression levels, sequences etc.) and they will be able to choose appropriate methods for their own research and application. On the other hand, we give brief notes on most popular tools/packages for statistical genetic analysis. This chapter aims to serve as a brief introduction to different kinds of statistical methods and also provides an extensive source of references.

Key words MicroRNA, MicroRNA data analysis, Gene expression analysis, Next-generation sequencing analysis, SNP–SNP interactions, Bioconductor

1 Introduction

MicroRNA profiling is an important task to understand miRNA functions and recent technologies such as microarray, single nucleotide polymorphism (SNP), quantitative real-time PCR (qPCR), and next-generation sequencing (NGS) have played a major role in their elucidation. In most of these analyses, classical statistical methods cannot be used because of the high-dimensionality of the datasets obtained from these technologies (curse of dimensionality; see below and Chapter 7 in this volume for more details). Most of the classical methods are based on the assumption that the number of dimensions should be less than the number of observations. However, it is not possible in these datasets and novel statistical methods are proposed to overcome this problem. Most useful approaches are "Machine Learning" and "Artificial Intelligence" (1–5, Chapter 7 in this volume). Despite the dimension problem,

Malik Yousef and Jens Allmer (eds.), *miRNomics: MicroRNA Biology and Computational Analysis*, Methods in Molecular Biology, vol. 1107, DOI 10.1007/978-1-62703-748-8_8, © Springer Science+Business Media New York 2014

it is possible to get good results from these methods which can then be used in a variety of computational biology problems, including the identification of genes related to diseases, the discovery of the gene groups having similar expressions, and estimation of expressions in response to drug doses. There are some considerations in making these analyses, and data pre-processing is one of the most important of them. Because deficiencies and errors affect the results obtained from the data and will not be useful for interpretations so it is essential to perform some pre-processing. The next thing to do after pre-processing and selecting the correct analysis is to do an ontological analysis. Many different tools and databases are available to do this.

The most used data types in genetic studies are sequencing and gene expression datasets [1, 2, 6, 7]. Sequencing data consisting of nucleotide sequences and they are very important for example to determine genomic regions related with diseases. In this context, the most common data types are SNP and NGS data. Determination of interactions, identification of genome regions, "alignment" and "mapping" are widely used analyses using these data formats. In addition to these, determining SNPs and their interactions is an important issue [8, 9]. If we can determine which SNPs are conjugated, this will provide necessary support to physicians' decisions. Researchers have been able to find solutions to many problems using these methods.

NGS data analysis is a very recent field in genetic research [8, 9]. In respect to the Sanger sequencing technique, this method is able to read more "short" reads in a very short time for example due to miniaturization and parallelization. This is important for both the cost and labor time and to speed up knowledge building necessary for decision making. Another difference to other sequencing methods is the "quality score" which is obtained for all sequenced reads. With this information, researchers can use the most informative reads. Knowledge of appropriate statistics is crucially important for analyzing these data types and trained staff will greatly benefit the work of the laboratory as well as experimental design. The experimental data will turn into knowledge quickly if statisticians are consulted at every stage of experiments.

Furthermore, differential expression (DE) analyses are essential tools for genomics in the detection of significant genes of diseases or conditions of interest. After, the DE genes can be used for disease diagnosis and treatment.

The paper is organized as following: Subheadings 2–5 address widely used methods of microarray, NGS, SNP, and qPCR analyses, respectively. The paper concludes in Subheading 6.

2 Microarray Analysis

2.1 Microarray Technology, Gene Expression, and Gene Expression Data

In microarrays thousands of probes are synthesized on a solid surface, which may be either glass or a silicon chip. The mRNA samples or probe targets are marked with fluorescent dyes and hybridized to their matching probes. The amount of hybridization estimates the respective target transcripts amounts, and is measured by the amount of fluorescent emission on their related spots. There are several microarray platforms which differ in array fabrication, probe length, number of fluorescent dyes being used, etc. An illustration of this process is shown in Fig. 1.

Experiments in genomics and gene expression studies can be used for predicting and diagnosing complex diseases by discovering new genes associated with pathways, expression markers or drug targets. It is feasible to include genes on an array which are entirely uncharacterized, since the arrays are able to be designed and made based upon only partial sequence information. The nature of this approach is similar to that of classical genetics in which mutations are random. Also selections are set up to identify mutants with interesting phenotypes [10].

To estimate the expression level, red and green point densities are obtained from image analysis and these densities are calculated using the differences between foreground and background of specific probes on a microarray chip. After that, $\log_2$ ratios of these densities are computed and further interpretations are made based on these ratios. A major advantage of this transformation is to produce a continuous mapping space for the expression values and treating up and down regulated genes in a similar and comparable fashion (Fig. 2). The gene expression level can be calculated as follows:

$$\textit{Density of Red} = \mathrm{R}_{fg} - \mathrm{R}_{bg}$$

$$\textit{Density of green} = \mathrm{G}_{fg} - \mathrm{G}_{bg}$$

$$fg = \textit{foreground}, \quad bg = \textit{background}$$

$$\textit{Expression Level} = \log 2\left(\textit{Density of Red} \,/\, \textit{Density of Green}\right).$$

The description of gene expression data is shown in Fig. 3 and the pipeline of gene expression analysis is shown in Fig. 4.

2.2 Pre-processing Gene Expression Data

The first step in a statistical analysis is the pre-processing of the data. That is one of the most important parts of an analysis because we cannot ensure reliable results with "raw data" which came from microarrays.

Pre-processing includes the following:

1. Log-transformation.
   Expression levels are expressed as ratios. Therefore, they can be scaled asymmetrically. High expressions will have larger

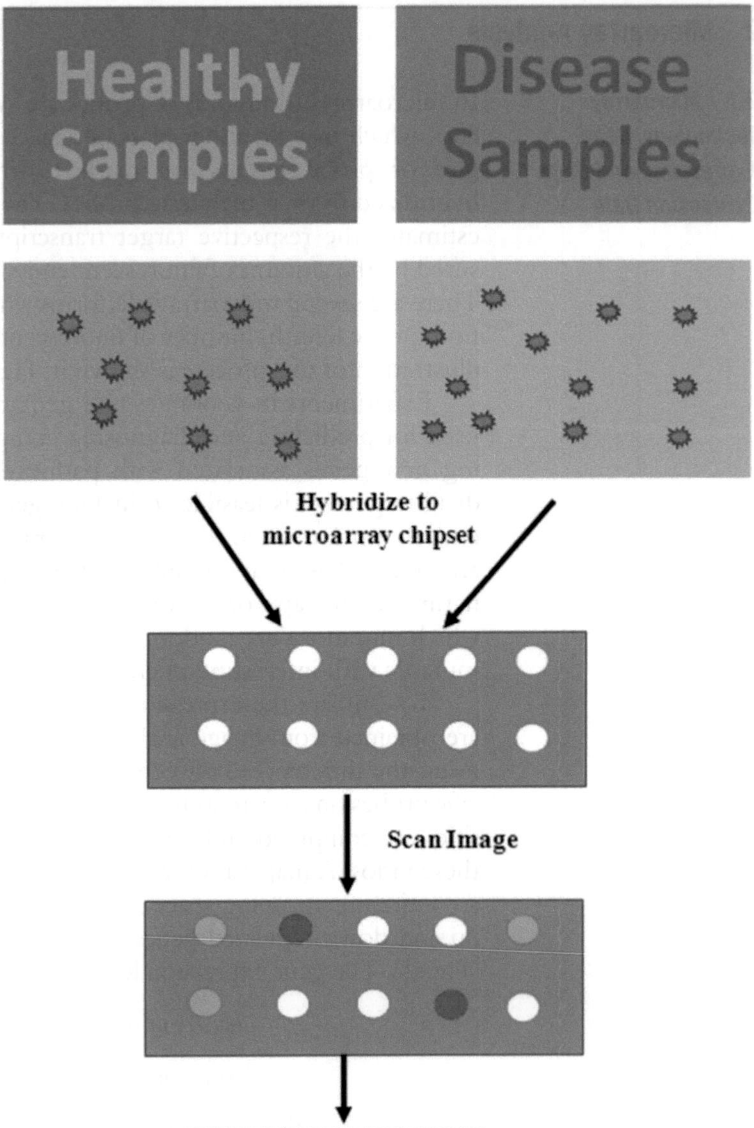

Gene Name	Healthy	Disease
Gene1	1	1
Gene2	2	0
Gene3	2	0
Gene4	2	2

Fig. 1 Preparation of target samples: the process from the cell samples to the microarray

values between 1 and ∞. On the other hand lower expressions will be between 0 and 1. We, however, need a balanced scale to be able compare these ratios. The most popular way to ensure balance is log-transformation.

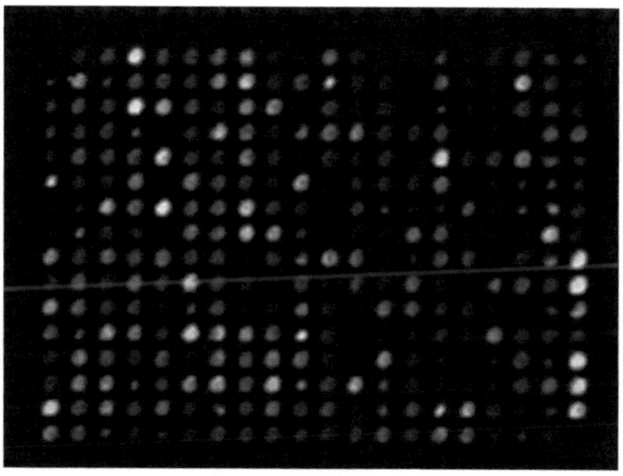

Fig. 2 An image result of microarray technology

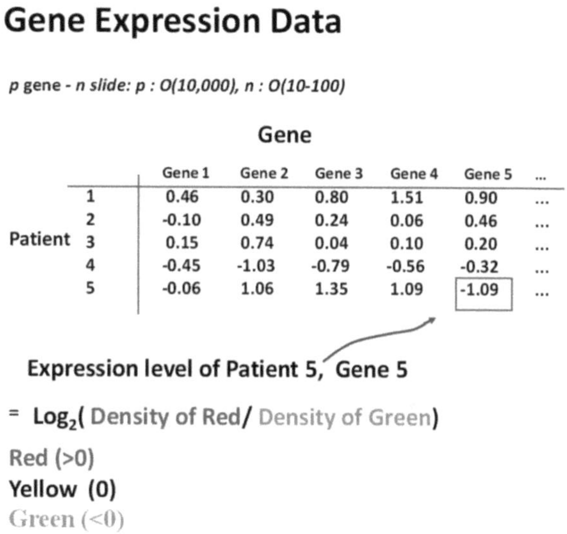

Fig. 3 Description of gene expression data

2. Gene replicate handling.
   In microarray experiments we may have several values for the same spot or gene (replicates). But we have to use only one result for downstream statistical calculations. To overcome this problem we first check the inconsistencies between replicates. Then, we use the average or median values of all replicates.
3. Management of missing values.
   This is a classical problem in microarray experiments. We can use different approaches for handling missing values. (1) filling blanks with "0", (2) use row median or average, and (3) use similarity distance measures (e.g., Euclidean distance).

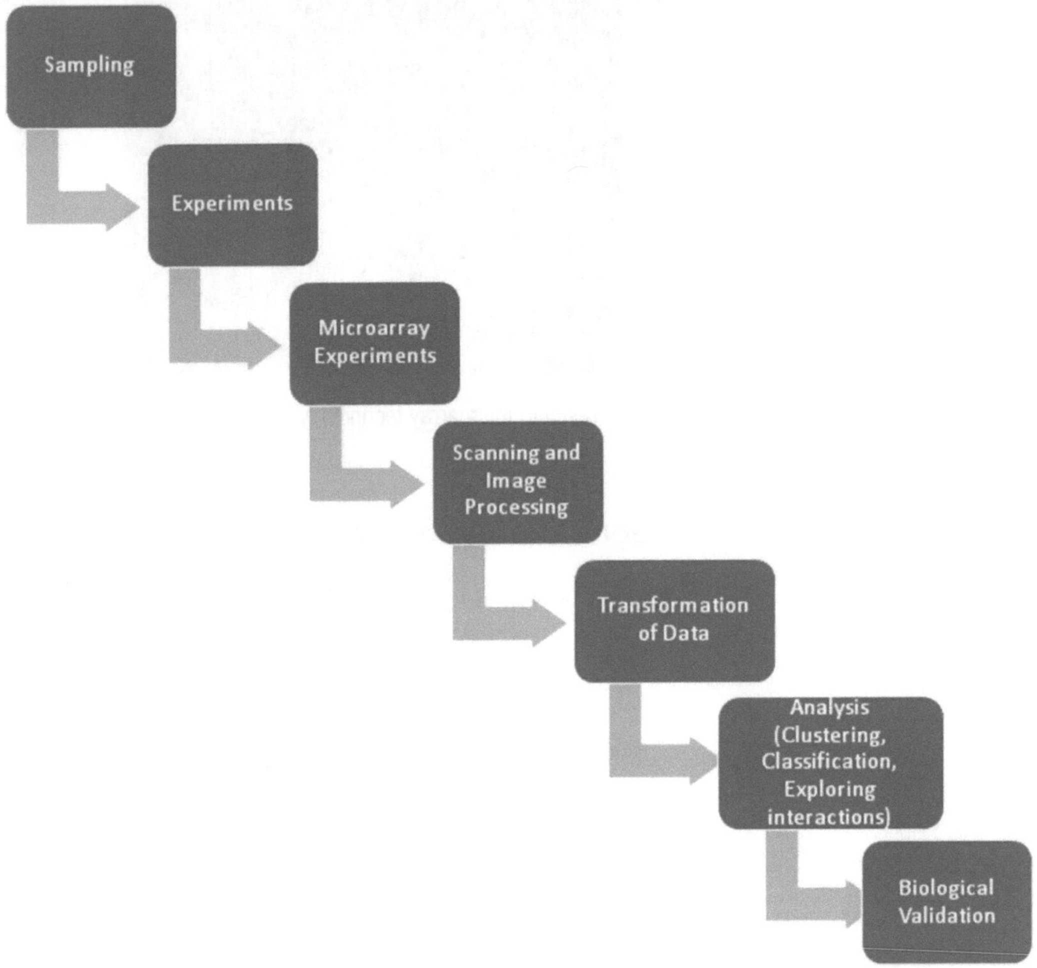

Fig. 4 Pipeline of gene expression analysis

4. Flat pattern filtering.
   It is hard to separate noise in non-flat datasets such as microarray data. We can discard this problem with using the root mean square, the number of peaks in the expression level, or by using standard deviation.
5. Pattern standardization.
   Pattern standardization is nothing but subtracting the average value of patterns from each value and dividing the result by the standard deviation [11]:

$$x_i^{standardized} = \frac{x_i - \overline{x}}{std.dev}.$$

2.3 Differential Expression Analysis

One of the major tasks in the statistical analysis of microarray data is to find DE genes between treatment groups or biological conditions.

Finding DE genes is helpful for understanding gene functions, regulations, and cellular processes. It is also a primary analysis for multivariate analysis such as clustering, classification, and gene set enrichment [12]. Microarrays measure continuous probe intensities, and thus methods proposed for this problem are based on continuous variable types [13]. Before DE analysis, the data must be pre-processed and p-values obtained by DE analysis should be adjusted with an appropriate multiple testing adjustment method.

Due to the small sample size problem in microarray data, parametric methods are widely used due to the powerlessness of nonparametric methods. The most widely used and powerful parametric methods for detecting DE genes based on microarray data are the following.

1. Fold-change analysis.
   Fold-change analysis finds DE genes by calculating the ratios (or log ratios) or differences between two conditions and considering genes that differ more than an arbitrary cut-off value [14]. Let x_{ij} and y_{ij} denote the expression levels of the ith gene in the jth replicate and $\overline{x}_{i.}$ and $\overline{y}_{i.}$ denote the mean expression values of the ith gene in the control and the treatment groups, respectively. Then, the fold-change values can be calculated for ratios and differences as follows [15]:

$$FC_i\left(ratio\right) = \log_2 \frac{\overline{x}_{i.}}{\overline{y}_{i.}}$$

$$FC_i\left(difference\right) = \log_2 \overline{x}_{i.} - \log_2 \overline{y}_{i.}.$$

   After calculating the fold-change values, suppose that we chose the cut-off values as twofold difference. Then the genes whose fold-change values are greater than 2 (or less than –2) will be identified as DE genes.

2. t test.
   The t test is one of the most frequently used statistical tests for DE analysis. We define the two-sample t statistic as:

$$t_i = \frac{\overline{x}_{i.} - \overline{y}_{i.}}{s_i}.$$

   where s_i is the standard error of the ith gene [14]. After determining the t statistic, a p-value can be calculated using the t probability distribution. Using these p-values, the best ranked k genes or the genes having p-values less than a predefined α threshold (e.g., $\alpha = 0.05$) are said to be DE.

3. Significance Analysis of Microarrays (SAM).
   SAM (also known as S test) is a modified version of the t test. When the number of replicates is small, the error variance will be hard to estimate and the ordinary t test will not give stable

results. In SAM, a positive small constant (s_0) is added to the denominator of t_i to minimize the coefficient of variation of t_i [14, 15]:

$$t_i = \frac{\overline{x}_{i.} - \overline{y}_{i.}}{s_0 + s_i}.$$

By adding the s_0 constant, genes having small fold changes will not be determined as DE.

4. Cyber-t test.
   This test is also a modification of the t test. It combines the t test and the global average variance estimates as follows:

$$t_i = \frac{\overline{x}_{i.} - \overline{y}_{i.}}{\sqrt{\frac{ws_g^2 + (n-1)s_i^2}{w+n-2}}}.$$

   n is the number of replicates for each condition, w is a weighting parameter and $s_g{}^2$ is the standard error calculated with combining the data across all genes [14].

5. Limma.
   Limma is a modified t test and is similar to SAM, but it uses a Bayesian approach to calculate the t statistic. In limma, each gene's variance is obtained from the weighted average of gene-specific and global variances [16]. Limma is also similar to the Cyber-t test in the calculation of the t statistic by Bayesian approach. They differ in the methods to estimate $s_g{}^2$ [17]. Limma can be found in the limma package available for the R/Bioconductor software [18].

6. ANOVA and mixed ANOVA models.
   The methods discussed above were for comparison of two conditions. When there are more than two conditions to compare, using multiple two condition tests will increase the chance of type I error (incorrect rejection of a true null hypothesis in statistics) occurrence. The ANOVA F test was proposed to compare multiple conditions. It generalizes the t test by comparing the variation among replicated samples within and between conditions using the F probability distribution. Furthermore, mixed ANOVA models can be used for multiple factor microarray designs. More details can be found in [14].

Moreover, MA and volcano plots can be used for graphical representation of DE analysis. Both plots are represented as scatter plots. The MA plot shows the fold-change variation between two expression profiles, and the volcano plot shows the fold-change variation between conditions and the associated confidence from replicate information simultaneously [19]. The detection of DE genes by these plots is exemplified in Fig. 5.

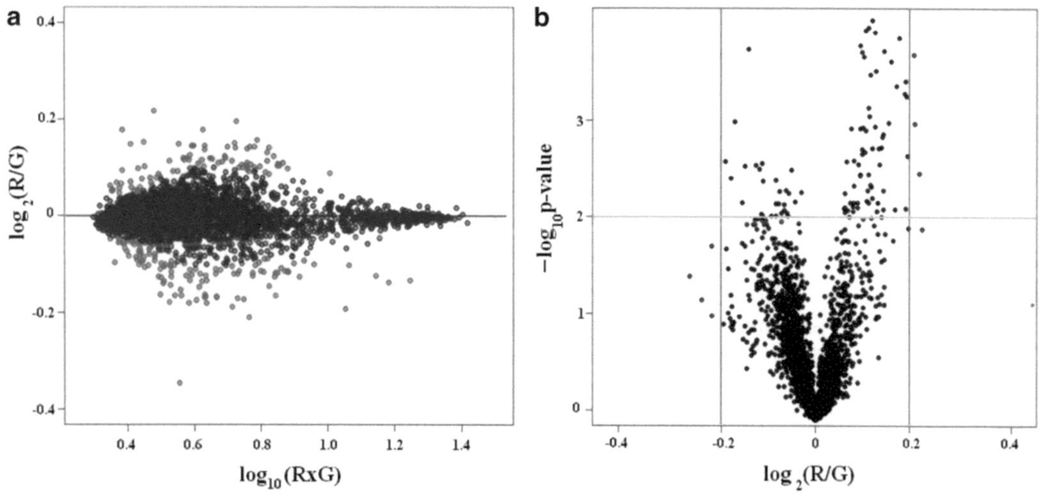

Fig. 5 Graphical demonstration of differential expression (**a**) MA plot (**b**) Volcano plot. Higher fold-change variated genes appear distant to center (zero) points and can be considered as differentially expressed

2.4 Multiple Testing Problem

A major problem of analyzing miRNA expression profiling data is that gene expression can be effected by confounding factors such as spatial heterogeneity and signal saturation [20]. Adding to this problem, these experiments investigate hundreds of genes at a time. Therefore, it is very difficult to find the most related gene for a disease. As a result, the chance for false discovery of genes perceived to be DE between comparison groups can be unacceptably high [21]. There clearly exists a need for statistical analysis and one is to find a reliable way to eliminate most of the unaffected genes from further study. There are many approaches that try to solve this problem and the most popular ones are:

1. Standardized *t*-statistics.
2. A permutation null distribution.
3. Bootstrapping approaches.

Further information can be found in [22].

2.5 Gene Regulatory Networks

Genes have a major role in effecting the cells' activity by creating mRNAs which instruct a ribosome to synthesize a protein in the cytoplasm. Some of the proteins generated are transcription factors that regulate the expression of one or several genes and this complexity in controlling gene expression can be described with gene regulatory networks (GRNs). In addition to this, microRNAs, often gene by-products or genes in their own right, also modulate gene expression and should thus be modeled in GRNs. GRNs have been studied for many years and novel methods were developed for modeling these networks from different disciplines. The most widely used ones can be classified based on the variables types allowed into discrete and continuous GRNs. Boolean networks, probabilistic Boolean networks, and Bayesian networks are good examples of discrete variable models as opposed to differential equation and neural network models which are examples of continuous variable models. Further details and recently proposed online tools based on these models (GenYsis, SIRENE, BANJO, etc.) can be found in [23].

2.6 Machine Learning

Machine learning methods can be classified in two broad categories as supervised (output labels are used) and unsupervised (output label are not used) learning methods. Using the right method in gene expression analysis is very important since a single method may not give the best performance for every dataset (there is no free lunch theorem). Thus, multiple models from different methods should be built and the method which gives the highest performance should be chosen. Moreover, hybrid methods are also used to combine the strengths of different methods. There are lots of methods to classify, cluster, or reduce the dimension of gene expression data. We aimed to show the most popular ones below:

1. Random Forest.
   A random forest (RF) is an ensemble decision tree classifier, where each tree is grown using randomization. RF has the ability to analyze high-throughput data with high training speeds on the basis of a classification and regression tree (CART) [3]. CART is an effective method which works well in large datasets by using recursive binary partitioning of the feature space (Fig. 6). However, problems such as noise and over fitting decrease the predictive ability of CART. RF overcomes this problem by working with variable subsets and when growing a tree.

   A bootstrap sample is drawn from the original samples for each classification tree [4]. Each tree casts a single vote and the final prediction is determined by the majority votes of all the trees in the random forest. As the bootstrap samples are selected with replacement, remaining samples are called out-of-bag (OOB) data which can be used for the estimation of the prediction error of the random forest. RF also has the ability to create a variable importance ranking by using the predictive importance of variables estimated from OOB cases [4, 5].
2. Boosted Classification Trees.
   The boosting tree (BT) algorithm evolved from the application of boosting methods to regression trees. The main idea of BT is to compute a set of single CARTs by building successive trees from the prediction residuals of preceding trees. BT constructs binary trees by splitting the data into two parts. The trees in each split can be defined in three nodes: a root node and two child nodes. In this way, best subsets of data can be determined and the residuals of each partition can be calculated in each step of the BT building. To reduce the residual variance, the next

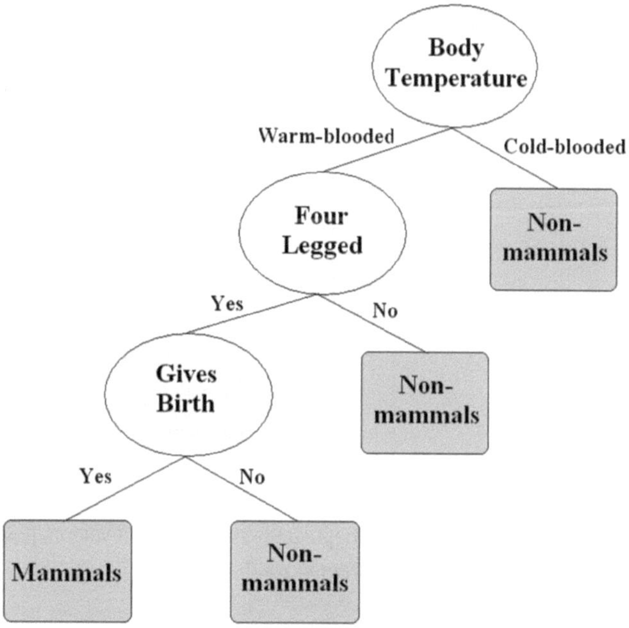

Fig. 6 An illustration of CART algorithm [24]. *Light grey* schemas indicate the input and *dark grey* schemas indicate the output variables. "A warm blooded, four-legged and birth giving animal is probably a mammalian." rule can be extracted from this example

three node tree is fitted to the residuals of the preceding set of trees. Even though there is a nonlinear and complex relationship between input and output variables, an additive weighted expansion of trees fits the predicted values perfectly to the observed values.

Additive models [25, 26] express $f(x)$ as a sum of basis functions $b(x; \gamma_m)$ as follows:

$$f(x) = \sum_m f_m(x) = \sum_m \beta_m b(x; \gamma_m).$$

For further information please refer to [27].

3. Support Vector Machines.
   Support vector machines (SVM) are widely used machine learning tools in many areas and have many advantages over other methods such as their strong mathematical background, their ability to analyze high dimensional complex datasets, and their accurate performance. The general idea of an SVM is to maximize the margin between two classes and minimize the total classification errors by placing a hyperplane or hyperplanes in high dimensional space [28]. The SVM finds $f(x) = b + w.h(x)$ by minimizing

$$\frac{1}{n}\sum_{i=1}^{n}\left[1 - y_i\left(b + w.h(x_i)\right)\right] + \lambda \,||\, w \,||^2,$$

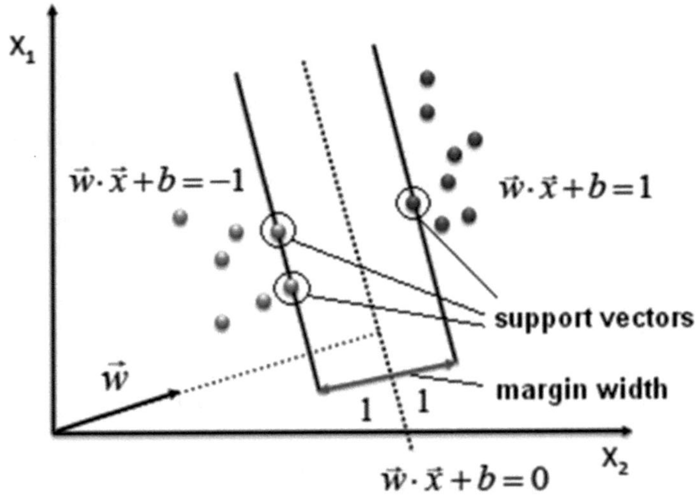

Fig. 7 Representation of support vectors [28]. These vectors are used to find maximum margin width and create the appropriate decision boundary for classification

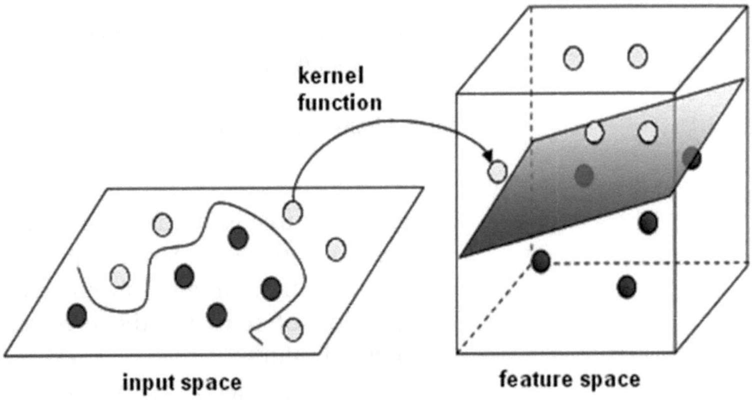

Fig. 8 The kernel trick of SVM [28]. The linearly inseparable data in two dimensions can be linearly separable in higher dimensions (three dimension in the plot) using kernel tricks

where w is the directional vector, b is the constant and $D = \{h_1(x), \ldots, h_p(x)\}$ is a library of basis functions [1].

In 1963, Vapnik proposed the original optimal hyperplane for linear classification problems (Fig. 7) and in 1992, he also suggested the use of kernel tricks for nonlinear classification problems [28, 29] (Fig. 8). Here are some of the most used kernel functions:

(a) Linear: $K(x_i, x_j) = x_i^{\mathrm{T}} x_j$.

(b) Polynomial: $K(x_i, x_j) = (x_i x_j)^{\mathrm{d}}$.

(c) Radial Based Function: $K(x_i, x_j) = \exp(-\gamma\|x_i - x_j\|^2), \gamma > 0$.

(d) Sigmoid: $K(x_i, x_j) = \tanh\,(k x_i x_j - \delta)$.

In recent years, multiple kernel learning (MKL) methods have been proposed to use multiple kernel functions instead of using the one explained above. MKL methods use a linear or nonlinear combination function to integrate single kernel functions. Details can be found in [30].

4. Relevance Vector Machines.
Relevance vector machines (RVMs) have recently attracted much interest in the research community because they provide a number of advantages [31]. RVMs are based on Bayesian formulations of linear models with suitable prior results in a sparse representation and it also assumes familiarity with vector representation of regression, matrix diorentiation, and kernel functions [32–34]. They areas are briefly covered in [35]. Assume that our dataset D, is comprised of l co-regulated genes:

$$D = \{(\hat{x}_i, t_i)\}_{i=1}^{l}, \hat{x}_i \varepsilon R^{d}, t_i \varepsilon \{-1, +1\}.$$

$\hat{x}_i$ represents the set of features defining the ith training pattern and t_i indicates whether the lth gene is up or down regulated ($t_i = \pm 1$). RVMs use the following logistic regression form for model building [32]:

$$p(t|\hat{x}) = \frac{1}{1 + \exp\{-f(\hat{x})\}}, \quad f(\hat{x}) = \sum_{i=1}^{l} \alpha_i X_i + \alpha_0.$$

5. Multivariate Adaptive Regression Splines.
Multivariate adaptive regression splines (MARS) is a nonparametric hybrid method that combines the strengths of recursive partitioning and spline fitting. It has many advantages [36]: (1) It does not require any assumptions about distribution or type of relationships (linear, logistic, etc.) between input and output variables, (2) both input and output variables may be either continuous or categorical, (3) it can work on missing data by using indicator functions to automatically impute the missing values, (4) it is able to work in high dimensional data, and (5) it has an automatic procedure for feature selection and transformation, and also can define the potential interactions. MARS builds models of the following form:

$$f(x) = \sum_{m=1}^{M} c_m B_m(x),$$

Where c_m is an intercept and $B_m(x)$ is the basis function which can be a constant, a hinge function ($max(0, x - c)$ or $max(x - c, 0)$) or a product of hinge functions [36]. The least squares method is used for parameter estimates. MARS has backward and forward model building procedure steps and uses generalized cross-validation in the backward step for model selection.

Recently, CMARS and RCMARS methods have been proposed. CMARS uses Tikhonov regularization instead of the backward stepwise procedure to estimate the model function of MARS and uses the conic quadratic programming framework and RCMARS has been proposed later to overcome some uncertainties in CMARS using robust optimization [37, 38].

6. Independent Component Analysis.
   The main objective of independent component analysis (ICA) is to get a linear representation of non-normal data in which the components are statistically independent [2]. In gene expression studies, n random variables $X_1, \ldots, X_n$ can be expressed as a linear combination of n random signals $s_1, \ldots, s_n$:

$$X_i = a_{i1}s_1 + a_{i2}s_2 + \ldots + a_{in}s_n, \quad \textit{for all } i = 1, \ldots, n.$$

   where $a_{ij}(i,j=1,\ldots,n)$ are some real coefficients and s_i is assumed to be statistically mutually independent using the definition in [39]. However, this assumption is impractical in many applications such as microarray experiments since any gene in the human genome is expressed entirely independently of other genes [40].

   Many algorithms have been proposed to improve the performance of ICA, i.e., minimum mutual information, FastICA and maximum non-Gaussianity and fastICA have been used in gene expression analysis. Further information can be found in [2].

7. Kohonen Map.
   An unsupervised learning method is Kohonen maps (KM) which are a special kind of neural network model [41]. It is used when the original structure of the groups are unknown at the beginning and to cluster datasets into different groups. The records are being grouped into different groups according to how similar or dissimilar they are [42].

   Kohonen first introduced the KM network between 1979 and 1982 based on the studies of Willshaw and Malsburg [6]. It was designed to search and find hierarchical structures of higher dimensional input spaces. During the learning stage, KM performs unsupervised training unlike most neural network applications. The input units in the network are being processed by KM to adjust their weights for potential lateral connections [7].

   KM algorithm steps are as follows [43]:

1. Initialization of weights to small random values.
2. Random selection of inputs from the datasets.
3. Computation of distances to all processing elements.
4. Choose winning processing element j with minimum distance.
5. Update weight vectors according to processing element j and its neighbors by using learning law. The learning law

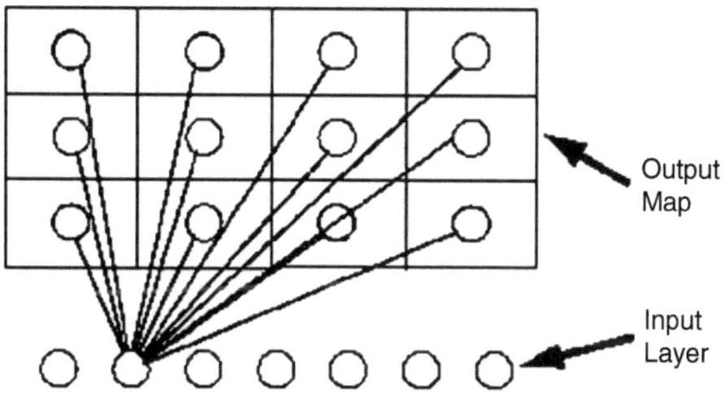

Fig. 9 Structure of Kohonen Map

(input–output vectors) moves weight vector toward input vector.

6. Go to **step 2** or stop iteration when enough inputs are presented.

The structure of KM is demonstrated in Fig. 9.

2.7 Gene Ontology

Once all the genes in an experiment have been analyzed, the next step is the biological interpretation of the statistics. The use of gene ontology programs is useful for the analysis of the gene lists identified by the experiment and to compare the patterns therein to the available literature. This can be explained as extraction of information about potentially important pathways affected by the experiment [44]. Different Pathway Analysis Tools are used to understand the biological meaning of genes. These are as follows:

1. Database for annotation, visualization, and integrated discovery (DAVID) annotation tool: annotates the gene lists by adding gene descriptions from public databases (http://david.abcc.ncifcrf.gov/) [45].
2. KeggCharts: assigns genes to the Kyoto encyclopedia of genes and genomes (KEGG) metabolic processes and enables users to view genes in the context of biochemical pathway maps (http://www.genome.jp/kegg/pathway.html) [46].
3. PATIKA: Pathway analysis tool for integration and knowledge acquisition: Patika is a multiuser tool that is composed of a server-side, scalable, object oriented database and a client-side (http://www.cs.bilkent.edu.tr/~patikaweb/) [47].
4. GoMiner: GoMiner maps lists of genes to functional categories using a tree view (http://discover.nci.nih.gov/gominer/) [48].
5. VANESA is a software solution to visualize and to examine networks in system biology applications. It addresses to biomedical case studies and is used to create and to model individual network systems and their details (http://vanesa.sourceforge.net/index.php) [49].

6. Blast2GO is an all in one functional annotation tool for (novel) sequences and the annotation data analysis (http://www.blast2go.com/b2ghome) [50].

2.8 Gene Set Enrichment Analysis

Gene set enrichment analysis (GSEA) is a statistical method that tests whether a predefined gene set is statistically significant on a biological condition or not. GSEA uses a Kolmogorov–Smirnov-like statistic to find the co-regulated gene groups. Three essential steps of GSEA are the following:

1. Calculation of enrichment score.
2. Estimation of enrichment score significance level.
3. Multiple testing adjustment.

More details about GSEA can be found in [51].

2.9 Public Microarray Data Repositories

Researchers always need original datasets to prove their approaches both statistically and biologically. The only way to do this is to use the publicly available datasets even though they are pre-processed in specific ways and despite the fact that they may not be reliable. Below are the most popular sources for public data:

1. Microarray gene expression data society (http://www.mged.org/).
2. Gene expression omnibus at NCBI (http://www.ncbi.nlm.nih.gov/gds).
3. ArrayExpress at EBI (http://www.ebi.ac.uk/arrayexpress/).
4. Center for information biology gene expression data-base at DDBJ (http://www.ddbj.nig.ac.jp/).

3 Next-Generation Sequencing Data Analysis

3.1 NGS for MicroRNAs

The sequencing instrumentation evolution has recently been very dynamic so that sequence throughput increases exponentially while the cost of sequencing a genome continues to fall [52]. NGS is a recent addition to the current miRNA expression profiling techniques. The millions of short sequence reads generated by NGS, like the SOLiD, ROCHE JS, and Illumina Genome Analyzer, are especially useful for small RNA transcription profiling. MicroRNA expression profiling is provided by NGS at an unusual sensitivity and resolution. NGS systems are not limited by a predetermined number of features, probe design, probe cross-hybridization, or array background issues, as compared to existing miRNA microarray platforms [53]. Classic sequencing steps of Illumina Genome Analyzer are:

1. Extraction of total RNA.
2. Size selection.

3. Adapter ligation and reverse transcription.
4. Size selection.
5. Sequencing.

One of the formats that NGS data is provided in is FASTQ which has four lines for each read. The first line contains the "@" symbol followed by a read identifier, and the second line contains the nucleotide sequence of each read; the third line also contains the read identifier, but this time preceded by a "+" symbol, and the fourth line contains quality scores [54].

In recent years, researchers prefer to compare miRNA sequencing with miRNA expression level. Expression levels are computed based on the read counts in each sequenced sample. The number of occurrences are calculated and normalized for each unique sequence between the reads against the total number of reads produced for the sample. A common way to do this is to compute the rpm (reads per million) value for each sequence *s* occurring in the sample according to the following formula [54]:

$$rpm = \frac{\textit{number of reads in the sample that are equal to } s}{\textit{total number of reads in the sample}} \times 1{,}000{,}000.$$

Bioinformaticians with statistics background should be careful with the analysis of NGS reads. Because NGS data is different than typical genomic data formats. Most important point is that it has quality scores of short-reads. This is a great idea for analysts. They are able to decide which method or pre-process steps are suitable for analysis. Motaneny et al. described the workflow of NGS analysis (Fig. 10) [54].

3.2 Differential Expression

The number of replicates is also small in NGS experiments and parametric methods are used for DE analysis (similar to microarrays). Since the variable type of NGS data is count (discrete); binomial, Poisson and negative binomial discrete probability functions are used and methods are proposed based on these distributions [12]. In early years of NGS studies, Poisson based methods were used to fit the count NGS data [55]. However, Poisson distribution has a single parameter which is equal to both mean and variance and the used statistical tests based on this distribution do not control the type I error by predicting smaller variations than the actual variance [56]. This problem is called over dispersion, and two approaches are used to deal with this problem. Quasi-likelihood, one of these approaches, uses a scaling factor to the variance to differ it from mean. The second approach is to use a negative binomial distribution instead of a Poisson distribution. The negative binomial distribution addresses the over dispersion problem by using a different parameter for variance [12]. Similarly to microarrays, the

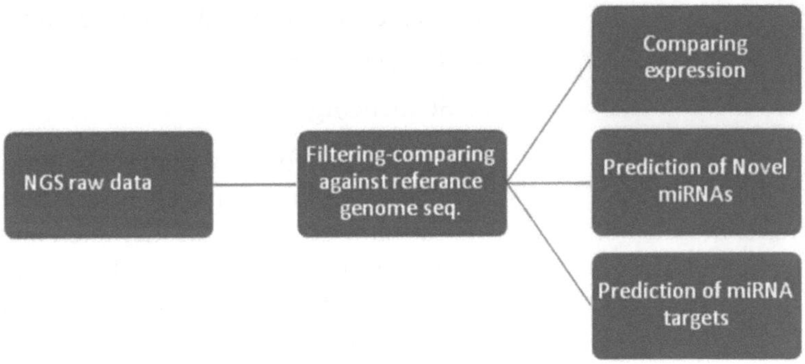

Fig. 10 The workflow of NGS analysis [54]

data must be pre-processed before DE and *p*-values can be obtained. DE analysis should be adjusted with an appropriate multiple testing adjustment method. Widely used methods for detecting DE genes based on NGS data are the following:

1. Two-Stage Poisson Model.
   The two-stage Poisson model (TSPM) uses Poisson or quasi Poisson approaches depending on the over dispersion of the data. In the first stage, they test if there is an over dispersion or not in data. In the second stage, the quasi Poisson approach is used if there is an over dispersion and the Poisson approach is used otherwise [12].
2. Generalized Linear Models.
   As the type of response variable is count, a log link function is mostly used for Poisson family:

$$\log(\mu_i) = \sum_j \beta_j x_{ij}.$$

   Based on generalized linear models, over dispersion can be calculated using likelihood ratios and Poisson or quasi Poisson based approaches can be used due to the over dispersion of data [12].
3. edgeR.
   The edgeR method is based on a negative binomial distribution and was initially proposed for serial analysis of gene expression (SAGE) data. One problem of using a negative binomial distribution for NGS data is the unreliable estimation of both parameters (mean and variance) due to the small number of replicates [56]. Robinson and Smyth assumed that one parameter estimation for each gene will be adequate using the formula below, and the other parameter can be estimated by finding the α, single proportionality constant from the data [56, 57]:

$$\sigma^2 = \mu + \alpha\mu^2.$$

edgeR uses qCML (quantile-adjusted conditional maximum likelihood) for parameter estimation and TMM (trimmed mean of *M* values) to normalize the data for DE [12]. More details about edgeR can be found in [57] and the method can be applied using the edgeR package of R/Bioconductor [58].

4. DESeq.
   Anders and Huber proposed the DESeq method, which is also based on a negative binomial distribution, to extend the edgeR method to a more general approach. They aimed to detect more balanced DE selections by fitting a data-driven relationship between mean and variance [56]. In the former case, the method was using a nonparametric regression to fit the model between mean and variance. The later version of the method makes a parametric fit using a gamma-family generalized linear model (Efficient experimental design and analysis strategies for the detection of DE using RNA-Sequencing). DESeq uses a median scaling for data normalization [12]. More details about DESeq can be found and applied in DESeq package of R/Bioconductor [56].
5. baySeq.
   The baySeq method is also based on a negative binomial distribution. This method uses an empirical Bayesian approach instead of quasi-likelihood methods for parameter estimation. The baySeq method uses the Bayesian approach for posterior probability estimation and ranks genes for DE [12]. More details about baySeq can be found and applied in baySeq R/Bioconductor package [59].

As in microarrays, volcano and smear plots (analogous to MA plot used in microarray analysis) can also be used for graphical display of DE analysis.

4 SNP Analysis

4.1 SNPs

The most common type of genetic diversity among people are single nucleotide polymorphisms (SNPs). Each SNP represents a nucleotide which is a difference in a single DNA building block. It has been found that some SNPs do not have a direct effect on human health or development. However, in the study of human health, some of these genetic differences can be very significant. SNPs may play an important role for the prediction of an individual's response to certain drugs and vulnerability to environmental factors such as toxins and risk of future diseases. Furthermore, SNPs can be used to find and track the inheritance of disease genes. It is possible to describe SNPs related with complex diseases such as heart disease, diabetes, and cancer [8]. SNPs are widely used in the functional analysis of miRNAs. For instance, Sun et al. [9]

examined the SNP effects on the generation and the function of mature miRNAs and concluded that naturally occurring SNPs are able to enhance or impair miRNA processing as well as alter their target sites.

4.2 SNP–SNP Interaction Analysis

In recent years, scientific research has focused on finding genetic factors that cause complex traits. The main aim is to find a gene or gene interactions which may be accountable for producing a specific trait. Classical statistical methods (i.e., logistics regression etc.) do not have a good performance finding interactions between genes. Especially, these methods are not good at identifying multi-locus effects in relatively small samples. Therefore, analysts have tried to use data mining methods for these kinds of problems. At this stage, the most widely used method is the multifactor dimensionality reduction (MDR) method. The MDR method is useful for finding and indicating high-order gene–gene and gene–environment interactions in case–control and discordant-sib-pair studies having small samples [60]. The main aim of the MDR method is to induce multi-locus genetic combinations of a set of genetic markers (e.g., SNPs) to two levels (i.e., high risk and low risk) of a variable [61]. After many studies, researchers have found the MDR method to be more advantageous than other methods.

In genetic association studies, some methods have been advised for the analysis of gene–gene interactions. Some of these are logistic regressions [62], the recursive partitioning method [63], logic regressions [64], neural networks [65], MDR method [60], focused interaction testing framework (FITF) [66], and grammatical evolution neural network (GENN) [67] interactions in a number of common diseases. A study by Motsinger-Reif et al. [67] presented the first example of using data mining methods for evaluating SNP data. In this study MDR, GENN, random forests, FITF, stepwise logistic regression, and explicit logistic regression were compared. As anticipated, the relative success of each method depends on the context. Another important characteristic of this study is that it demonstrates the strengths and weaknesses of each method and illustrates the importance of continued method development. Therefore, researchers are able to understand the framework of data mining usage.

In the field of data mining, finding SNP–SNP interactions is a highlighted research topic. Especially, tree-based methods have been widely used for finding SNP–SNP interactions. One of the early studies was performed by Meng et al. [68]. They tested the performance of random forest as a screening procedure for the identification of small numbers of risk-associated SNPs from among large numbers of unassociated SNPs using complex disease models with up to 32 loci, incorporating both genetic heterogeneity and multi-locus interaction. They compared their approach with Fisher's exact test. They reported that keeping other factors constant, if risk SNPs interact, the random forest importance measure significantly outperforms the Fisher exact test as a screening tool. If the number of interacting level increases, the improvement in performance of random forest analysis relative to Fisher exact test for screening also increases. Random forest performs similarly to the univariate Fisher exact test as a screening tool when SNPs in the analysis do not interact. Bureau et al. [69] illustrated the application of random forest with a dataset of asthma cases and unaffected controls genotyped for 42 SNPs in ADAM33, a previously identified asthma susceptibility gene. SNPs and SNP pairs highly associated with asthma tend to have the highest importance index value, but predictive importance and association are not always conclusive.

Chang et al. [70] studied a real dataset about Glioma which is a complex disease. They selected 10 pathways potentially involved in gliomagenesis that had SNPs represented on the panel. They performed random forest (RF) analyses of SNPs within each pathway group and logistic regression to assess interaction among genes in the one pathway for which the RF prediction error was better than chance and the permutation $p < 0.10$. They reported that RF analysis identified one important biological pathway and several SNPs potentially associated with the development of glioblastoma.

Garcia-Magarinos et al. [71] aimed to analyze the ability of logistic regression (LR) and two tree-based supervised learning methods, classification and regression trees (CART) and random forest (RF), to detect epistasis. They modeled interactions under different scenarios of sample size, missing data, minor allele frequencies (MAF), and several penetrance models: three involving both (indistinguishable) marginal effects and interaction, and two simulating pure interaction effects. They simulated 99 different scenarios. Although CART, RF, and LR yielded similar results in terms of detection of true association, CART and RF performed better than LR with respect to the classification error. This was an important result for researchers who used data mining methods for analyzing GWAS data. Jiang. et al. [72] worked on the detection of epistatic in case–control studies. For this purpose, they first ran a random forest with all SNPs to obtain the Gini importance of each SNP and then used a sliding window sequential forward feature selection (SWSFS) algorithm to select a subset of SNPs that can minimize the classification error. Since this subset typically contains only a small number of SNPs (e.g., ~100), it is possible to enumerate all k-way ($k = 1,2,3$) interactions of the candidate SNPs and to test them for their statistical significance and their association with disease risk. In recent studies, tree based methods have been widely used for pre-selection of SNPs.

De Lobel et al. [73] used a Random Forest (RF) based prescreening method, before executing MDR, to improve its performance. They found that the power of MDR increases when noisy SNPs are first removed, by creating a collection of candidate markers with RFs.

4.3 Multifactor Dimensionality Reduction

With rapidly advancing technology, research in the area of genetic epidemiology has been increasing day by day. As a result, it leads to new analytical challenges to identify genetic risk factors for disease in high dimensional data of huge diversity [74]. Moreover, it can be claimed that complex diseases may be the result of an interaction between multiple genetic and environmental factors and gene-gene and gene- environment cooperation, or epistasis and they play a significant role in the etiology of these types of diseases [75]. Multifactor dimensionality reduction which is a common data mining approach to assess potential gene–gene interactions is designed to accurately choose potentially interacting genetic variables among the most associated ones with disease and case–control studies [60].

MDR has been successful in identifying a number of interactions in real data applications including multiple sclerosis, breast cancer, and HIV immunogenetics [76].

In summary, MDR can be explained with the following steps:

1. When used with k-fold cross-validation (CV), MDR randomly splits the data into k-folds. The optimality functioning condition of the CV is between 5 and 10 intervals, with lower values of k optimized for computation time.
2. The case/control ratio at all multilocus genotypes within a combination of loci are created in $(k-1)/k$ of the data.
3. A binary variable is being formed from the multilocus genotype combinations that summarize the risk for each multilocus comparison such as all high risk genotypes are in one group and low risk genotypes are in another group. Balanced accuracy (BA) is defined as (sensitivity + specificity)/2, where sensitivity is true positives/total sample size and specificity is true negatives/total sample size is then computed. Then, BA is used to choose models from each order of the comparison, or number of loci, for testing. To determine the model's ability to predict outcomes in independent datasets the model with the highest BA in the remaining $1/k$ of the data is tested. In test sets, k models will be tested for k CV intervals.
4. The steps mentioned before are repeated k times. To select the final models maximized average predicted BA and maximized cross-validation consistency over the k-fold cross-validation procedure are used. The tie breaker is higher BA, among all models with the highest observed CV consistency. According to

the principle of statistical parsimony the model with the fewest loci is selected if these two criteria will not fit together [77].

In spite of all of its advantages, MDR has two main disadvantages: First, cells in high-dimensional tables will often be empty; these cells cannot be labeled based on the case/control ratio. Second, the binary assignment (high risk/low risk) is highly unstable when the proportions of cases and controls are similar [78].

5 Quantitative Real Time PCR Analysis

Classical methods such as northern blot or Southern blot are not reliable for the quantification of gene expression. Quantitative real time PCR (qPCR) is a more powerful and recent technique that is used for the amplification and quantification of a targeted DNA molecule at the same time. Relative quantification and absolute quantification are the two widely used methods for quantification of qPCR [79].

Performing some preliminary analysis for qPCR is important. One of these analyses is the transformation of the number of cycles (Cq) at a fluorescence threshold level to a relative quantity (RQ) of an input template $RQ = 1/E^{Cq}$, E is 2 for optimum situation). Next, the calculated RQ is normalized to the total amount of cDNA used in the reaction, and the normalized relative quantity (NRQ) values are obtained. Finally, a last transformation, which is usually a base 2 logarithmic transformation, is needed before starting statistical analysis [80].

After pre-processing the qPCR data, traditional statistical methods (*t*-test, ANOVA, etc.) can be used for further analysis. If a researcher is working with a huge number of miRNAs (hundreds, thousands), high-dimensional statistical analysis, which is discussed in Subheading 1, should be used.

6 Conclusion

In this chapter, we have outlined introductory statistical methods of miRNA analysis and explained the major steps of microarray, NGS, SNP, and qPCR technologies. For microarrays and NGS, we explained the commonly used pre-processing methods.

DE is one of the major tasks in genomics. We reviewed the most powerful DE methods for both microarray and NGS data. For microarrays, modified versions of *t*-tests and fold-change analysis are suggested [15]. NGS analysis methods are still emerging and methods used in DE analysis are well established as of now. There are several studies comparing the results of these methods. Kvam and Liu [12] have found that the baySeq method performs

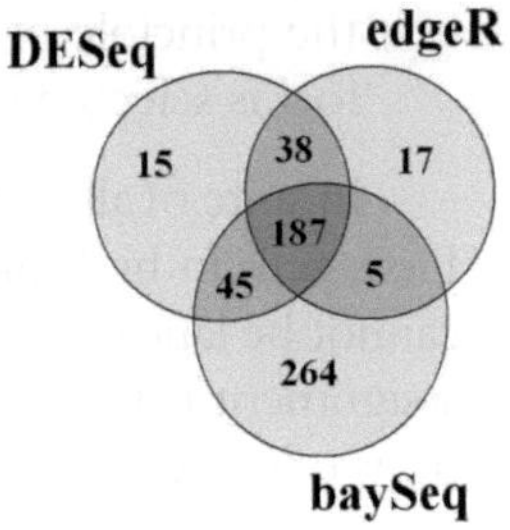

Fig. 11 Venn diagrams to combine results of DE methods. In this example, 187 genes are commonly identified as differentially expressed using DESeq, edger, and baySeq methods

best based on simulation results. Robles et al. [81] have shown that DESeq method performs well for DE based on less sensitive changes in replication. More research is needed to define the appropriate method for DE analysis of NGS data. We advise to use combined approaches such as Venn diagrams which have also been widely used by researchers [82, 83]. By these diagrams, they applied various methods for DE analysis and detected DE genes which were commonly detected by all methods (Fig. 11). Same approach can be applied for DE analysis of microarray data. Also, it has been used to combine the results of different genetic technologies [84].

In machine learning, clustering, classification, and dimensionality reduction analysis are mostly used and we reviewed the most powerful machine learning methods used for this purpose. We described ICA for dimension reduction, Kohonen maps for clustering and SVMs, RVMs, MARS, random forest, and boosted classification trees for classification problems.

Identifying SNP–SNP interactions is very significant for finding genetic factors which lead to complex traits and classical statistical methods do not perform well. Here, we described multifactor dimensionality reduction which is a very successful data mining method to identify these interactions. We also briefly mentioned the analysis of qPCR data.

For the application of methods described above, we advise the R/Bioconductor open source software [85], which has more than 600 software packages and an active community. The pre-processing and DE analysis of microarrays can be applied in affy, limma, SAM packages, machine learning methods can be applied in rpart, e1071, MDA, ICA, MLInterfaces, Random Forest, ada, kohonen, and rvmbinary packages. GSEA can be applied in GSEABase and gene ontology methods can be applied in GOstats, annotate, topGO packages. NGS analysis can be applied in Biostrings, ShortRead, edger, DESeq, and baySeq packages. For SNP–SNP interactions, multifactor dimensionality reduction method can be applied in MDR package.

References

1. Zhang HH, Ahn J, Lin X et al (2006) Gene selection using support vector machines with non-convex penalty. Bioinformatics 22(1): 88–95
2. Kong W, Vanderburg CR, Gunshin H et al (2008) A review of independent component analysis application to microarray gene expression data. Biotechniques 45(5):501–520
3. Ko BC, Kim SH, Nam JY (2011) X-ray image classification using random forests with local wavelet-based CS-local binary patterns. J Digit Imaging 24(6):1141–1151
4. Breiman L (2001) Random forests. Mach Learn 45(1):5–32
5. Chen CC, Schwender H, Keith J et al (2011) Methods for identifying SNP interactions: a review on variations of logic regression, random forest and Bayesian logistic regression. IEEE/ACM Trans Comput Biol Bioinform—TCBB 8(6):1580–1591
6. Wilk MB, Shapiro SS (1968) The joint assessment of normality of several independent samples. Technometrics 10(4):825–839
7. Kohonen T (1984) Self-organization and associative memory. Springer, Berlin
8. http://ghr.nlm.nih.gov/handbook/genomicresearch/snp (09.20.2012)
9. Sun G, Yan J, Noltner K et al (2009) SNPs in human miRNA genes affect biogenesis and function. RNA 15(9):1640–1651
10. Simon ML, Kimberly FJ (2002) Methods of microarray data analysis II. Kluwer Academic Publishers, Boston
11. Herrero J, Diaz-Uriarte R, Dopazo J (2003) Gene expression data preprocessing. Bioinformatics 19(5):655–656
12. Kvam VM, Liu P, Si Y (2011) A comparison of statistical methods for detecting differentially expressed genes from RNA-seq data. Am J Bot 99(2):248–256
13. White Paper (2011) RNA-Seq Data Comparison with Gene Expression Microarrays. by: Illumina
14. Cui X, Churchill GA (2003) Statistical tests for differential expression in cDNA microarray experiments. Genome Biol 4(210):1–10
15. Witten D, Tibshirani R (2007) A comparison of fold-change and the t-statistic for microarray data analysis. Stanford University, Technical Report
16. Dziuda DM (2009) Data mining for genomics and proteomics: analysis of gene and protein expression data. Wiley, New Jersey
17. Kooperberg C, Aragaki A, Strand AD et al (2005) Significance testing for small microarray experiments. Stat Med 24:2281–2298
18. Smyth GK (2004) Linear models and empirical Bayes methods for assessing differential expression in microarray experiments. Stat Appl Genet Mol Biol 3(1):1–25
19. Krawetz S (2009) Bioinformatics for system biology. Springer, New York
20. Delongchamp RR, Bowyer JF, Chen JJ et al (2004) Multiple-testing strategy for analyzing cDNA array data on gene expression. Biometrics 60(3):774–782
21. Craig BA, Black MA, Doerge RW (2003) Gene expression data: the technology and statistical analysis. J Agric Biol Environ Stat 8:1–28
22. Pollard KS, Dudoit S, van der Laan MJ (2004) Multiple testing procedures: R multtest package and applications to genomics. http://works.bepress.com/mark_van_der_laan/115. Accessed 9 Nov, 2012
23. Lee WP, Tzou WS (2009) Computational methods for discovering gene networks from expression data. Brief Bioinform 10(4): 408–423
24. Tan PN, Steinbach M, Kumar V (2006) Introduction to data mining. Pearson Education, Boston
25. Hastie T, Tibshirani R (1990) Generalized additive models. Chapman and Hall, CRC
26. Hastie T, Tibshirani R, Friedman J (2001) The elements of statistical learning: data mining, inference and prediction. Springer Verlag, New York
27. De'ath G (2007) Boosted trees for ecological modeling and prediction. Ecology 88(1): 243–251
28. Zararsiz G, Elmalı F, Öztürk A (2012) Bagging support vector machines for leukemia classifications. Int J Comput Sci 9(6):355–358
29. Aizerman MA, Braverman EM et al (1964) Theoretical foundations of the potential function method in pattern recognition learning. Autom Remote Control 25: 821–837
30. Gönen M, Alpaydın E (2011) Multiple kernel learning algorithms. J Mach Learn Res 12: 2211–2268
31. Daniela M (2012) New approaches to open problems in gene expression microarray data. Bologna, Marzo 2008. http://amsdottorato.cib.unibo.it/842/1/Tesi_Marconi_Daniela.pdf. Accessed 13 Feb, 2012
32. Cawley GC, Talbot NLC (2006) Gene selection in cancer classification using sparse logistic regression with Bayesian regularization. Bioinformatics 22(19):2348–2355
33. Tzikas DG, Wei L, Likas A, et al. A tutorial on relevance vector machines for regression and classification with applications. http://citeseerx.ist.psu.edu/viewdoc/summary?doi=10.1.1.99.3559
34. Tipping ME (2001) Sparse Bayesian learning and the relevance vector machine. J Mach Learn Res 1:211–244

35. Fletcher T (2008) Relevance vector machines explained (Tutorial Paper—PhD 2008), http://www.tristanfletcher.co.uk. Accessed 10 Feb, 2012
36. Friedman JH (1991) Multivariate adaptive regression splines. Ann Stat 19(1):1–141
37. Weber GW, Batmaz İ, Köksal G et al (2011) CMARS: a New contribution to nonparametric regression with multivariate adaptive regression splines supported by continuous optimisation. Inverse Probl Sci Eng 20(3):371–400
38. Özmen A, Weber GW, Batmaz İ et al (2011) RCMARS: Robustification of CMARS with different scenarios under polyhedral uncertainty set. Commun Nonlinear Sci Numer Simulat 16(12):4780–4787
39. Hyvärinen A, Karhunen J, Oja E (2001) Independent component analysis. John Wiley & Sons, Inc, New York
40. Comon P (1994) Independent component analysis—a new concept? Signal Process 36: 287–314
41. Kohonen T (2001) Self-organizing maps, 3rd edn. Springer, Berlin
42. Clementine® 12.0 Algorithms Guide, Copyright © 2007 by Integral Solutions Limited
43. Lippmann RP (1987) An introduction to computing with neural nets. IEEE Acoust Speech Signal Processing Magazine 4(2):4–22
44. Michael J (2007) Microarray data analysis: methods and applications. Humana Press Inc., Totowa, NJ
45. Huang DW, Sherman BT, Lempicki RA (2007) Systematic and integrative analysis of large gene lists using DAVID Bioinformatics Resources. Nature Protoc 4(1):44–57
46. Kanehisa M, Goto S, Kawashima S et al (2004) The KEGG resource for deciphering the genome. Nucleic Acids Res 32:D277–D280
47. Dogrusoz U, Erson EZ, Giral E et al (2006) PATIKAweb: a Web interface for analyzing biological pathways through advanced querying and visualization. Bioinformatics 22(3):374–375
48. Zeeberg BR, Feng W, Wang G et al (2003) GoMiner: a resource for biological interpretation of genomic and proteomic data. Genome Biol 4(4):R28
49. Janowski J (2008) An integrative bioinformatics solution to visualize and examine biological networks (MSc. thesis sup: Hofestadt R., Willassen N.P.). Bielefeld Univ.
50. Conesa A, Götz S, Garcia-Gomez JM et al (2005) Blast2GO: a universal tool for annotation, visualization and analysis in functional genomics research. Bioinformatics 21:3674–3676
51. Subramanian A, Tamayo P, Mootha VK (2005) Gene set enrichment analysis: a knowledge-based approach for interpreting genome-wide expression profiles. PNAS 102(43):15545–15550
52. Brent GR, David PS (2009) Managing and analyzing next-generation sequence data. PLoS Comput Biol 5(6)
53. Buermans HPJ, Ariyurek Y, van Ommen G et al (2010) New methods for next generation sequencing based microRNA expression profiling. BMC Genomics 11(716)
54. Motameny S, Wolters S, Nurnberg P et al (2010) Next generation sequencing of miRNAs – strategies, resources and methods. Genes 1(1):70–84
55. Bullard JH, Purdom E, Hansen KD et al (2010) Evaluation of statistical methods for normalization and differential expression in mRNA-Seq experiments. BMC Bioinforma 11(94):1–13
56. Anders S, Huber W (2010) Differential expression analysis for sequence count data. Genome Biol 11(R106):1–12
57. Robinson MD, Smyth GK (2008) Small-sample estimation of negative binomial dispersion, with applications to SAGE data. Biostatistics 9(2):321–332
58. Robinson MD, McCarthy DJ, Smyth GK (2010) edgeR: a Bioconductor package for differential expression analysis of digital expression data. Bioinformatics 26:139–140
59. Hardcastle TJ, Kelly KA (2010) BaySeq: empirical Bayesian methods for identifying differential expression in sequence count data. BMC Bioinforma 11:422
60. Ritchie MD, Hahn LW, Roodi N et al (2001) Multifactor-dimensionality reduction reveals highorder interactions among estrogen-metabolism genes in sporadic breast cancer. Am J Hum Genet 69:138–147
61. Namkung J, Kim K, Yi S et al (2009) New evaluation measures for multifactor dimensionality reduction classifiers in gene–gene interaction analysis. Bioinformatics 25(3):338–345
62. Cordell H, Clayton D (2002) A unified stepwise regression procedure for evaluating the relative effects of polymorphisms within a gene using case/control or family data: application to HLA in type 1 diabetes. Am J Hum Genet 70(1):124–141
63. Zhang HP, Bonney G (2000) Use of classification trees for association studies. Genet Epidemiol 19:323–332
64. Kooperberg C, Ruczinski I (2005) Identifying interacting SNPs using Monte Carlo logic regression. Genet Epidemiol 28(2):157–170
65. Sherriff A, Ott J (2001) Applications of neural networks for gene finding. Adv Genetics 42:287–297, Genetic Dissection of Complex Traits DC Rao, MA Province (eds.) Academic Press
66. Millstein J, Conti DV, Gilliland FD et al (2006) A testing framework for identifying susceptibility genes in the presence of epistasis. Am J Hum Genet 78:15–27

67. Motsinger-Reif AA, Dudek SM, Hahn LW et al (2008) Comparison of approaches for machine-learning optimization of neural networks for detecting gene-gene interactions in genetic epidemiology. Genet Epidemiol 32: 325–340
68. Meng YA, Yu Y, Cupples LA et al (2009) Performance of random forest when SNPs are in linkage disequilibrium. BMC Bioinforma 10(78)
69. Bureau A, Dupuis J, Falls K et al (2005) Identifying SNPs predictive of phenotype using random forests. Genet Epidemiol 28(2): 171–182
70. Chang JS, Yeh RF, Wiencke JK et al (2008) Pathway analysis of single-nucleotide polymorphisms potentially associated with glioblastoma multiforme susceptibility using random forests. Cancer Epidemiol Biomarkers Prev 17:1368–1373
71. García-Magariños M, De-UllibarriI L, Cao R et al (2009) Evaluating the ability of tree-based methods and logistic regression for the detection of SNP-SNP interaction. Ann Hum Genet 73(3):360–369
72. Jiang R, Tang W, Wu X et al (2009) A random forest approach to the detection of epistatic interactions in case–control studies. BMC Bioinforma 10(1)
73. Lobel LD, Geurts P, Baele G et al (2010) A screening methodology based on random forests to improve the detection of gene–gene interactions. Eur J Hum Genet 18:1127–1132
74. Winham S, Wang C, Motsinger-Reif AA (2011) A comparison of multifactor dimensionality reduction and L1-penalized regression to identify gene-gene interactions in genetic association studies. Stat Appl Genet Mol Biol 10:1–4
75. Moore JH (2003) The ubiquitous nature of epistasis in determining susceptibility to common human diseases. Hum Hered 56:73–82
76. Haas SL, Jesnowski R, Steiner M et al (2006) Expression of tissue factor in pancreatic adenocarcinoma is associated with activation of coagulation. World J Gastroenterol 12: 4843–4849
77. Edwards TL, Pericak-Vance M, Gilbert JR et al (2009) An association analysis of Alzheimer disease candidate genes detects an ancestral risk haplotype clade in ACE and putative multilocus association between ACE, A2M, and LRRTM3. Am J Med Genet B Neuropsychiatr Genet 150(5):721–735
78. He H, Oetting WS, Brott MJ et al (2009) Power of multifactor dimensionality reduction and penalized logistic regression for detecting gene-gene Interaction in a case–control study. BMC Med Genet 10(127)
79. Dhanasekaran S, Doherty TM, Kenneth J (2010) Comparison of different standards for real-time PCR-based absolute quantification. Immunol Methods 354:34–39
80. Rieu I, Powers SJ (2009) Real-time quantitative RT-PCR: design, calculations, and statistics. Plant Cell 21:1031–1033
81. Robles JA, Qureshi SE, Stephen SJ et al (2012) Efficient experimental design and analysis strategies for the detection of differential expression using RNA-Sequencing. BMC Genomics 13(484)
82. Cantu D, Pearce SP, Distelfeld A (2011) Effect of the down-regulation of the high grain protein content (GPC) genes on the wheat transcriptome during monocarpic senescence. BMC Genomics 12(492):1–17
83. Wang Y, Wu QF, Chen C et al (2012) Revealing metabolite biomarkers for acupuncture treatment by linear programming based feature selection. BMC Syst Biol 6(1)
84. Marioni JC, Mason CE, Mane SM (2008) RNA-seq: An assessment of technical reproducibility and comparison with gene expression arrays. Genome Res 18:1509–1517
85. Gentleman RC, Carey VJ, Bates DM et al (2004) Bioconductor: open software development for computational biology and bioinformatics. Genome Biol 5(10)

Chapter 9

Computational and Bioinformatics Methods for MicroRNA Gene Prediction

Jens Allmer

Abstract

MicroRNAs (miRNAs) have attracted ever-increasing interest in recent years. Since experimental approaches for determining miRNAs are nontrivial in their application, computational methods for the prediction of miRNAs have gained popularity. Such methods can be grouped into two broad categories (1) performing ab initio predictions of miRNAs from primary sequence alone and (2) additionally employing phylogenetic conservation. Most methods acknowledge the importance of hairpin or stem–loop structures and employ various methods for the prediction of RNA secondary structure. Machine learning has been employed in both categories with classification being the predominant method. In most cases, positive and negative examples are necessary for performing classification. Since it is currently elusive to experimentally determine all possible miRNAs for an organism, true negative examples are hard to come by, and therefore the accuracy assessment of algorithms is hampered. In this chapter, first RNA secondary structure prediction is introduced since it provides a basis for miRNA prediction. This is followed by an assessment of homology and then ab initio miRNA prediction methods.

Key words miRNA, Secondary structure prediction, Homology-based prediction, Ab initio prediction, miRNA prediction accuracy, Multiple sequence alignment-based prediction

1 Introduction

Noncoding RNAs (ncRNAs) represent a large portion of the transcriptome and have recently received much attention [1] although the term ncRNA may not have been chosen well since many so-called ncRNAs also lead to mRNAs [2]. These ncRNAs have been grouped into families [3, 4], one of which, microRNAs (miRNAs), is the focus of this book. MicroRNA can originate from any part of a genome [5] and can lead to silencing of transcripts originating from anywhere in the genome. MicroRNA genes' presence or their effects have been shown in many species, and even viruses make use of miRNAs to regulate host- and virus-encoded genes [6].

Malik Yousef and Jens Allmer (eds.), *miRNomics: MicroRNA Biology and Computational Analysis*, Methods in Molecular Biology, vol. 1107, DOI 10.1007/978-1-62703-748-8_9,

There are at least two computational challenges: (1) the prediction of miRNAs in a genome and (2) the mapping of the miRNAs to likely targets. This chapter focuses on the prediction of miRNAs within a genome. Computational miRNA gene prediction can be grouped into several approaches. Generally, either homology modeling or machine learning is applied to extract likely miRNAs from a genome. Although homology modeling can glean information from already successfully established miRNAs from a related organism's genome, it is also limited since completely novel miRNAs cannot be determined in this way. Furthermore, miRNAs evolve quickly, and very close homology is thus needed for successful miRNA gene prediction [7]. Another approach, machine learning, is hampered in a similar manner but assumes that the examples for learning are derived from the organism in question. In the following, first miRNA gene prediction will be further explored followed by a brief discussion of RNA secondary structure prediction, a process vital to all approaches in miRNA gene prediction. Then homology-based miRNA gene prediction and ab initio gene prediction are discussed.

2 MicroRNA Gene Prediction

Identification of miRNA genes is computationally challenging since a genome can be divided into millions of putative miRNAs of appropriate sequence length (e.g.: 80–200 nucleotides for pre-miRNAs). Folding all these sequences in silico increases the complexity and may only be practical for small genomes. Furthermore, many hairpin structures can be found in the predictions and will thus lead to an abundance of putative miRNAs, many of which may represent false positive results. An inherent problem to the experimental validation of miRNAs occurs because their expression may only happen in response to specific signals or at certain developmental stages [8]. *See* Chapters 13 and 14 in this volume or Bentwich 2005 for more details on miRNA gene validation [9]. In order to decrease the number of false positive results many filtering strategies have been developed and are discussed later in this chapter. Since both homology-guided detection algorithms and ab initio miRNA gene prediction algorithms rely on the prediction of RNA secondary structure, a number of such tools shall be introduced first before the two miRNA gene prediction approaches are discussed in more detail.

2.1 RNA Secondary Structure Prediction

Prediction of RNA secondary structure is integral to many algorithms trying to find hairpins, also known as stem–loop structures and pre-miRNAs, which may give rise to miRNAs. In general, the prediction of secondary structure is much easier for shorter sequences, which means that the longer the sequence becomes the more difficult the prediction which is further reflected in exponentially increasing algorithm run time. Therefore, most algorithms which use secondary structure prediction resort to merely predicting the hairpin structure which is always contained in a sequence of less than 500 nucleotides which can successfully be folded in a short time. There are a number of algorithms which can be used for RNA secondary structure prediction (Table 1). The table is sorted by usage statistics not by successfulness of the algorithm. A recent paper has shown, however, that in the realm of predicting the secondary structure of short nucleotide sequences RNAfold seems to be most successful [10].

For both methods, the homology-based prediction of miRNA genes and their ab initio prediction, RNA structure prediction is vital. One feature of miRNAs is the stem–loop structure which seems to be important for processing of the pre-miRNA into a mature miRNA with Drosha and Dicer. The homology-based prediction of miRNA genes is inherently simpler than their ab initio prediction and shall thus be discussed first.

2.2 Homology-Based miRNA Gene Prediction

In contrast to ab initio gene prediction where miRNA genes need to be found without additional knowledge, homology-based mapping methods can build on available and experimentally validated miRNAs and find similar structures and sequences in related species.

All software that enable mapping of a known miRNAs to homologous genomes take sequence similarity as well as RNA secondary structure into account (Table 2). The assumption is that a mature miRNA derives from a hairpin structure formed by folding its pre-miRNA. The approach taken by one of the most recent developments, MapMi [21], first scans the miRNA sequences against the target genome and then creates two potential pre-miRNAs from it. The ViennaRNA package [13] is used to fold the extracted RNA sequences. Finally, the results are scored, ranked, and displayed. Both a Web service with rich display facilities and a downloadable, local, version are freely available for this program which as the authors report achieves 92 % sensitivity at 98 % specificity.

Although mapping by homology is a straightforward approach, it can only reproduce results and cannot find new miRNA genes. Since many miRNAs are species specific these will always be missed by this method, and therefore other strategies need to be used in tandem. Additionally, miRNA genes evolve very rapidly which further limits the applicability of homology-based methods [37, 38].

A recent study by Keshavan and colleagues pointed out that it is important to make sanity checks when constructing a computational pipeline for miRNA gene detection since in their case the temperature at which *Ciona intestinalis* operates is only 18 °C while most folding programs default to 37 °C [39]. They were able

Table 1
Non-comprehensive list of programs predicting the secondary structure from primary RNA sequence

Name	Summary	Systems	Availability	Reference
Dynalign	Aligns two nucleotide sequences and predicts their common structure	ANSI C++ code, Part of RNAstrucuture (MS Windows)	rna.chem.rochester.edu, Open Source	[11]
Unnamed	Predicts RNA secondary structure using covariational and free energy methods	–	–	[12]
RNAfold	Predicts RNA secondary structure using minimum free energy	Web service, local installation	http://rna.tbi.univie.ac.at/cgi-bin/RNAfold.cgi	[13]
RNAHybrid	Finds the minimum free energy hybridization of two RNAs	Web service, limited local installation	http://bibiserv.techfak.uni-bielefeld.de/rnahybrid/	[14]
RNAStructure	Determines secondary structure using dynamic programming with free energy minimization	MS Windows, C++	rna.chem.rochester.edu	[15]
mfold	Determines secondary structure using dynamic programming with free energy minimization	Fortran, C, UNIX	www.ibc.wustl.edu/~zuker/rna/form1.cgi	[16]
RNADistance	Calculates the distance among structures based on string editing and base pair distance	Local installation	http://www.tbi.univie.ac.at/~ivo/RNA/man/RNAdistance.html	[17]
ViennaRNA	Unified access to various RNA tools of the Vienna package	Web service, software package	rna.tbi.univie.ac.at	[13]
taveRNA	A package containing secondary structure prediction, RNA–RNA interaction, and a database pruning algorithm	Web service	compbio.cs.sfu.ca/taverna	[18]
RNAShapes	Predicts secondary structure by evaluating promising shapes with Boltzman probabilities	Web service, local installation	http://bibiserv.techfak.uni-bielefeld.de/rapidshapes/submission.html	[19]
UNAFold	Simulates folding, hybridization, and melting pathways for up to two sequences	Local installation	http://mfold.rna.albany.edu/	[20]

The rows are sorted decreasingly by average citation count per year

Table 2
Non-comprehensive selection of software that allows homology mapping of miRNAs to the source genome or to related species

Name	Summary	Clade	URI	Reference
MicroHarvester	BLAST search for candidates filtered by structural features specific to plant miRNAs	Plant	http://www-ab.informatik.uni-tuebingen.de/brisbane/tb/index.php?view=microharvester2	[22]
miRNAminer	BLAST search for homologs with filtering by minimum free energy and alignment conservation	Animal	http://groups.csail.mit.edu/pag/mirnaminer/	[23]
miROrtho	Homology: extended alignments of known miRBase families and putative miRNA families using SVM and orthology	Animal	http://cegg.unige.ch/mirortho	[24]
CoGemiR	Sequence similarity and secondary structure analysis similar to miRNAminer but with a larger number of species	Animal	http://cogemir.tigem.it/	[25]
MapMi	Maps miRNAs within species and across species using sequence homology and structure	Any	http://www.ebi.ac.uk/enright-srv/MapMi/	[26]
MiRscan	Trained on examples conserved between two closely related species derived from a fold-first find-homologs later strategy	Worms	http://genes.mit.edu/mirscan/	[27]
MiRscanII	Supersedes MiRscan, adds conservation of miRNA gene flanking regions and a conserved motif	Worms	http://genes.mit.edu/burgelab/MiRscanII/	[28]
ProMiR II	Integrative approach using several databases and criteria as well as several custom modules	Animal	http://cbit.snu.ac.kr/~ProMiR2/introduction.html	[29]
Unnamed	Homologous miRNA genes among primates used to determine general characteristics of miRNA genes in vertebrates	Vertebrates	Not associated website allowing phylogenetic shadowing: http://eshadow.dcode.org/	[30]

(continued)

Table 2
(continued)

Name	Summary	Clade	URI	Reference
MiRFinder	Based on pairwise genome searches for shRNA using SVM for filtering, introduces mutation model for hairpins	Any	http://www.bioinformatics.org/mirfinder/	[31]
Unnamed	Homology between Arabidopsis and Oryza; approach also takes target information into account	Plant	–	[32]
Unnamed	Homology between Arabidopsis and Oryza; approach also takes target information into account	Plant	–	[33]
Unnamed	Exploit clustering of miRNAs to filter miRNA predictions	Mammals	–	[21]
Unnamed	GSS, EST versus known miRNAs and proteins with subsequent feature-based filtering	Plant	–	[34]
RNAz	Detects thermodynamically stable and evolutionarily conserved ncRNA secondary structures in MSA	Any	http://rna.tbi.univie.ac.at/cgi-bin/RNAz.cgi	[35]
QRNA	Uses comparative genome sequence analysis to detect conserved ncRNA secondary structures	Any	http://selab.janelia.org/software.html	[36]

The rows are sorted decreasingly by average citation count per year

to confirm about half of their predictions by either microarray analysis or by the fact that the predicted hairpins were already in other databases.

The two aforementioned studies are just a small selection of the large amount of available studies, but the following section aims to briefly summarize common approaches among different studies.

2.2.1 Methods Used in Homology-Based miRNA Detection

There are many ways to detect and filter hairpin structures and miRNAs. The list below is separated into two sections, the first one showing methods for hairpin/miRNA detection and the second one listing methods used to filter/remove false-positive identifications. The methods below are a non-comprehensive list, and some methods may not be used synergistically while others can be combined. In general any algorithm used for homology detection of hairpins or miRNAs uses a combination of some of the methods in the list, but no algorithm has been proposed that integrates most of the detection and filtering methods below.

Methods for Initially Detecting miRNAs

- Difference in evolutionary conservation.
  - Coding arm, noncoding arm, seed region.
  - Loop and stem flanking regions.
  - Effect on secondary structure.
- Scanning for hairpin structures conserved in closely related species.
  - Sliding window (70–110), folding sequences for each window.
  - Level of expected similarity can be adjusted.
- Windows with high sequence conservation (sometimes higher than for coding sequences) flanked by windows with high sequence variation.
- Homology of the miRNA targets among genomes.

Since studies have shown that excessive number of conserved hairpin structures can be found [40], additional criteria for their filtering need to be established, some of which are listed below.

Methods for Filtering Detected Hairpins

- Varying level of sequence conservation within stem structure.
- Using general properties of hairpin structures that can be learned from examples.
- Repetitively detected structures are generally discarded.
- Minimum free energy.
- Length of stem–loop structure.
- If matching to certain annotation of a genome (e.g.: coding sequence) the detections may be discarded.

- Base composition.
- miRNA gene clustering.
- Upstream and downstream conserved regions surrounding miRNA genes.
- Sequence entropy.
- Identity with a multiple sequence alignment.
- Position of mature sequence within hairpin structure.
- Maximal internal loop and bulge sizes.
- EST sequences can confirm that sequences are transcribed.
- Text mining.

2.2.2 Accuracy of Homology-Based miRNA Prediction

MirScan [27] has been applied to *Caenorhabditis elegans*, and the predictions were validated experimentally setting the sensitivity to 0.50 at a specificity of 0.70 [27]. The same study has also shown that many miRNAs are present at high levels, between 1,000 and 50,000 molecules per cell. Another study, which also validated the predictions experimentally, studied the conservation among ten primate species and found that sequences representing stem–loops are conserved whereas flanking regions and loop region are highly variable [30]. The sensitivity of the method was reported at 0.83, but the specificity was not given. It may be rather low due to their prediction of 976 putative miRNAs where 179 were confirmed in miRNA databases and only 16 out of 69 predicted miRNAs have been confirmed via Northern blot analysis. A study using two Drosophila species had a similar sensitivity (0.75) to the other studies presented above, but no value for the specificity was presented. 24 new miRNAs were, however, confirmed by Northern blotting [40]. Huang and colleagues presented MirFinder which on their training and test set achieved an accuracy of 99.6 % and had an area under the receiver operator characteristic (ROC) curve of almost 1 [41]. They compared their ROC curve to that from other studies, but this may be at best misleading due to the usage of completely different training and test datasets. Artzi and colleagues set their filter specificity to 95 %; they estimated the sensitivity of their algorithm at 88 % (85–94 % on seven mammalian species) [23]. The content of the miROrtho database has been constructed with a hairpin prediction accuracy of 95 %, yielding a sensitivity of 84 % at 97 % specificity [24]. They then filtered these hairpins by homology with an independent accuracy of 91 %, but they do not report the overall accuracy measures. MapMi reports a sensitivity of 92 % at a specificity of 98 %. Wang and colleagues did not explicitly report on the accuracy of their algorithm but were able to confirm 67 % of their predicted miRNAs by Northern blotting [33].

That the reported accuracies have to be viewed critically can be seen in a study by Leung and colleagues who found quite different sensitivities for ProMirII and miR-abela and then the ones reported in their respective publications [21]. They also report that they were able to increase the positive prediction value by more than 15 % at high sensitivity. Since all accuracy measures reported above are derived from different studies, using different datasets, they are not integrated into a table for easy comparison since that would be misleading. In fact, the measures reported above can hardly be compared and are most likely highly optimistic. A study independently comparing these measures objectively needs yet to be done. Experimental validation may seem to actually proof the existence and effect of a miRNA, but the opposite is not true so that these approaches can only be used to confirm the existence but never to prove the absence of a miRNA.

Two examples of algorithms for homology-based miRNA gene prediction are presented as anecdotes in the next section.

2.2.3 Selected Examples Performing Homology-Based miRNA Gene Detection

Due to the large number of available miRNA gene prediction algorithms only the most cited ones, MiRscan [27] and ProMiR II [29], are discussed in some more detail followed by a more general statement about prediction accuracy.

ProMiR II

ProMiR was first introduced in 2005 as an algorithm that simultaneously considers structure and sequences of pre-miRNAs [42]. A machine learning approach was used with positive examples from known human miRNAs and negative examples extracted quasi-randomly from the human genome. Their hidden Markov model includes both sequence and structure and predicts for each element of the sequence whether it is part of a pre-miRNA or not. The predicted pre-miRNAs are then further evaluated in regard to their minimum free energy and their conservation among vertebrates.

ProMiR II extends ProMiR by adding knowledge about miRNA gene clustering, G/C ratio conservation, and entropy of candidate sequences [29]. Another improvement of ProMiR II is that different criteria are now implemented in modules making the approach very extensible. In addition to that, several databases are integrated into the analysis without the need for user intervention. ProMiR II is a web server available at http://cbit.snu.ac.kr/~ProMiR2/introduction.html. No values for specificity and sensitivity are reported, but the provided ROC curve seems to have an area under the curve somewhere between 80 and 95 % which is similar to other algorithms (*see* Table 2).

MiRscan II

MiRscan was first introduced in 2003 to find miRNA genes conserved between two species [27]. Initially, a screen for hairpin structures conserved between two genomes is performed; afterward the hairpin structures are evaluated in respect to their features.

Among these features, which are used to discriminate between true and false miRNA genes, are stringent base pairing in the miRNA:mRNA target duplex seed region, less stringent base pairing in the remaining structure, sequence bias in the first five bases, loop symmetry, and bulges.

MiRscan II [28] extends MiRscan by including the genomic sequence upstream of the miRNA gene into the analysis algorithm. In addition to general conservation for the miRNA gene flanking regions, at about 200 bp a conserved motif was observed. These findings and orthology of host genes for intronic RNA were incorporated into the new program. The new version supersedes MiRScan and is the one referenced on the web server (http://genes.mit.edu/burgelab/MiRscanII/).

2.3 Ab Initio miRNA Gene Prediction

Ab initio miRNA gene prediction needs no other information than the primary sequence in order to determine whether it is a true miRNA. Two possible modes of operation are possible with one using multiple sequences and the other based on single sequences.

2.3.1 Multiple Sequence-Based miRNA Gene Prediction

RNAmicro is an SVM-based classifier that enables detection of hairpin structures in multiple sequence alignments [43]. The approach tries to balance sensitivity and specificity unlike most other approaches in miRNA detection which try to minimize the number of false positives. In their initial tests they achieved a sensitivity of 91 % at a specificity of 99 % which as they point out cannot be achieved in a real dataset due to limitations of RNAz which places an optimistic upper bound of 80 % sensitivity at 99 % specificity on experiments with real data.

2.3.2 Single Sequence-Based miRNA Gene Prediction

One important field of research is the detection of novel miRNA genes. While there are experimental methods (*see* Chapters 1–3 and 6 in this volume) to perform this task like forward genetic screens and identification in small RNA libraries [44] as well as deep sequencing methods [45], also refer to [46] for approaches to identify ncRNAs [47]. These methods are time consuming, inefficient, or expensive. Therefore, it is necessary to also develop computational methods to predict miRNA genes that can be used in tandem with experimental strategies. Some of the current approaches are listed in Table 3. In general a mature miRNA should be derived from the stem part of a short hairpin RNA (shRNA) which should form a large number of Watson–Crick pairs and few internal loops and bulges (cf. Fig. 1). Other criteria are, for instance, that the mature miRNA is conserved in closely related species (*see* Subheading 2.2). Presence of Drosha and Dicer in the organism and accumulation of relevant product in deficient mutants is an experimental validation for a miRNA. Thermodynamic stability of hairpins and similarity to known miRNAs can also serve as supporting evidence when predicting new miRNAs. This can,

Table 3
Non-comprehensive list of software that allows the ab initio prediction of miRNA genes

Program	Summary	Clade	URL	Reference
miRseeker	First homologous miRNA gene fishing and then structure and nucleotide sequence divergence filtering	Flies	Not functional: http://www.fruitfly.org/seq_tools/miRseeker.html	[40]
PalGrade	Hairpin structural and sequence characteristic model with subsequent experimental validation	Human	–	[49]
Dynalign	Finds ncRNAs by optimizing total free energy between RNA sequences, alternative fast SVM classification	Any	http://rna.urmc.rochester.edu/dynalign.html	[11]
MiRenSVM	Employs multiple targeted SVM to model different types of miRNAs	Any	–	[50]
MiR-abela	Assumes miRNA gene clustering and searches for new genes in proximity of known genes	Human, mouse, rat	http://www.mirz.unibas.ch/cgi/pred_miRNA_genes.cgi	[3]
Triplet-SVM	Forms structure sequence triplets from hairpins and classifies them using an SVM	Any	http://bioinfo.au.tsinghua.edu.cn/mirnasvm/	[51]
RNAmicro	First structure of shRNAs (RNAz) and then SVM filtering of MSAs	Any	http://www.tbi.univie.ac.at/~jana/software/RNAmicro.html	[43]
miPred	Introduces a new machine learning approach, random forest, and improves upon Triplet-SVM	Any	http://www.bioinf.seu.edu.cn/miRNA/	[52]
miPred	Uses SVM classification without homology and defines 29 parameters to describe hairpin structures	Any	Not functional: http://web.bii.a-star.edu.sg/~stanley/Publications	[53]
NovoMir	Uses a series of filter steps and statistical models to determine pre-miRNAs in a plant genome	Plant	www.biophys.uni-duesseldorf.de/~teune/Data/novomir-2010-10-10.tgz	[54]
SplamiR	Predicts miRNAs which derive from spliced transcripts	Plant	www.uni-jena.de/SplamiR.html	[55]
MiRPara	Predicts miRNAs from high-throughput sequencing data using an SVM	Any	www.whiov.ac.cn/bioinformatics/mirpara	[56]

Rows are sorted by number of citations

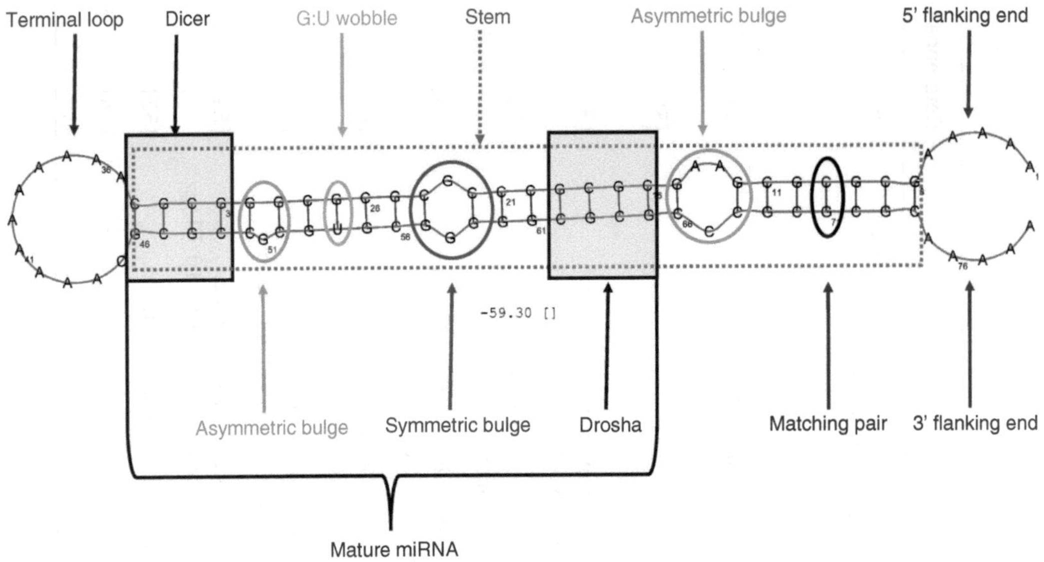

Fig. 1 The primary sequence for this hairpin was manually designed such that the selected elements were guaranteed to be present in one hairpin. The sequence was folded using RNAShapes

for example, be done by defining features of known miRNAs and training a classifier such as a support vector machine (SVM). Clustering of miRNA genes in a genomic locus can further support the validity of miRNA genes [48].

In a more systems-driven approach the predicted mature miRNAs can be validated further by looking for targets in, for example 3′UTRs, and by evaluating the multiplicity of targets per miRNA and target sites per regulated mRNA (*see* Chapters 12–14 in this volume). The non-comprehensive list of miRNA gene prediction programs and web servers in Table 3 contains algorithms using different strategies which are summarized in the following section for single-sequence miRNA gene prediction.

Methods Used in Single Sequence-Based miRNA Gene Prediction

- Proximity to known miRNA genes since miRNAs sometimes reside in clusters.
- Varying level of sequence conservation within stem structure.
- Using general properties of hairpin structures that can be learned from examples.
- Minimum free energy threshold.
- Length of stem–loop structure threshold.
- Base composition.
- Local contiguous substructures paired with central sequence information of the substructure.
- *P*-value derived from the predicted structure compared with randomized structures of the same sequence.

Filtering Strategies

Obviously, it would be beneficial to include as much information as possible for discriminating false-positive identification to increase prediction accuracy although the use of too many parameters can lead to over-training (*see* Chapters 7 and 10 in this volume). For instance, sequence, structure, and homology information can be used in tandem. Some of the information that can be used to distinguish true from false-positive miRNA gene predictions are given below:

- miRNA genes are small noncoding genes (<150 nt).
  - miRNA length.
    - Varies between plants and animals.
- Originates from pre-miRNA (80–120 nt).
  - Forms a characteristic hairpin structure.
  - Low free energy.
  - Sequence composition.
    - G/C composition varies between plants and animals.
- Sequence conservation by homology.
  - Sequence.
    - Different for plants and animals.
  - Stem–loop structure.
    - Varies between plants and animals.
- Clustering of multiple miRNAs in a genome locus.
- Each miRNA needs a target with sufficient complementarity.
- Location of miRNA and target.
  - Origin (intron, exon, intergenic).
  - Target (exon, 3′UTR).

Methods for Filtering Detected Hairpins

Whether a computationally detected hairpin is truly interesting and whether it affords spending time and money on follow-up experimental research are not always clear. Some filtering can be performed to narrow down the number of putative miRNAs to an amount that is suitable for budget and time constraints.

- Varying level of sequence conservation within stem structure (for homology-based predictions or post-filtering for ab initio approaches).
- Using general properties of hairpin structures that can be learned from examples.
- Repetitively detected structures are generally discarded.
- Minimum free energy threshold filtering.

- Length of stem–loop structure threshold filtering.
- If matching to certain annotation of a genome (e.g.: coding sequence) the detections may be discarded.
- Base composition.
- miRNA gene clustering.
- Upstream motif about 200 nucleotides before miRNA genes.
- Text mining.

Other information that could be included is, for example, the existence of a cap and a poly-A tail for pri-miRNAs that are often found in experimentally validated miRNAs.

Although the annotation of the genomic region has been used for filtering, it is clear that miRNAs can come from any region of a genome [5], and this filtering can thus only be used for reducing computational complexity and not for a biological valid reason.

Selected Examples Performing Ab Initio miRNA Gene Prediction

Due to the large number of available miRNA gene prediction algorithms only two of them, miR-abela [3] and MiPred [53], are discussed in some more detail followed by a more general statement about prediction accuracy.

MiR-Abela

The approach for ab initio prediction of miRNAs by Sewer et al. assumes that miRNAs cluster and that they may be co-transcribed [3]. Therefore, they restrict the search of novel miRNAs to areas having close proximity to already known miRNAs. For determining miRNAs, they first check for robust stem–loop structures in the area around known miRNAs because they state that the structure is important for recognition and processing by Drosha and Dicer. For this, the similarity to known stem–loops is calculated using an SVM based on weighted sequence and structural features. Overall they describe 40 features for pre-miRNA determination with 16 features describing stem–loop structures, 10 features for symmetrical regions of a stem–loop, 11 features with relaxed symmetry constraints, and 3 features in respect to mature miRNA-sized portions of a hairpin.

When using their method to predict hairpins in the proximity of known hairpins from the Rfam database in human, mouse, and rat, they were able to achieve a sensitivity of about 89 % for their hairpin detection in these species and for their artificial negative examples they achieved a false-negative discovery rate of 29 % and a sensitivity of 71 % with only 3 % false positives.

MiPred

Ng and Mishra proposed an SVM-based ab initio prediction method for finding miRNAs in 2007 [53]. In this study, 29 features have been employed to describe a hairpin at the dinucleotide, folding, thermodynamic, and topological levels. Ng and Mishra trained the classifier on human pre-miRNAs and later used the model to predict miRNAs for human with high sensitivity and specificity. When they used the same model to test the generalization ability of MiPred, an average sensitivity of 88 % at an average specificity of 98 % on a variety of species was achieved. They also compared their method with other existing predictors and found that their method and RNAmicro [43] perform similar, both outperforming the other tools tested, by large. While RNAmicro employs multiple sequences for the prediction, MiPred only uses a single sequence which makes these programs not directly comparable. Thus, according to the authors, MiPred is the most successful quasi ab initio miRNA predictor for single sequences among the methods tested in their assessment.

Accuracy of miRNA Gene Prediction

It is hard to assess the specificity and sensitivity of algorithms in the absence of at least one fully annotated genome; therefore, this section does not compare the accuracy of existing algorithms. The reported values from different publications are listed, but the reader should be aware of the fact that these values cannot be compared and may even be misleading (*see* Subheading 2.2.2).

NovoMir, software for plant miRNA gene prediction, achieved a sensitivity of 80 % at a specificity of 99 % [54]. MiRenSVM, an algorithm combining three SVM, achieved a sensitivity of 93 % at a specificity of 97 % [50].

Xue and colleagues trained an SVM to distinguish between real and pseudo pre-miRNAs which achieved about 90 % accuracy within human, from which the training data were derived, but interestingly also achieved high accuracies of up to 90 % in other species [51]. On human data they achieved a sensitivity of about 93 % at a specificity of about 88 %.

A study by Jiang and colleagues [52] which reused the same approach as Xue and colleagues [51], but added *P*-value and minimum free energy to the classification parameters and also used random forest, a different classification algorithm, achieved a sensitivity of 95 % at a specificity of 98 %.

A recent study by Zeller and co-workers first extracted all shRNAs from the *Ciona intestinalis* genome and filtered the results by structure/sequence conservation, homology to known microRNAs, and phylogenetic footprinting. For all 458 putative miRNAs predicted in this way a microarray was designed [39]. They were able to identify 100 of these using the microarray and 170 as homologous in the small RNA database for *C. intestinalis* [57].

Many algorithms for miRNA gene prediction are based on machine learning strategies. In general, these algorithms need a sufficient number of positive as well as negative examples. Although many miRNA genes seem to be unique in any organism, positive training examples can easily be found, whereas negative examples are hard to come by. They are also difficult to be established experimentally since an mRNA needs to be expressed in order to be affected by a miRNA which may only be possible in some specific developmental stadium. Some negative examples that were picked in studies like mRNA sequences [3] are dubious since to our current knowledge miRNAs can originate from any part of an mRNA. Therefore, one-class classifiers which do not need negative examples may be of help in the future [58].

Without an encompassing knowledge of miRNA genesis only a systems approach can increase the accuracy of current methods. To the best of my knowledge, there is no existing systems approach that evaluates all initially introduced descriptors and discriminators for miRNA genes and further validates them with additional discriminating information such as transcription factor-binding sites, expression assays using microarrays, and many others. Usage of several of these features in tandem is obligatory since when scanning the genome for putative miRNAs the number is enormous and thus it needs to be strictly scrutinized.

3 Methods for Filtering of Predicted miRNAs

Rfam is a database grouping noncoding RNAs, from over 200 complete genome sequences, into families aiming to facilitate the identification and classification of new noncoding RNA sequences [4, 59]. This resource can help assessing whether a predicted miRNA actually fits to the miRNA family and thus aid in deciding whether it should be retained or removed from the predictions. Further databases such as UCbase [60] and others can provide supporting information for confirmation of potential miRNAs.

Our recent assessments of miRBase, however, contradict the above statement since we found many sequences which were labeled as miRNA but obviously must use a different mechanism since they do not fit to the current definition of miRNAs and their genesis or to the proposed processing pathway via Drosha, Dicer, and RISC [61].

4 Conclusion

Today's databases contain many miRNAs. At least one study suggests that these miRNAs may only represent abundant variants [40]. This was confirmed in an independent study which found that the miRNAs, they were able to confirm experimentally, also turned out to be quite abundant [27]. Therefore, there is a large need for ab initio prediction of miRNAs in addition to homology detection. Ab initio prediction of genes has been discussed in this chapter, but despite many approaches (Table 3) there is no user-friendly software which would allow the ab initio prediction of miRNAs from sequences.

5 Outlook

Future miRNA gene prediction approaches should take a systems approach and evaluate all parts of the system here, for instance, miRNA genesis and miRNA targeting at the same time. This can raise the confidence in individual predictions and reduce the number of false predictions [62, 63]. They could further include text mining [64], gene ontologies and networks [65], and promoter sequences [66].

Integrative approaches like MMIA [62], which uses multiple miRNA target prediction algorithms in parallel, will also enhance prediction coverage and accuracy in the future.

Besides miRNAs, very similar structures adjacent to them, termed moRs, have been shown to induce gene silencing [57] which shows that we have not yet seen all biological regulatory options.

Strategies that make use of experimental data, such as deep sequencing data, for miRNA prediction [67] will in the future be more abundant and likely lead to detection of new miRNAs which do not closely resemble currently known miRNAs.

Other new findings, like spliced miRNAs [55], may be found in the future, further complicating the already complex prediction of miRNAs.

Acknowledgements

I would like to thank Müşerref Duygu Saçar for preparing Fig. 1. This study was in part supported by an award received from the Turkish Academy of Sciences for outstanding young scientists (TUBA GEBIP, http://www.tuba.gov.tr).

References

1. Soldà G, Makunin IV, Sezerman OU et al (2009) An Ariadne's thread to the identification and annotation of noncoding RNAs in eukaryotes. Brief Bioinform 10:475–489
2. Dinger ME, Pang KC, Mercer TR et al (2008) Differentiating protein-coding and noncoding RNA: challenges and ambiguities. PLoS Comput Biol 4:e1000176
3. Sewer A, Paul N, Landgraf P et al (2005) Identification of clustered microRNAs using an ab initio prediction method. BMC Bioinformatics 6:267
4. Griffiths-Jones S, Moxon S, Marshall M et al (2005) Rfam: annotating non-coding RNAs in complete genomes. Nucleic Acids Res 33:D121–D124
5. Rodriguez A, Griffiths-Jones S, Ashurst JL et al (2004) Identification of mammalian microRNA host genes and transcription units. Genome Res 14:1902–1910

6. Pfeffer S, Zavolan M, Grässer FA et al (2004) Identification of virus-encoded microRNAs. Science 304:734–736
7. Fahlgren N, Jogdeo S, Kasschau KD et al (2010) MicroRNA gene evolution in arabidopsis lyrata and arabidopsis thaliana. Plant Cell 22:1074–1089
8. Aravin A, Tuschl T (2005) Identification and characterization of small RNAs involved in RNA silencing. FEBS Lett 579:5830–5840
9. Bentwich I (2005) Prediction and validation of microRNAs and their targets. FEBS Lett 579:5904–5910
10. Janssen S, Schudoma C, Steger G et al (2011) Lost in folding space? Comparing four variants of the thermodynamic model for RNA secondary structure prediction. BMC Bioinformatics 12:429
11. Mathews DH, Turner DH (2002) Dynalign: an algorithm for finding the secondary structure common to two RNA sequences. J Mol Biol 317:191–203
12. Juan V, Wilson C (1999) RNA secondary structure prediction based on free energy and phylogenetic analysis. J Mol Biol 289: 935–947
13. Hofacker IL (2003) Vienna RNA secondary structure server. Nucleic Acids Res 31: 3429–3431
14. Krüger J, Rehmsmeier M (2006) RNAhybrid: microRNA target prediction easy, fast and flexible. Nucleic Acids Res 34:W451–W454
15. Reuter JS, Mathews DH (2010) RNAstructure: software for RNA secondary structure prediction and analysis. BMC Bioinformatics 11:129
16. Zuker M (2003) Mfold web server for nucleic acid folding and hybridization prediction. Nucleic Acids Res 31:3406–3415
17. Shapiro BA (1988) An algorithm for comparing multiple RNA secondary structures. Comput Appl Biosci 4:387–393
18. Aksay C, Salari R, Karakoc E et al (2007) taveRNA: a web suite for RNA algorithms and applications. Nucleic Acids Res 35:W325–W329
19. Janssen S, Giegerich R (2010) Faster computation of exact RNA shape probabilities. Bioinformatics 26:632–639
20. Markham NR, Zuker M (2008) UNAFold: software for nucleic acid folding and hybridization. In: Keith JM (ed) Bioinformatics: structure, function and applications. Humana Press, Totowa, NJ, pp 3–31
21. Leung W-S, Lin MCM, Cheung DW et al (2008) Filtering of false positive microRNA candidates by a clustering-based approach. BMC Bioinformatics 9(Suppl 12):S3
22. Dezulian T, Remmert M, Palatnik JF et al (2006) Identification of plant microRNA homologs. Bioinformatics 22:359–360
23. Artzi S, Kiezun A, Shomron N (2008) MiRNAminer: a tool for homologous microRNA gene search. BMC Bioinformatics 9:39
24. Gerlach D, Kriventseva EV, Rahman N et al (2009) miROrtho: computational survey of microRNA genes. Nucleic Acids Res 37:D111–D117
25. Maselli V, Bernardo DD, Banfi S (2008) CoGemiR: a comparative genomics microRNA database. BMC Genomics 9:457
26. Guerra-Assunção JA, Enright AJ (2010) MapMi: automated mapping of microRNA loci. BMC Bioinformatics 11:133
27. Lim LP, Lau NC, Weinstein EG et al (2003) The microRNAs of Caenorhabditis elegans. Genes Dev 17:991–1008
28. Ohler U, Yekta S, Lim LP et al (2004) Patterns of flanking sequence conservation and a characteristic upstream motif for microRNA gene identification. RNA 10:1309–1322
29. Nam J-W, Kim J, Kim S-K et al (2006) ProMiR II: a web server for the probabilistic prediction of clustered, nonclustered, conserved and nonconserved microRNAs. Nucleic Acids Res 34:W455–W458
30. Berezikov E, Guryev V, van de Belt J et al (2005) Phylogenetic shadowing and computational identification of human microRNA genes. Cell 120:21–24
31. Huang T-H, Fan B, Rothschild MF et al (2007) MiRFinder: an improved approach and software implementation for genome-wide fast microRNA precursor scans. BMC Bioinformatics 8:341
32. Bonnet E, Wuyts J, Rouzé P et al (2004) Detection of 91 potential conserved plant microRNAs in Arabidopsis thaliana and Oryza sativa identifies important target genes. Proc Natl Acad Sci U S A 101:11511–11516
33. Wang X-J, Reyes JL, Chua N-H et al (2004) Prediction and identification of Arabidopsis thaliana microRNAs and their mRNA targets. Genome Biol 5:R65
34. Lang Q, Jin C, Lai L et al (2011) Tobacco microRNAs prediction and their expression infected with cucumber mosaic virus and potato virus X. Mol Biol Rep 38:1523–1531
35. Gruber AR, Findeiß S, Washietl S et al (2010) Rnaz 2.0: improved noncoding RNA detection. Pac Symp Biocomput 15:69–79
36. Rivas E, Klein RJ, Jones TA et al (2001) Computational identification of noncoding RNAs in E. coli by comparative genomics. Curr Biol 11:1369–1373
37. Liang H, Li W-H (2009) Lowly expressed human microRNA genes evolve rapidly. Mol Biol Evol 26:1195–1198
38. Lu J, Shen Y, Wu Q et al (2008) The birth and death of microRNA genes in Drosophila. Nat Genet 40:351–355

39. Keshavan R, Virata M, Keshavan A et al (2010) Computational identification of Ciona intestinalis microRNAs. Zoolog Sci 27:162–170
40. Lai EC, Tomancak P, Williams RW et al (2003) Computational identification of Drosophila microRNA genes. Genome Biol 4:R42
41. Huang JC, Morris QD, Frey BJ (2007) Bayesian inference of MicroRNA targets from sequence and expression data. J Comput Biol 14:550–563
42. Nam J-W, Shin K-R, Han J et al (2005) Human microRNA prediction through a probabilistic co-learning model of sequence and structure. Nucleic Acids Res 33:3570–3581
43. Hertel J, Stadler PF (2006) Hairpins in a Haystack: recognizing microRNA precursors in comparative genomics data. Bioinformatics 22:197–202
44. Berezikov E, Cuppen E, Plasterk RHA (2006) Approaches to microRNA discovery. Nat Genet 38(Suppl):2–7
45. Hafner M, Landthaler M, Burger L et al (2010) Transcriptome-wide identification of RNA-binding protein and MicroRNA target sites by PAR-CLIP. Cell 141:129–141
46. Vogel J, Sharma CM (2005) How to find small non-coding RNAs in bacteria. Biol Chem 386:1219–1238
47. Hüttenhofer A, Vogel J (2006) Experimental approaches to identify non-coding RNAs. Nucleic Acids Res 34:635–646
48. Lau NC, Lim LP, Weinstein EG et al (2001) An abundant class of tiny RNAs with probable regulatory roles in Caenorhabditis elegans. Science 294:858–862
49. Bentwich I (2008) Identifying human microRNAs. Curr Top Microbiol Immunol 320: 257–269
50. Ding J, Zhou S, Guan J (2010) MiRenSVM: towards better prediction of microRNA precursors using an ensemble SVM classifier with multi-loop features. BMC Bioinformatics 11(Suppl 1):S11
51. Xue C, Li F, He T et al (2005) Classification of real and pseudo microRNA precursors using local structure-sequence features and support vector machine. BMC Bioinformatics 6:310
52. Jiang P, Wu H, Wang W et al (2007) MiPred: classification of real and pseudo microRNA precursors using random forest prediction model with combined features. Nucleic Acids Res 35:W339–W344
53. Ng KLS, Mishra SK (2007) De novo SVM classification of precursor microRNAs from genomic pseudo hairpins using global and intrinsic folding measures. Bioinformatics 23: 1321–1330
54. Teune J-H, Steger G (2010) NOVOMIR: De Novo Prediction of MicroRNA-Coding Regions in a Single Plant-Genome. J Nucleic Acids 2010. doi: 10.4061/2010/495904, Pubmed: 20871826
55. Thieme CJ, Gramzow L, Lobbes D et al (2011) SplamiR-prediction of spliced miRNAs in plants. Bioinformatics (Oxford, England) 27:1215–1223
56. Wu Y, Wei B, Liu H et al (2011) MiRPara: a SVM-based software tool for prediction of most probable microRNA coding regions in genome scale sequences. BMC Bioinformatics 12:107
57. Shi W, Hendrix D, Levine M et al (2009) A distinct class of small RNAs arises from pre-miRNA-proximal regions in a simple chordate. Nat Struct Mol Biol 16:183–189
58. Yousef M, Jung S, Showe LC et al (2008) Learning from positive examples when the negative class is undetermined–microRNA gene identification. Algorithms Mol Biol 3:2
59. Gardner PP, Daub J, Tate JG et al (2009) Rfam: updates to the RNA families database. Nucleic Acids Res 37:D136–D140
60. Taccioli C, Fabbri E, Visone R et al (2009) UCbase & miRfunc: a database of ultraconserved sequences and microRNA function. Nucleic Acids Res 37:D41–D48
61. Saçar MD, Hamzeiy H, and Allmer J (2013) Can MiRBase provide positive data for machine learning for the detection of MiRNA hairpins?. J Integr Bioinform 10:215
62. Cakir MV, Allmer J (2010) Systematic computational analysis of potential RNAi regulation in Toxoplasma gondii. Health Informatics and Bioinformatics (HIBIT), 2010 5th International Symposium on, pp. 31–38 IEEE, Ankara, Turkey
63. Nam S, Li M, Choi K et al (2009) MicroRNA and mRNA integrated analysis (MMIA): a web tool for examining biological functions of microRNA expression. Nucleic Acids Res 37:W356–W362
64. Naeem H, Küffner R, Csaba G et al (2010) miRSel: automated extraction of associations between microRNAs and genes from the biomedical literature. BMC Bioinformatics 11:135
65. Backes C, Meese E, Lenhof H et al (2010) A dictionary on microRNAs and their putative target pathways. Nucleic Acids Res 38: 4476–4486
66. Long Y-S, Deng G-F, Sun X-S et al (2011) Identification of the transcriptional promoters in the proximal regions of human microRNA genes. Mol Biol Rep 38:4153–4157
67. Hendrix D, Levine M, Shi W (2010) miRTRAP, a computational method for the systematic identification of miRNAs from high throughput sequencing data. Genome Biol 11:R39

Chapter 10

Machine Learning Methods for MicroRNA Gene Prediction

Müşerref Duygu Saçar and Jens Allmer

Abstract

MicroRNAs (miRNAs) are single-stranded, small, noncoding RNAs of about 22 nucleotides in length, which control gene expression at the posttranscriptional level through translational inhibition, degradation, adenylation, or destabilization of their target mRNAs. Although hundreds of miRNAs have been identified in various species, many more may still remain unknown. Therefore, discovery of new miRNA genes is an important step for understanding miRNA-mediated posttranscriptional regulation mechanisms. It seems that biological approaches to identify miRNA genes might be limited in their ability to detect rare miRNAs and are further limited to the tissues examined and the developmental stage of the organism under examination. These limitations have led to the development of sophisticated computational approaches attempting to identify possible miRNAs in silico. In this chapter, we discuss computational problems in miRNA prediction studies and review some of the many machine learning methods that have been tried to address the issues.

Key words Machine learning, miRNA gene prediction, miRNA gene detection, Classification, Test data, Examples

1 Introduction

Current attempts to distinguish microRNA (miRNA) genes have led to the detection of thousands of miRNAs in various species, but many may remain undiscovered [1]. These efforts, mainly based on experimental methods such as directional cloning of endogenous small RNAs, are time consuming, expensive, and work intensive [2]. Inadequacy of experimental approaches can be showcased by the fact that miRNAs are expressed in specific cell types, at low levels, or only in a specific condition which complicates their experimental detection. To overcome these problems several computational methods have been designed and applied to miRNA gene detection.

Numerous approaches for the in silico prediction of miRNAs have been created so far. These programs commonly regard the hairpin secondary structure of the miRNA precursor as the most important characteristic of a miRNA gene [3, 4]. RNA secondary structure prediction algorithms such as RNAfold [5] are used to

Malik Yousef and Jens Allmer (eds.), *miRNomics: MicroRNA Biology and Computational Analysis*, Methods in Molecular Biology, vol. 1107, DOI 10.1007/978-1-62703-748-8_10, © Springer Science+Business Media New York 2014

predict the secondary structure and thermodynamic stability of RNA hairpin structures. Existing bioinformatics methods for the prediction of miRNA usually consist of (1) genome-wide estimation of hairpin structures, (2) filtering or scoring of those hairpins based on their similarity in structure and sequence to known miRNA hairpins, and (3) experimental confirmation of putative candidates [3]. In order to extract possible miRNAs from a genome, either homology modeling or ab initio methods are used.

2 Homology-Based MicroRNA Gene Prediction

Homology-based miRNA gene mapping methods can build on available, experimentally validated, miRNAs and find similar structures and sequences in related species. The idea is that if a miRNA is identified in one genome then its homologs can be possibly found in other species [6]. Since conservation indicates a function, it is assumed that conserved candidates are more likely to be miRNAs. Although it has been shown that for noncoding RNAs absence of conservation does not inevitably mean lack of function [7], searching for homologs especially in newly annotated genomes may be a beneficial approach. Software facilitating mapping of known miRNAs to homologous genomes take both sequence similarity and miRNA secondary structure information into account. The theory is based on derivation of mature miRNAs from hairpin structure formed by folding its pre-miRNA. The approach taken by one of the most recent developments, MapMi [8], first scans the miRNA sequences against the target genome and then creates potential pre-miRNAs from them. In the end, the results are scored, ranked, and displayed. The scoring function considers both the quality of the sequence match (match, mismatch, perfect match) and the predicted structure of hairpins [8]. The best candidate is chosen according to the calculated score, and candidates are further filtered based on a score threshold which is either user defined or selected from suggested thresholds [8]. All candidates above threshold are displayed with their related scores and other relevant information. The Web version of MapMi provides more detailed analysis including the generation and display of maximum likelihood phylogenetic trees, multiple sequence alignments, and RNA structural logos [8].

Although homology modeling can gather information from already successfully established miRNAs of a related organism's genome, it is also limited since completely novel miRNAs cannot be determined in this way. First attempts in this approach have mainly relied on identifying close homologs of published pre-miRs, i.e., let-7 [9]. This method might seem as straightforward as aligning sequences through NCBI BlastN [10], but it can only reproduce results and cannot find new miRNA genes. Since many miRNAs are species specific, they will always be missed by this method and therefore other strategies need to be used in tandem. Additionally, miRNA genes evolve very rapidly which further limits the applicability of homology-based methods [11]. A powerful approach developed for genome-wide screening of phylogenetically well-conserved pre-miRNAs between closely related species is cross-species sequence conservation based on computationally intensive multiple genome alignments. However, it also suffers lower sensitivity especially for more divergent evolutionary distances [12, 13]. Moreover, identifying pre-miRNAs that differ significantly or undergo rapid evolution at the sequence level while keeping their characteristic evolutionarily conserved hairpin structures may also pose problems [2]. Another important issue is that non-conserved pre-miRNAs with genus-specific patterns are likely to escape detection [2].

There are various homology-based miRNA gene prediction software such as MirScan [14], miROrtho [15], miRNAminer [16], and ProMiR II [17]. Also, some of these software use machine learning approaches such as ProMir [18], which uses hidden Markov models, and MirFinder [19]. MirFinder is designed for genome-wide, pair-wise sequences from two chosen species and includes two key steps: (1) genome-wide searching of hairpin candidates and (2) elimination of the non-robust structures based on 18 features analyzed by support vector machine (SVM) classification [19]. The tool was tested on chicken/human and *Drosophila melanogaster*/*D. pseudoobscura* pair-wise genome alignments. The results showed that the proposed method can be used for genome-wide pre-miRNA predictions [19].

3 Ab Initio-Based MicroRNA Gene Prediction

While homology-based methods mainly use comparative genomics, ab initio miRNA gene prediction needs no information other than the primary sequence for the prediction of miRNAs (albeit ab initio methods require negative and positive datasets, which is conceptually similar to homology-based approaches). Nonetheless, ab initio methods may enable the identification of new miRNAs which have no close homologs [20]. The main difficulty of ab initio methods is choosing proper parameters that allow determining a given sequence to be a miRNA based on its properties. For instance, hairpin structure and minimum free energy (MFE) are widely used features [21] of miRNAs in prediction tools such as miPred [2]. If the chosen parameters do not provide good specificity, it would not be very informative and might increase the potential to produce false positive results. This would lead to a decrease in the accuracy of the miRNA prediction method and make validating the results of predictions in the wet-lab much more elaborate, time consuming, and expensive. The main problem is that, although precursor miRNAs should possess an evolutionarily conserved hairpin structure which is critical for the early stages of the mature miRNA biogenesis, the hairpin shape is not unique to miRNAs and is found in many other noncoding RNAs [22]. For instance, all translational RNAs contain multiple hairpins. It has been estimated that there are millions of hairpin-like structures in the human genome, and differentiating the millions of hairpins from the few true miRNAs is the grand challenge [23].

There are many programs using ab initio methods with machine learning approaches including Triplet-SVM [24], MiRenSVM [22], miPred (SVM) [2], MiPred (random forest) [25], and MiRPara (SVM) [26].

4 Machine Learning and MicroRNA Gene Prediction

Next to defining proper features that allow differentiation between true and false miRNAs the selection of training data for machine learning algorithms is crucial for prediction success. Therefore, we will shortly comment on training and test data used in machine learning for miRNA prediction.

4.1 Learning and Test Data

4.1.1 Positive Data

Usually positive data for miRNA gene predictions are obtained from miRBase [27]. However, there are some entries in miRBase which are suggested as miRNAs but are not fulfilling the necessary properties to be classified as miRNAs such as having more than one loop. It was shown that reference set of positive controls taken from miRBase requires additional improvement to create a high-confidence set proper for use as positive controls [28]. We recently elaborated on this and found that prediction accuracy can be improved upon filtering of unlikely miRNAs from miRBase [29]. Except for these minor problems, in miRNA gene prediction studies, it is usually uncomplicated to select positive examples (e.g., using the known miRNAs), while it is challenging to create negative samples [6].

4.1.2 Negative Data

The collection of an appropriate negative dataset is vital for many machine learning algorithms to produce a well-trained classifier. If the sequences are too artificial, then there is a high probability that the machine learning method will not be trained adequately to differentiate between true miRNAs and non-miRNA sequences [26]. On the other hand, if the negative dataset is very similar to the positive dataset, the machine learning approach will be incapable of distinguishing between these two datasets [26].

One of the criteria for a small RNA sequence to be classified as a miRNA is that it should be recognized and processed by the enzyme Dicer. While defining a negative dataset, this criterion should be used efficiently so that selected negative controls are not recognized by Dicer [28]. The negative dataset sequences should be composed of transcripts that are expressed in the same cellular location as true miRNAs but are not recognized by Dicer. Since this is a very complicated way to generate negative samples, instead of this, in most of the algorithms random genomic sequences or exonic sequences are used [24, 28]. These sequences are very weak negative controls because there is no confirmation that these transcribed small RNAs would not be recognized by Dicer and other components of miRNA biogenesis pathway (i.e., Drosha, RISC) and processed into functional mature miRNAs [28]. On the contrary, there is evidence that miRNAs can stem from any region of a genome (see other chapters in this volume) so that the assumption that hairpins from exons are good negative data is quite dangerous.

A well-known negative dataset for miRNA gene prediction consists of 8,494 pseudo hairpins from human RefSeq genes [30] which have been selected such that they do not undergo any alternative splicing events [2].

4.2 Algorithms for Machine Learning

Machine learning is used in many bioinformatics applications and studies (Fig. 1). The quickly increasing amount of data, created by modern molecular biology techniques, has caused the need for accurate classification and prediction algorithms since handling it with traditional methods is not feasible anymore [31]. There are numerous biological fields where machine learning methods are applied for knowledge extraction from data such as genomics, system biology, evolution, microarray, and proteomics [32].

Machine learning algorithms are different from the rule-based miRNA prediction algorithms since the rules to decide whether a given sequence is a miRNA are not manually created; instead, these rules are fit, trained, or learned from examples [32]. Usually machine learning-based methods start with the learning process of sequence, structure, or thermodynamic characteristics of miRNAs. Next, a classifier is formed to decide whether unknown sequences are true miRNAs based on the information gained through positive and negative datasets. Normally, the parameters are a set of numerical features defining a candidate miRNAs such as MFE of folding, and the results would be true or false indicating whether the candidate is a miRNA or not.

However, there are two main weaknesses with the existing machine learning-based miRNA gene identification methods. The first one is the imbalance between positive and negative examples. Since the exact number of real miRNAs in any genome is unknown, it is supposed that there are few miRNA precursors in a genome [22] so that any arbitrarily selected hairpin extracted from the genome is unlikely to be a pre-miRNA. Also, the number of positive examples is significantly smaller than that of generated negative examples (note

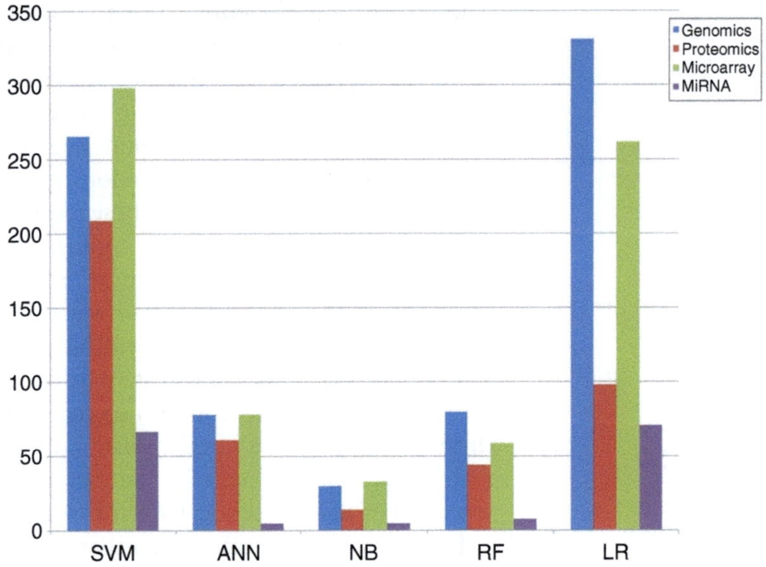

Fig. 1 Fields in biology where machine learning methods are applied (the number of publications (*y*-axis) is calculated by searching PubMed with machine learning approach and the field name as key words)

caution in Subheading 4.1.2). For instance, one of the commonly used negative datasets for miRNA prediction algorithms consists of approximately 9,000 pseudo hairpins, while the number of human miRNAs that can be obtained from miRBase is less than 1,500 [2]. The imbalance problem between positive and negative datasets can significantly reduce the performance of current machine learning approaches [22]. The other problem is that most of the current machine learning-based algorithms make assumptions such as length of the stem, loop size, and MFE of the data. Thus, sequences outside of these predetermined borders are not considered as a true miRNA and cannot be predicted by those methods, which may reduce the prediction performance and accuracy [22].

To our knowledge, there is no published study that uses unsupervised machine learning approaches for miRNA gene prediction. On the other hand, there are many studies using supervised machine learning algorithms such as SVM, neural networks (NN), hidden Markov models (HMM), and Naive Bayes (NB) (for more details on these algorithms *see* Chapter 7 in this volume).

4.2.1 Supervised MicroRNA Gene Prediction Approaches (Classification)

Machine learning for miRNA gene prediction is almost exclusively based on supervised learning in which an algorithm is trained to learn, approximating a function that maps input data to required outputs [33]. Usually, the inputs are a set of parameters designating a candidate (e.g., MFE, number of dinucleotides, length of stem), and the output would be miRNA or non-miRNA. While the

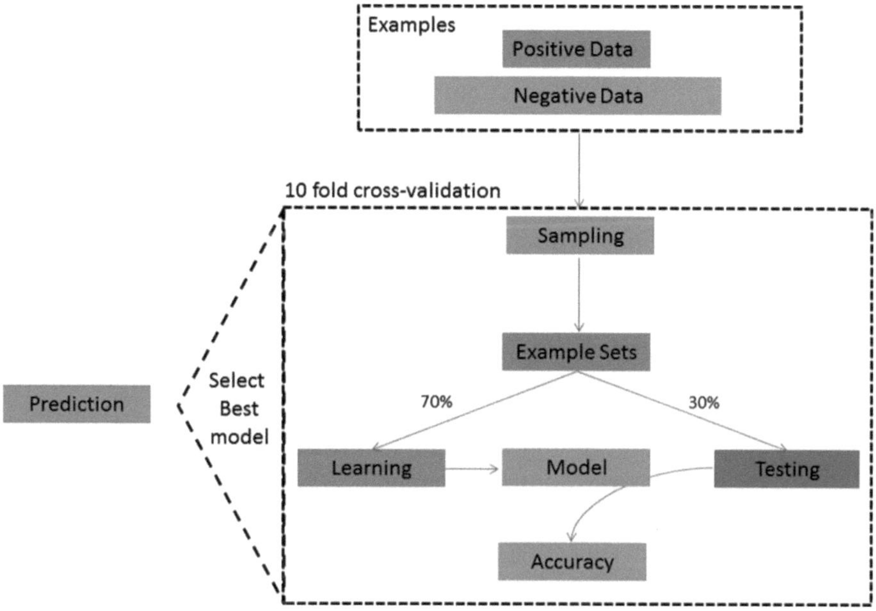

Fig. 2 General work-flow of machine learning algorithms for miRNA gene prediction

anticipated output is unknown, the machine is trained by input. The main idea of the process is that the machine learner should be capable of simplifying from these examples (input data; positive and negative examples) and properly classify candidates [6]. The most important factor influencing the accuracy of the results is the choice of features since parameterization of the examples into features is not performed automatically [6, 22]. To test the accuracy and precision of the machine learning process, a system called cross-validation is used. Cross-validation is important to prevent Type III errors (as put by Mosteller: "correctly rejecting the null hypothesis for the wrong reason" [34]), particularly in situations where further samples are dangerous or expensive to obtain. One round of cross-validation includes dividing a sample of data into corresponding subsets, performing the analysis on one subset (the training or the learning set), and validating the analysis on the test set (Fig. 2). The example sets can be divided in defined percentages (e.g., 70 % of samples included in learning set, remaining 30 % included in testing set; *see* Fig. 2), but the essential point is that these datasets must not have shared examples. After cross-validation the best model is selected and applied to perform predictions.

One of the initial works in the field by Sewer et al. (2005) assembled 40 different sequence and structural parameters to label a candidate as pre-miRNA. The SVM classifier model was trained using 178 known human pre-miRNAs as positive examples and 5,395 random sequences obtained from tRNA, rRNA, and mRNA genes as negative examples (in reality, there is no guarantee that these RNAs would not contain any functional miRNAs, *see* Subheading 4.1.2). As a result of huge difference between the number of positive and negative samples, their results have high specificity (91 %) and low sensitivity (71 %) for their dataset.

ProMiR was introduced in 2005 as an algorithm that uses an HMM and simultaneously takes into account structure and sequences of pre-miRNAs (Nam et al. 2005). A machine learning approach was used with positive examples from known human miRNAs and negative examples obtained arbitrarily from the human genome. The predicted pre-miRNAs are further assessed according to their MFE and searched to find out whether they are conserved among vertebrates. ProMiR II includes additional features than ProMiR such as addition of knowledge about miRNA gene clustering, G/C ratio conservation, and entropy of candidate sequences (Nam et al. 2006).

MatureBayes is a probabilistic algorithm developed by Gkirtzou et al., which uses a Naive Bayes classifier to characterize potential mature miRNAs [35]. Similar to previous approaches, it also performs classification based on sequence and secondary structure information of miRNA precursors.

4.2.2 One-Class Classification

The major challenge of classification is appointing a new object to one of a set of classes which are defined in advance. This classification process is performed by using the learned rules based on a number of examples. Differing from other classification approaches, in one-class classification it is supposed that only information of one of the classes, also known as the target class, is accessible. Hence, since there is no information apart from the examples of the target class, the distinction between the two classes has to be assessed from data of only the real class [36].

Defining the negative class is the most difficult challenge to overcome in developing machine learning algorithms for miRNA identification. Therefore, machine learning approaches have been proposed for identifying miRNAs without the requirement of a negative class. Yousef and colleagues performed a study using one-class machine learning approach for miRNA gene prediction by using only positive data to construct the classifier [37]. Although one-class method is less complex to implement which makes it easier to handle, the two-class procedures generally seem to be superior. Moreover, there are additional problems due to some characteristic properties of miRNAs; e.g., pre-miRNAs must fold in a hairpin structure, but not all the hairpins in the genome are miRNA sequences [38].

5 Conclusion

The biggest challenge for miRNA gene prediction is that most eukaryotic genomes include vast numbers of inverted repeats (IR), so the transcripts of these IRs can form strong hairpins [6]. Without considering phylogenetic conservation it has been shown that about ≈11 million hairpins can be found in the human genome [1]. These hairpins can have various origins and take part in numerous processes, one of which might be miRNA-mediated posttranscriptional regulation [6]. Since not all hairpins are miRNAs, identifying the hairpins which would become functional miRNAs is a very difficult task. Moreover, the big number of possible hairpins makes reducing the false-positive rate and increasing the accuracy of the prediction a difficult task.

Machine learning approaches have become popular for miRNA gene prediction studies. Since there are known miRNAs either experimentally validated or discovered through bioinformatics tools, positive datasets which is a necessity for machine learning methods, are available for miRNA precursors. Moreover, there are also some rules defining a sequence as a miRNA (e.g., recognition and being processed by miRNA biogenesis pathway enzymes such as Dicer and Drosha), so the sequences that do not pass this criteria can be used as negative datasets. However, it is important to keep in mind that the quality of these datasets will affect the sensitivity and specificity of the designed programs (*see* Subheading 4.1). Still, in order to overcome the difficult issue of creating appropriate negative datasets one-class classification method can be applied to the miRNA gene prediction problem. The abundance of machine learning methods employed for miRNA gene prediction shows that these approaches are deemed to be suitable to deal with this problem.

Acknowledgements

This study was in part supported by an award received from the Turkish Academy of Sciences for outstanding young scientists (TUBA GEBIP, http://www.tuba.gov.tr).

References

1. Bentwich I, Avniel A, Karov Y et al (2005) Identification of hundreds of conserved and nonconserved human microRNAs. Nat Genet 37:766–770
2. Ng KLS, Mishra SK (2007) De novo SVM classification of precursor microRNAs from genomic pseudo hairpins using global and intrinsic folding measures. Bioinformatics 23: 1321–1330
3. van der Burgt A, Fiers MWJE, Nap J-P et al (2009) In silico miRNA prediction in metazoan genomes: balancing between sensitivity and specificity. BMC Genomics 10:204
4. Janssen S, Schudoma C, Steger G et al (2011) Lost in folding space? Comparing four variants of the thermodynamic model for RNA secondary structure prediction. BMC Bioinformatics 12:429

5. Hofacker IL (2003) Vienna RNA secondary structure server. Nucleic Acids Res 31: 3429–3431
6. Lindow M, Gorodkin J (2007) Principles and limitations of computational microRNA gene and target finding. DNA Cell Biol 26:339–351
7. Pang KC, Frith MC, Mattick JS (2006) Rapid evolution of noncoding RNAs: lack of conservation does not mean lack of function. Trends Genet 22:1–5
8. Guerra-Assunção JA, Enright AJ (2010) MapMi: automated mapping of microRNA loci. BMC Bioinformatics 11:133
9. Pasquinelli AE, Reinhart BJ, Slack F et al (2000) Conservation of the sequence and temporal expression of let-7 heterochronic regulatory RNA. Nature 408:86–89
10. McGinnis S, Madden TL (2004) BLAST: at the core of a powerful and diverse set of sequence analysis tools. Nucleic Acids Res 32: W20–W25
11. Liang H, Li W-H (2009) Lowly expressed human microRNA genes evolve rapidly. Mol Biol Evol 26:1195–1198
12. Berezikov E, Guryev V, van de Belt J et al (2005) Phylogenetic shadowing and computational identification of human microRNA genes. Cell 120:21–24
13. Boffelli D, McAuliffe J, Ovcharenko D et al (2003) Phylogenetic shadowing of primate sequences to find functional regions of the human genome. Science (New York, NY) 299: 1391–1394
14. Lim LP, Lau NC, Weinstein EG et al (2003) The microRNAs of Caenorhabditis elegans. Genes Dev 17:991–1008
15. Gerlach D, Kriventseva EV, Rahman N et al (2009) miROrtho: computational survey of microRNA genes. Nucleic Acids Res 37: D111–D117
16. Artzi S, Kiezun A, Shomron N (2008) MiRNAminer: a tool for homologous microRNA gene search. BMC Bioinformatics 9:39
17. Nam J-W, Kim J, Kim S-K et al (2006) ProMiR II: a web server for the probabilistic prediction of clustered, nonclustered, conserved and nonconserved microRNAs. Nucleic Acids Res 34: W455–W458
18. Nam J-W, Shin K-R, Han J et al (2005) Human microRNA prediction through a probabilistic co-learning model of sequence and structure. Nucleic Acids Res 33:3570–3581
19. Huang T-H, Fan B, Rothschild MF et al (2007) MiRFinder: an improved approach and software implementation for genome-wide fast microRNA precursor scans. BMC Bioinformatics 8:341
20. Brameier M, Wiuf C (2007) Ab initio identification of human microRNAs based on structure motifs. BMC Bioinformatics 8:478
21. Allmer J, Yousef M (2012) Computational methods for ab initio detection of microRNAs. Front Genet 3:209
22. Ding J, Zhou S, Guan J (2010) MiRenSVM: towards better prediction of microRNA precursors using an ensemble SVM classifier with multi-loop features. BMC Bioinformatics 11(Suppl 1):S11
23. Bentwich I (2008) Identifying human microRNAs. Curr Top Microbiol Immunol 320: 257–269
24. Xue C, Li F, He T et al (2005) Classification of real and pseudo microRNA precursors using local structure-sequence features and support vector machine. BMC Bioinformatics 6:310
25. Jiang P, Wu H, Wang W et al (2007) MiPred: classification of real and pseudo microRNA precursors using random forest prediction model with combined features. Nucleic Acids Res 35:W339–W344
26. Wu Y, Wei B, Liu H et al (2011) MiRPara: a SVM-based software tool for prediction of most probable microRNA coding regions in genome scale sequences. BMC Bioinformatics 12:107
27. Kozomara A, Griffiths-Jones S (2011) miRBase: integrating microRNA annotation and deep-sequencing data. Nucleic Acids Res 39: D152–D157
28. Ritchie W, Gao D, Rasko JEJ (2012) Defining and providing robust controls for microRNA prediction. Bioinformatics (Oxford, England) 28:1058–1061
29. Saçar MD, Hamzeiy H, Allmer J (2013) Can MiRBase provide positive data for machine learning for the detection of MiRNA hairpins? J Integr Bioinform 10(2):215
30. Pruitt KD, Tatusova T, Maglott DR (2005) NCBI reference sequence (RefSeq): a curated non-redundant sequence database of genomes, transcripts and proteins. Nucleic Acids Res 33: D501–D504
31. Bhaskar H, Hoyle DC, Singh S (2006) Machine learning in bioinformatics: a brief survey and recommendations for practitioners. Comput Biol Med 36:1104–1125
32. Larrañaga P, Calvo B, Santana R et al (2006) Machine learning in bioinformatics. Brief Bioinform 7:86–112
33. Zhang Y-Q, Rajapakse JC, Zhang B-T et al (2008) Supervised learning methods for MicroRNA studies., *machine learning in bioinformatics*. Wiley, New York, p 339
34. Mosteller F (1948) A k-sample slippage test for an extreme population. Ann Math Stat 19:58–65

35. Gkirtzou K, Tsamardinos I, Tsakalides P et al (2010) MatureBayes: a probabilistic algorithm for identifying the mature miRNA within novel precursors. PloS one 5:e11843
36. Tax DMJ (2001) One-class classification. ISBN: 90-75691-05-x
37. Yousef M, Jung S, Showe LC et al (2008) Learning from positive examples when the negative class is undetermined–microRNA gene identification. Algorithms Mol Biol 3:2
38. Bentwich I (2005) Prediction and validation of microRNAs and their targets. FEBS Lett 579: 5904–5910

Chapter 11

Functional, Structural, and Sequence Studies of MicroRNA

Chanchal K. Mitra and Kalyani Korla

Abstract

In this review, current knowledge and ideas regarding several important functional-, structural-, and sequence-related aspects of microRNAs (miRNAs) are summarized. The current research on structural and functional aspects of miRNAs is rapidly growing, and new information appears regularly. Well-established information from literature useful for researchers working in this area has been compiled in this work. Although details of the methodology have not been elaborated, the outline should be useful for a broad and general understanding. The current information is highly interdisciplinary including molecular biology and bioinformatics; we have tried to bring both sides together. Little is known about the 3-D structure of miRNAs and its significance. miRNAs are usually active in conjunction with some proteins, and the complex is responsible for the intended task. Little information is available in the literature about the protein–miRNA interactions and their nature and properties. Nonetheless, we believe that the review will be useful to both bioinformaticians and molecular biologists.

Key words miRNA, Sequence similarity, miRNA function

1 Introduction

MicroRNAs (miRNAs) are ~22-nucleotide-long regulatory RNA molecules present in eukaryotes and other realms of life (*see* Chapter 1 in this volume). Hundreds of miRNA genes have been found in various species and even more are predicted to be present. Computational analysis suggests that approximately 30 % of protein-coding genes are regulated by miRNAs and each miRNA can target a number of genes [1]. Nearly all biological processes have been found to be influenced by miRNAs, and their ubiquitous functional occurrence suggests a new perspective to the whole gene regulation scenario. Some of the miRNAs have been found to be highly conserved across species.

With increasing data accumulating on novel miRNAs from various organisms, the need to identify and more importantly to distinguish miRNAs from other classes of small RNAs became evident. Criterion for a small RNA to be called miRNA may include

Malik Yousef and Jens Allmer (eds.), *miRNomics: MicroRNA Biology and Computational Analysis*, Methods in Molecular Biology, vol. 1107, DOI 10.1007/978-1-62703-748-8_11, © Springer Science+Business Media New York 2014

the following experimental verification including their expression-, structural-, and biogenesis-related features [2]:

- Its expression should be confirmed by hybridization to a size-fractionated RNA sample, preferably by northern blotting. Other detection methods may be used like PCR after reverse transcription of RNA (RT-PCR), primer extension analysis, RNase protection assay, and microarray hybridization. Northern blotting is the preferred choice for the confirmation of miRNAs, because the blot usually shows both the mature miRNA (an ~22-nucleotide band) and precursor-miRNA (pre-miRNA) (an ~70-nucleotide band).
- The small RNA sequence should be present in one arm of the hairpin precursor (pre-miRNA), which lacks large internal loops or bulges. The precursors are usually ~60–80 nucleotides in animals, but the lengths may be more variable in plants.
- The small RNA sequences should be phylogenetically conserved. The sequence conservation should also be seen in the precursor hairpin, usually to a lesser extent than in the mature miRNA segment.
- The evidence can be strengthened if the precursor accumulates in the nucleus in the presence of reduced Dicer function.

The last point is not mandatory, and there can be other technical reasons for reduced Dicer function. Usually the first three or the combination of first, second, and fourth are adequate to annotate a candidate gene product as novel miRNA.

1.1 Location of miRNA

miRNAs may be derived from the exonic or the intronic regions of transcription units (or processed transcripts) but are also found in intergenic clusters. There are also mixed miRNA genes which can be classified to one of the above groups depending on the variations of the splicing pattern. miRNAs produced via this route (from exonic or intronic regions) are likely to be effective in regulating the translation process for the particular protein via a possible feedback mechanism. Messenger RNA silencing by miRNAs has already been reported experimentally. Several miRNAs are present in close proximity to other miRNAs, forming a cluster of 2–7 miRNA genes [3–5]. This suggests the possibility of their transcription from a single polycistronic transcription unit. Such miRNAs may act generically on a number of related pathways for transcription and translation and may act as enhancer. It has been reported that miRNAs transcribed from a single-gene cluster are co-expressed during embryogenesis indicating the cluster to be functionally cooperative. On the contrary, it has been reported that miRNAs from a single cluster may be differentially regulated (*see* Chapter 18 in this volume).

In the following sections we briefly outline the canonical biogenesis of miRNAs and touch on some alternative strategies.

2 The Molecular Biology of MicroRNA

2.1 Biogenesis in Animals

The experimental evidences suggest that miRNAs are mostly transcribed by RNA polymerase II (pol II) [6]. The evidences that support this fact include the following:

1. Primary-microRNAs (pri-miRNAs) were shown to contain both cap structures and poly(A) tails [6, 7] which are unique properties of pol II gene transcripts.
2. miRNA transcription activity was found sensitive to α-amanitin at a concentration that specifically inhibits pol II, but not pol I or pol III [6], and this was observed as an effect of decreased level of pri-miRNA.
3. Chromatin immunoprecipitation analyses have demonstrated physical association of pol II with the promoter of miR-23a, miR-27a, and miR-24-2 [6]. Transcription by pol II facilitates the control of miRNA gene transcription by various pol II-associated regulatory factors so as to express a specific set of miRNAs at a given stage in the development and in certain specified cell types in a particular cellular condition. This would also assist in coordinating expression of both miRNA and protein-coding regions, when both of them reside in the same transcript region.

RNA polymerase III (pol III) also transcribes some miRNAs, especially those with upstream Alu sequences, transfer RNAs (tRNAs), and mammalian wide interspersed repeat (MWIR) promoter units [8]. Such repeats are considered to have the sequences important for pol III activity.

Sometimes, a single RNA transcript can contain both a protein-coding region and miRNA sequences (miRNAs in protein-coding transcription units) [9, 10]. It has recently been shown that a single transcription unit including both luciferase cDNA and miR-21 precursor sequences can produce both luciferase protein and miR-21 RNA [7]. This pri-miR-21 gene sequence is flanked on the 5′ end by a promoter element able to transcribe heterologous mRNAs and on the 3′ end by a consensus polyadenylation sequence. Nuclear processing of pri-miRNAs was found to be efficient, thus largely preventing the nuclear export of full-length pri-miRNAs. However, the deciding mechanism which directs the primary transcript to take one of the two synthesis pathways—either the miRNA pathway or the mRNA pathway—is still unknown.

It has been observed experimentally that miRNAs are transcribed as long primary transcripts (pri-miRNA) that are first trimmed into the hairpin intermediates (pre-miRNAs) and subsequently cleaved into mature miRNAs. The catalytic activities for the first and the second processing are compartmentalized into the nucleus and the cytoplasm, respectively. Transcription of miRNA genes yields primary transcripts, pri-miRNAs that contain a local

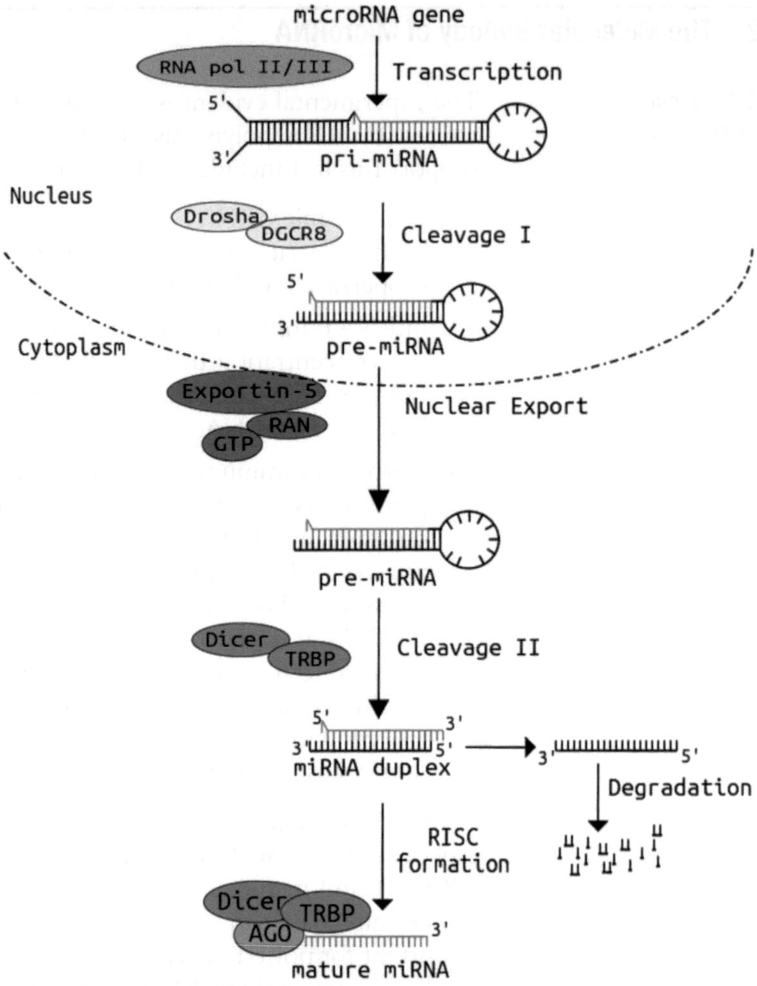

Fig. 1 Biogenesis of miRNA (redrawn from Winter, J., Jung, S., Keller, S., Gregory, R.I., Diederichs, S. (2009) Many roads to maturity: microRNA biogenesis pathways and their regulation. Nature Cell Biology 11, 228–234)

hairpin structure (Fig. 1). These pri-miRNAs undergo two-step processing, the first one in the nucleus followed by export to the cytoplasm and then the final processing step in the cytoplasm to yield a mature miRNA.

Inside the nucleus, the stem–loop structure of the pri-miRNA is recognized by the DiGeorge syndrome critical region 8 (DGCR8 or Pasha in invertebrates), which is a double-stranded RNA-binding protein [11]. DGCR8 is an essential cofactor of Drosha, and together they are referred to as microprocessor complex. DGCR8 is capable of binding to the pri-miRNA and is essential for proper processing by Drosha [12]. DGCR8 facilitates the release of hairpins from pri-miRNAs by cleaving RNA about 11 nucleotides from the hairpin base using the RNase III domain of Drosha [13]. The Drosha processing leaves it with a two-nucleotide overhang at the 3′ end, a 3′ hydroxyl and 5′ phosphate group. It is often termed as a pre-miRNA.

Drosha can form two different complexes, a small microprocessor complex that contains only Drosha and DGCR8 or a larger complex that contains RNA helicases, double-stranded RNA-binding proteins, heterogeneous nuclear ribonucleoproteins, and Ewing's sarcoma proteins [14].

Not all miRNAs are produced according to the canonical biogenesis pathway as we show in the next section.

2.2 Mirtrons

Mirtrons are the class of miRNAs which are located in the intron region of an mRNA-encoding gene. Splicing out process of these mirtrons replaces Drosha cleavage, if the released miRNA has the appropriate size to form a hairpin resembling a pre-miRNA. It is then directly processed by Dicer in cytoplasm [15, 16] *upon export into the cytoplasm*. Such miRNAs have been found also in mammals, *D. melanogaster*, and *C. elegans*, and recent studies indicate that they also exist in plants.

pre-miRNAs formed from the processing of pri-miRNAs are then exported to the cytoplasm for further processing. The two-step processing of pri-miRNA by two different RNase III enzymes is compartmentalized, and therefore nuclear export is crucial for miRNA biogenesis. The export is mediated by Exportin-5 (Exp-5), a nucleocytoplasmic shuttle. In reduced concentration of Exp-5, the pre-miRNA, and consequently the product, its mature miRNA, levels are also reduced in the cytoplasm. There was no accumulation of pre-miRNA in the nucleus, indicating relative instability of pre-miRNA in the nucleus. Alternatively, it has been suggested that they are stabilized by Exp-5 binding. Exp-5 recognizes the two-nucleotide overhang left by the RNase III enzyme Drosha at the 3′ end of the pre-miRNA hairpin, releases it from Drosha, and exports it to cytoplasm using energy from Ras-related nuclear protein-GTP (RanGTP) [17].

Once the pre-miRNAs are exported to the cytoplasm, Dicer, another RNase III enzyme, gets involved in the processing of pre-miRNAs. Dicer cleaves the ~70-nucleotide-long stem–loop pre-miRNAs (Fig. 2) to a ~22-nucleotide-long mature miRNA duplex [18]. The endoribonuclease interacts with the 3′ end of the stem–loop and cleaves the loop joining the 3′ and 5′ arms, resulting into an imperfect miRNA:miRNA*—a duplex ~22 nucleotides in length. The ~22-nucleotide miRNA duplex formed initially has a short life. One of the strands (miRNA*; also called the passenger strand) of this duplex is digested rapidly, and the other strand (miRNA) persists as mature miRNA. Knocking down of Dicer results, as expected, in the accumulation of pre-miRNA and depletion of mature miRNA. Dicers consist of two RNase III domain (RIIID) and a double-stranded RNA-binding domain (dsRBD). The N-terminal segment of Dicer contains a helicase domain along

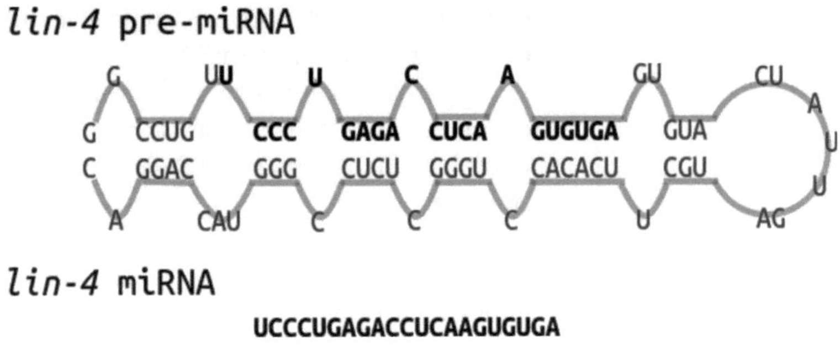

Fig. 2 Structure of pre-miRNA and miRNA (redrawn from He, L. and Hannon, G.J. (2004) MicroRNAs: Small RNAs with a big role in gene regulation. Nature Reviews Genetics, 5, 523)

with DUF283 and PAZ domains. The PAZ (Piwi/Argonaute/Zwille) domain is said to bind the 3′ end of small RNAs. Dicer may interact with several other proteins such as RDE-4, R2D2, FMR1, and others, which are not involved in the cleavage process directly but may have their roles in the miRNA stability and effector complex formation as well as miRNA action [19].

The miRNA biogenesis pathway in animals as described above is slightly different from the one in plants which will be described in the following.

2.3 Biogenesis in Plants

miRNA biogenesis in plants is also a stepwise process but differs from biogenesis in animal in the steps of nuclear processing and export. The maturation of miRNA from pri-miRNA in plants is not carried out by two different enzymes, but by a Dicer-like protein DCL1, which is said to be a Dicer homolog. There are no homologs for Drosha or DGCR8, and it is DCL1 which processes pri-miRNA to pre-miRNA. This DCL1 is expressed in the nucleus of plant cells [20]. The pre-miRNA thus formed is further processed to yield miRNA:miRNA* duplex. HYL1, which is a dsRNA-binding protein, plays an important role in mi*RNA* processing by DCL1 [21]. The duplex before transportation from the nucleus is methylated by an RNA methyltransferase protein called Hua-Enhancer (HEN1) at the 3′ overhang. HEN1 consists of a dsRNA-binding motif and a C-terminal methyltransferase domain. It is found to be highly selective for its substrate and can recognize the structure of the duplex. This step is undertaken to restrict the action of enzymes such as ligases, polymerases, and terminal transferases, which can otherwise attack the 3′-hydroxyl group. The miRNA:miRNA* duplex is then transported from the nucleus by the plant homolog of Exp-5, Hasty (HST) [22], where the miRNA is selectively loaded (perhaps with different preferences) into the RNA-induced silencing complex (RISC) and miRNA* is subsequently degraded.

miRNA biogenesis ends in general with the incorporation of the mature miRNA into RISC which then performs the targeting using the miRNA as a key to guide its target selection.

2.4 The RNA-Induced Silencing Complex

The activation of RISC starts with the loading of processed mature miRNA duplex which is essentially an ~22-nucleotide-long dsRNA fragment with a two-nucleotide overhang on both 3′ ends. Of the two strands one strand with lesser thermodynamic stability at the 5′ end is selected by the argonaute protein and is then integrated into RISC and is called the guide strand. The other strand, called the passenger strand, is then degraded [23]. The guide strand then recognizes the target mRNA based on sequence complementarity, which either results in complete endonucleolytic cleavage of targeted mRNA or leads to translational repression. Once the cleaved mRNA is released, RISC is ready to catalyze the cleavage of another target RNA. This process indicates functional coupling between pre-miRNA processing by Dicer and the activity of RISC. RISC assembly thus undertakes pre-miRNA processing, strand separation, and multiple rounds of mRNA cleavage without utilizing ATP, for any of the steps [23].

Now that the biogenesis of miRNAs has been delineated, we will discuss structure and function of miRNAs in the following sections.

3 Structure and Function of MicroRNAs: A Short Review

3.1 Structure

The two processing steps undertaken during miRNA biogenesis: First remove the stem of the primary precursor hairpin to give the semi-processed duplex containing the functional miRNA strand and the nonfunctional RNA strand which is later degraded. This semi-processed duplex RNA directs it toward the RISC for the degradation of the non-miRNA (passenger) strand. The stepwise processing from pri-miRNA to miRNA by Drosha and Dicer indicates the importance of proper recognition and the precise cleavage. Thus the analysis of structural pattern of miRNA seems beneficial for increased understanding of the biogenesis and functions of miRNA.

A study conducted on ten human miRNA precursors indicated a few conserved features among them which are stated below [24]:

- Hairpin structure of pre-miRNA (*see* Fig. 2) consists of regions (bulges and loops) with varying stability at certain intervals, and these have both structural and functional implications.
- The base of the hairpin stem (the open end) which constitutes the precursor processing site is expected to have both structural and functional determinants (*see* Fig. 1).

- The sixth nucleotide in this region seems to be conserved. In the study of 140 miRNA precursor sequences, U was found to be present in 50 % of cases and a pyrimidine (U or C) in 70 % of the cases.
- The nucleotides spanning from third to seventh position almost consistently include mismatches (usually the G-G, U-U, G-A, and A-G) resulting in a distorted duplex, i.e., with moderate destabilization.
- The -1 (one base to the left of the first base of the mature miRNA) position has G or C in 70 % of the cases, and position 1 (the first base of mature miRNA) has A/U in 72 % of the cases depending on the arm of the precursor where the mature miRNA resides, i.e., A is mostly present in pre-miRNAs with miRNA on the 3′ side and those with miRNAs on the 5′ side have U in the first position. No prominent base/sequence preference was observed at the cleavage site near the hairpin's terminal loop.

3.2 Function

Advancements in scientific techniques used to study the effect of miRNAs have revealed their surprising roles in a wide range of biological processes. miRNAs have been shown to regulate gene expression either positively or negatively at various stages including both transcription and translation. miRNAs participate in a variety of biological and pathological processes such as embryonic development, cell proliferation, cell differentiation, cell fate determination, organ development, hematopoietic lineage differentiation, apoptosis, signal transduction, virus–host interactions, insulin secretion, and carcinogenesis [25]. Recent studies indicate that approximately one-third of human genes are potentially regulated by miRNAs and each miRNA can target more than 200 genes on an average. A single miRNA can interact and regulate many different mRNA targets, and conversely, several different miRNAs can cooperatively interact and control a single mRNA target (also see other chapters in this volume).

The most widely studied aspect of miRNA function is gene silencing, either via mRNA degradation or by repression of translation. It has been experimentally demonstrated that complete complementarity between miRNA and mRNA sequence results in the direct mRNA degradation. In case of partial complementation, translation is either repressed or completely inhibited.

miRNAs occasionally also cause histone modification and DNA methylation of promoter sites, which ultimately affects the expression of target genes. Recent evidences indicate that the expression pattern of miRNAs varies in different differentiation stage of cells and tissues [26].

Some of the miRNAs are enriched in the nucleus [27]. It has been reported that some miRNAs are imported to nucleus through a process directed by a hexanucleotide sequence present in the miRNA [28]. miRNAs and pre-miRNAs are found to induce gene expression by targeting promoter sequences in a sequence-specific manner [29]. miRNAs may also direct epigenetic gene silencing pathway in the nucleus [30]. The role of miRNAs in transcriptional gene regulation is little studied, but the studies indicate both activating and silencing effect of miRNAs. Besides, various studies have attributed a wide variety of functions for miRNAs inside the nucleus including RNAi-directed DNA methylation (RdDM), RNAi/miRNA-mediated chromatin remodelling, miRNA-guided target RNA cleavage in nucleus, and transcriptional gene activation and silencing [31].

Recently, another miRNA, miR-181c, which is encoded in the nucleus and assembled in the cytoplasm, is found to be finally translocated into the mitochondria of cardiac myocytes [32].

It is important that this knowledge about structure and function is incorporated into bioinformatics analyses in respect to miRNA gene prediction and targeting. In the following some tools that may be helpful in this regard are discussed.

4 Bioinformatics: Tools to Predict Structural Similarity

Prediction of folding structure and sequence similarity between the miRNAs and their targets is an important tool to decipher the mode of action of miRNAs. The structure of RNA mainly determines its type and function. miRNA sequences are usually ~22 nt long with a sequence match of 2–8-nt-long seed sequence. It has been suggested that the miRNA action would depend not solely on the seed sequence match but also on structural features. It has been indicated that the secondary structure of the miRNA gene is more conserved than the primary structure, i.e., the nucleotide sequence [33]. In such a scenario, the low conservation of the miRNA sequences limits the use of sequence alignment-based search of miRNA genes and targets. With this approach the tools for gene and target prediction have evolved to use the information of secondary structures along with the sequence match. It is also difficult to understand the exact role of structure for such a small sequence.

Various computational algorithms designed to predict the miRNA genes or their target focus on both sequence and structural similarity. This is based on the fact that miRNA acts on its targets by complementary base pairing. The surprising fact is that the small size of miRNA (~22 nt) indicates that the matching sequence present in the target would be no longer than 20 nt. Further, analysis of the genome (say human) indicates that those sequences complementary to miRNA sequences occur with high frequency, which raises a question on the selective and specific activity of miRNA.

Some structural aspects of pri- and pre-miRNA have already been discussed in detail in other chapters of this book. We list some of the software (Table 1) which use structural features of miRNA in their algorithms. The server links, if available, are also given.

Table 1
List of some of the software (Table 1) which use structural features of miRNA in their algorithms

miRD	Detects secondary structure in a given sequence and decides whether the given sequence is a candidate pre-miRNA	http://mcg.ustc.edu.cn/rpg/mird/mird.php	[34]
miRAlign	Detect miRNAs in animals based on both sequence and structure alignment. It assigns similarity score to a given pre-miRNA using secondary structure prediction, pairwise alignment, miRNA's position on the stem–loop structure, and RNA secondary structure (of the target) alignment	http://bioinfo.au.tsinghua.edu.cn/miralign	[33]
miRFinder	The software uses features such as the local secondary structure differences of the stem region of miRNA and non-miRNA hairpins and correlations between different types of mutations and secondary structures of pre-miRNAs to predict pre-miRNAs	http://www.bioinformatics.org/mirfinder/	[35]
miRPred	A method for ab initio prediction of miRNAs by genome scanning based on characteristic motifs of (predicted) secondary structure. Used to differentiate miRNA precursors from other similar segments of the human genome	Not given	[36]
NOVOMIR	A program for the identification of plant miRNA genes. It uses features such as relative thermodynamic stability of their structure, length of helices, and number and size of loops to discriminate a pre-miRNA from other RNAs	http://www.biophys.uni-duesseldorf.de/novomir/	[37]
PmirP	This pre-miRNA prediction method is based on structure-sequence hybrid features using left-triplet method, the free nucleotides, the minimum free energy of secondary structure, and base-pairing features	http://ccst.jlu.edu.cn/ci/bioinformatics/MiRNA	[38]
microRNAfold	This software aids in miRNA secondary structure prediction based on modified NCM model with thermodynamics-based scoring strategy	Not given	[39]
RISCbinder	The software can be used to predict guide strand of miRNAs based on its sequence and secondary structure. A web server has been developed based on SVM models described in this study	http://crdd.osdd.net:8081/RISCbinder/	[40]
Micro-HARVESTER	The program identifies candidate miRNA homologs in any set of sequences, based on a sequence similarity search step followed by a set of structural filters	http://www-ab.informatik.uni-tuebingen.de/software/microHARVESTER	[41]
miRacle	Server for miRNA target prediction incorporating RNA secondary structure	miracle.igib.res.in/miracle/	[42]
Polycistronic transcriptional units	The algorithm is able to produce the optimal secondary structure of polycistronic miRNAs based on master slave architecture	Not given	[43]
Network based	The pre-miRNA stem–loop secondary structure is translated to a network, and the network parameters effectively characterize pre-miRNA secondary structure, thus contributing to both prediction ability and computation efficiency	Not given	[44]

5 Bioinformatics: Tools to Study Sequence Similarity

Sequence-specific activity of miRNAs has been reported for pre-mRNA and also for action on its splicing. However, details of selection and action of miRNA in the above roles are not clear.

The sequence-specific activity of miRNA points toward a large number of possible mechanisms of action. Besides their posttranscriptional activity, recent studies have shown their activity in the transcription process as well. Considering these aspects various modes of action can be attributed to miRNAs:

- It is possible that miRNAs bind to complementary sequences in the DNA and cause the chromatin-associated changes required for activating that gene.
- miRNAs can bind to the complementary sequences of promoters and create motifs which are useful for the binding of transcription factors (presently a hypothetical scenario).
- miRNAs can bind to complementary regions in active sites of transcription factor or in the promoter region of DNA and can inhibit the binding of the transcription factor to the promoter region.
- miRNAs can bind to complementary sequences on newly synthesized regions of RNA and cause changes required for the inhibition of the translation.

In one of our studies, we have taken ten TLR genes from human and downloaded their promoter regions (simplified, the stretch of 1,000 nucleotides upstream of the transcription start site, TSS) from the ENSEMBL database. The sequences of these ten promoter regions (and their complementary strands) were compared with mature miRNA sequences (1921- obtained from mirbase.org) to find 6-nt (6-nucleotides) matches between them. Numerous such common sequences were found as a result. These 6-nt matches were taken as seeds and were extended on both sides, considering various combinations, so as to obtain 7-nt matches, which were extended to obtain 8-nt matches and further extended in a similar fashion gradually up to 12-nt matches.

The length of the human genome is estimated to be around 3.4 billion base pairs, so a unique 12-nt-long sequence is expected to occur approximately 200 ($3.4 \times 10^9/4^{12}$) times by chance. Similarly a 13-nt-long sequence is expected to be present 50 times in the whole genome. A 14-nt-long sequence can occur around 12 times. The result has shown some 13- and 14-nucleotide-long matching sequences. Also, a 19-nt match (hsa-miR-1273 g-3p) was found as seen in Table 2, and the miRNA itself is 21 nt long; such a match is very rare.

32 sequences from the TLR promoter regions of length 12 nt and more were found to be present in mature miRNA also, as shown in Table 2. A study of location of these sequences showed that 18 out of 32 matching sequences lie within 500 bp from the

Table 2
The sequence found common between human mature miRNA and the promoter regions, their expected frequency, and the frequency obtained through exact match search

TLR	Matching sequence	miRNA name	Frequency	
			Expected	Observed
TLR1	CATTTATTGTGT	hsa-miR-545-3p	200	1,752
	CCTTTTTGTTTT	hsa-miR-4668-3p	200	4,924
	CTAAGAAGAAAA	hsa-miR-4659a-5p	200	2,755
TLR2	GCACTGAAACAA	hsa-miR-635	200	550
	CAAATGTGTCTTG	hsa-miR-642a-5p	51	192
TLR3	GTCTAGCTGAAGCT	hsa-miR-3157-3p	13	16
	GAAGCTGGAGCC	hsa-miR-1254	200	717
TLR4	AGGTAGAATGAGG	hsa-miR-4650-3p	51	191
	GGCTATTTTAAT	hsa-miR-3606	200	932
	ATGTTGGATTAGGG	hsa-miR-3923	13	131
	TGAGGTAGAGGG	hsa-miR-3138	200	486
TLR5	ACCTTCCTCTCC	hsa-miR-3667-3p	200	1,163
	ATAGAACTTTCC	hsa-miR-625-3p	200	590
	GGAAAGTTCTAT	hsa-miR-625-5p	200	590
TLR6	AAACTGCAATTA	hsa-miR-548ae	200	2,438
		hsa-miR-548aq-3p		
		hsa-miR-548x-3p		
		hsa-miR-548aj-3p		
	ACCCTTTTCCCCAG	hsa-miR-1227	13	110
	TAATTGCAGTTT	hsa-miR-570-5p	200	2,438
		hsa-miR-548x-5p		
		hsa-miR-548aj-5p		
		hsa-miR-548ar-5p		
		hsa-miR-548ai		
		hsa-miR-548 g-5p		
TLR7	CCCTGGAGTTTC	hsa-miR-4308	200	639
TLR8	CCACAATTATGT	hsa-miR-548as-3p	200	529
	CTGGAGGAGGCA	hsa-miR-3911	200	1,791
	TGATGAAACTCA	hsa-miR-4517	200	687
	GTCTCCCTCCCA	hsa-miR-5196-3p	200	1,055
	TGGGTCTCCCTC	hsa-miR-615-3p	200	602
TLR9	CACCCAAGGCTT	hsa-miR-532-3p	200	361
	GTTGGCCATCTG	hsa-miR-571	200	499
	CTCCTTTGCCTG	hsa-miR-4742-3p	200	1,040
	TCCAGTACAGTG	hsa-miR-141-5p	200	493
TLR10	ACAAAGTGAGACC	hsa-miR-1285-3p	51	6,175
	ACCACTGCACTCCAGCCTG	hsa-miR-1273 g-3p	0.01	89,048
	GAGTGCAGTGGTG	hsa-miR-3135b	51	120,786
	CTGGGACTACAGG	hsa-miR-5585-3p	51	230,602
	TAGAGACGGGGT	hsa-miR-1303	200	157,757

TSS. Even the 19-nt match lies at just 250 bp upstream from the TSS on the template strand.

A study of expected and observed frequency of these matching sequences indicated that the occurrence of such sequences is far from random distribution of A, C, G, T. This explains the importance of those sequences and also indicates that the miRNA may be regulating various other regions of the genome as well, where the same sequence is present.

It is important to note that the binding by nucleotide complementarity is fundamentally different from protein ligand binding. In case of the binding of a ligand to a protein molecule, the overall geometry of the ligand is important and a small change in geometry can have a large effect on the binding. But, in nucleotide sequence hybridization, a small mismatch of a few nucleotides causes only a proportional loss of binding energy. This contributes to the lack of specificity of nucleotide sequence matches compared to protein ligand complexes. Recent prediction algorithms are also using structural aspects of miRNA and pre-miRNA as discussed in the previous section.

Considering sequence similarities including exact and near-exact matches has given way to a number of algorithms for the prediction of miRNA genes and their probable targets.

6 Current Scenario: Modern Trends in MicroRNA Research

Recent developments in miRNA prediction and validation resulted in many software which incorporates strategies based on characteristic features of miRNAs, such as sequence similarity, structural similarity, phylogenetic conservation, and thermodynamic stability (high minimal folding free energy index—MFEI) either alone or in various combinations (*see* Chapters 9 and 10 in this volume). Computational methods can form the preliminary stage before initiating experimental identification of novel miRNAs, as they can locate/predict potential candidates which can be verified through experiments. The earliest members of the miRNA family were however discovered using genetic experiments, but screening the whole genome would be time consuming and cost inefficient.

The regulation involving miRNAs has been found to influence not only mRNA or promoters but also a wide range of factors influencing various metabolic and developmental pathways. One such example is miR-133 which was found to down-regulate an important splicing factor during muscle development [45]. Similarly miRNAs have also been attributed with the regulation of pre-miRNA splicing. Another miRNA, miR-124, was found to promote neuronal differentiation and partly regulates the network of brain-specific alternative pre-mRNA splicing [46, 47]. This is both important and interesting as miRNAs always work together with another protein or factor and most likely help it to locate some important region on the nucleic acid.

Localization of miRNAs is an important aspect to be studied in detail by experimental methods. Both pri- and pre-miRNA are produced in the nucleus and transported to the cytoplasm where they are processed on demand to produce mature miRNAs. It is believed that only mature miRNAs can have selective regulatory roles. It has been reported that some mature miRNAs are transported back to the nucleus where they are expected to have some regulatory role. It is important to identify the particular class of miRNAs that are transported back to the nucleus and their regulatory roles.

The miRNAs present in the cytoplasm are most likely regulating the transcriptional process and the splicing selection process. Exact details also need to be explored. During embryogenesis a number of miRNAs are involved in the regulation of several pathway enzymes and these may be active in the nucleus.

7 Conclusion

The pri-miRNAs generated by RNA pol II or pol III consist of long stem–loop structures, which are processed by Drosha in the nucleus to form ~70-nt hairpin pre-miRNAs. As mentioned earlier, this processing defines the 3′ end of the pre-miRNA by leaving a 2-nt overhang. Furthermore, these pre-miRNAs are transported to the cytoplasm by Exp-5, which is suggested to recognize the 3′ overhang. It was reported that Exp-5 effectively binds to the 3′ overhang, whereas a 5′ overhang was found to be inhibitory [48]. Also, binding to Exp-5 prevents the nuclear degradation of pre-miRNA. In the cytoplasm, Dicer processes pre-miRNAs to form an imperfect duplex (miRNA/miRNA*), thus defining the other end of miRNA. This imperfect duplex paves the way for the disposal of the non-miRNA strand from the RISC. These details essentially suggest that the structures of the pri- and pre-miRNA are important for the efficient processing and thus for the formation of mature miRNAs.

A study of the secondary structure motifs which destabilize and distort the structure of the stem of pre-miRNAs has shown that the sequences were given a minor preference in the course of precursor recognition. A study indicated that the protein complexes involved in miRNA biogenesis rely more on structure than on sequence for recognition [49].

A number of structural parameters have been used to develop algorithms for miRNA prediction. They include general parameters such as the length of the longest fully base-paired part of the stem, terminal loop size, number of nucleotides in symmetric and asymmetric bulges, as well as more specific structural characteristics such as frequency of triplet structure elements. Other highly significant features of pre-miRNAs' structure that might facilitate miRNA gene prediction include the prominent overrepresentation of bulges in the 5′ arm of the pre-miRNA, opposite polarity of symmetric and asymmetric motif distribution along the hairpin stem, and increased contribution of asymmetric motifs when the total number of stem structure-destabilizing motifs in pre-miRNAs increases [49].

Softwares used for predicting RNA structures can also be used for the prediction of miRNA structure. Incorporating the trivial algorithms with the structural parameters discussed above has been practiced. This resulted in more accurate results.

miRNAs involved in regulating biological processes are formed from long pri-miRNAs, and during the maturation process a substantial part of initial pri-miRNA is lost to yield a small ~22-nucleotide sequence. The relevance of such a long process to ultimately discard major parts of the source RNA is still unexplained. Furthermore, the intermediate product, pre-miRNAs, was found to be highly structurally conserved across species. This information suggests that the stepwise biogenesis of miRNA has evolved so as to provide accuracy and precision to the ultimate regulation activity.

Regulatory molecules in the cell are usually present at a basal level, which may accumulate according to the requirements of the cell. miRNAs are also regulatory molecules derived from a precursor (pre-miRNA), which may be present at a basal level, ready to be converted to their active form depending upon a signal. This paints a picture similar to proteolytic enzymes which usually exist in the cell in their zymogenic form ready for activation. Such a hypothesis is likely to be true, since pre-miRNAs being a longer molecule are expected to have a longer half-life than miRNAs which are less than half their length. The discovery of a family of exoribonucleases which degrade miRNA in *A. thaliana* [50] also strengthens this hypothesis. This points toward a profound regulatory mechanism regulating the stepwise processing of miRNAs.

8 Outlook

The function of miRNAs has been subjected to extensive study, and several modes of action are now well established. The literature already contains references about transcription and translation being regulated by miRNAs. However, detailed mechanisms are not yet clear, and several miRNAs act as enhancers or silencers. Most likely the regulation by miRNA is achieved by binding to a suitable complementary sequence on DNA or mRNA.

miRNAs are produced in the nucleus where they have several roles, but the final mature miRNA is processed in the cytoplasm. A number of miRNAs are reportedly transported from the cytoplasm to the nucleus where they perform their regulatory role. There are miRNA transporters reportedly present on the nuclear membrane. The exact nature and selectivity of these transporters are not yet fully known.

In the miRBase database for miRNAs, we found more than 2,000 mature miRNA sequences for humans. It is very likely that the database is incomplete and the expected number of miRNAs may be ~10,000. Comparing with other species, we find that the number of miRNA for humans is relatively high. This is likely because (1) of greater interest in miRNA for humans for medical and other reasons, (2) miRNAs were found during the course of experiments in unrelated studies, and (3) of wide availability of computational tools for sequence prediction and target validation.

Because miRNAs are relatively small molecules their isolation, characterization, and regulation properties are difficult to study experimentally. However, the mature miRNA is expected to be a linear molecule ready to pair with a suitable complementary sequence. On the other hand, the pri- and pre-miRNAs are relatively longer molecules with a well-defined three-dimensional folded structure. This folded structure is recognized by the enzymes (Drosha and Dicer) which process to produce the mature miRNA.

Acknowledgements

One of the authors (KK) gratefully acknowledges the financial support as a Junior Research Fellowship from the University Grants Commission, Government of India.

References

1. Lewis BP, Burge CB, Bartel DP (2005) Conserved seed pairing, often flanked by adenosines, indicates that thousands of human genes are microRNA targets. Cell 120:15–20
2. Ambros V, Bartel B, Bartel DP et al (2003) A uniform system for microRNA annotation. RNA 9:277–279
3. Lau NC, Lim LP, Weinstein EG et al (2001) An abundant class of tiny RNAs with probable regulatory roles in *Caenorhabditis elegans*. Science 294:858–862
4. Lagos-Quintana M, Rauhut R, Lendeckel W et al (2001) Identification of novel genes coding for small expressed RNAs. Science 294: 853–858
5. Mourelatos Z, Dostie J, Paushkin S et al (2002) miRNPs: a novel class of ribonucleoproteins containing numerous microRNAs. Genes Dev 16:720–728
6. Lee Y, Kim M, Han J et al (2004) MicroRNA genes are transcribed by RNA polymerase II. EMBO J 23:4051–4060
7. Cai X, Hagedorn CH, Cullen BR (2004) Human microRNAs are processed from capped, polyadenylated transcripts that can also function as mRNAs. RNA 10:1957–1966
8. Borchert GM, Lanier W, Davidson BL (2006) RNA polymerase III transcribes human microRNAs. Nat Struct Mol Biol 13:1097–1101
9. Smalheiser NR (2003) EST analyses predict the existence of a population of chimeric microRNA precursor–mRNA transcripts expressed in normal human and mouse tissues. Genome Biol 4:403
10. Rodriguez A, Griffiths-Jones S, Ashurst JL et al (2004) Identification of mammalian microRNA host genes and transcription units. Genome Res 14:1902–1910
11. Denli AM, Tops BB, Plasterk RH et al (2004) Processing of primary microRNAs by the Microprocessor complex. Nature 432:231–235
12. Han J, Lee Y, Yeom KH et al (2006) Molecular basis for the recognition of primary microRNAs by the Drosha-DGCR8 complex. Cell 125:887–901
13. Lee Y, Ahn C, Han J et al (2003) The nuclear RNase III Drosha initiates microRNA processing. Nature 425:415–419
14. Gregory RI, Yan K-P, Amuthan G et al (2004) The Microprocessor complex mediates the genesis of microRNAs. Nature 432:235–240
15. Okamura K, Hagen JW, Duan H et al (2007) The mirtron pathway generates microRNA-class regulatory RNAs in Drosophila. Cell 130:89–100
16. Ruby JG, Jan CH, Bartel DP (2007) Intronic microRNA precursors that bypass Drosha processing. Nature 448:83–86

17. Okada C, Yamashita E, Lee SJ et al (2009) A high-resolution structure of the Pre-microRNA nuclear export machinery. Science 326:1275–1279
18. Knight SW, Bass BL (2001) A role for the RNase III enzyme DCR-1 in RNA interference and germ line development in Caenorhabditis elegans. Science 293:2269–2271
19. Kim VN (2005) microRNA biogenesis: coordinated cropping and dicing. Nat Rev Mol Cell Biol 6:376–385
20. Papp I, Mette MF, Aufsatz W et al (2003) Evidence for nuclear processing of plant micro RNA and short interfering RNA precursors. Plant Physiol 132:1382–1390
21. Kurihara Y, Takashi Y, Watanabe Y (2006) The interaction between DCL1 and HYL1 is important for efficient and precise processing of pri-miRNA in plant microRNA biogenesis. RNA 12:206–212
22. Bollman KM, Aukerman MJ, Park MY et al (2003) HASTY, the Arabidopsis ortholog of exportin 5/MSN5, regulates phase change and morphogenesis. Development 130:1493–1504
23. Gregory RI, Chendrimada TP, Cooch N et al (2005) Human RISC couples microRNA biogenesis and posttranscriptional gene silencing. Cell 123:631–640
24. Krol J, Sobczak K, Wilczynska U et al (2004) Structural features of microRNA (miRNA) precursors and their relevance to miRNA biogenesis and small interfering RNA/short hair-pin RNA design. J Biol Chem 279:42230–42239
25. Huang Y, Shen XJ, Zou Q et al (2011) Biological functions of microRNAs: a review. J Physiol Biochem 67:129–139
26. Hawkins PG, Morris KV (2008) RNA and transcriptional modulation of gene expression. Cell Cycle 7:602–607
27. Liao J-Y, Ma L-M, Guo Y-H et al (2010) Deep sequencing of human nuclear and cytoplasmic small RNAs reveals an unexpectedly complex subcellular distribution of miRNAs and tRNA 39 trailers. PLoS ONE 5:e10563
28. Hwang HW, Wentzel EA, Mendell JT (2007) A hexanucleotide element directs MicroRNA nuclear import. Science 315:97–100
29. Place RF, Li LC, Pookot D et al (2008) MicroRNA-373 induces expression of genes with complementary promoter sequences. Proc Natl Acad Sci U S A 105:1608–1613
30. Kim DH, Sætrom P, Snøve O Jr et al (2008) MicroRNA-directed transcriptional gene silencing in mammalian cells. Proc Natl Acad Sci U S A 105:16230–16235
31. Tang R, Zen K (2011) Gold glitters everywhere: nucleus microRNAs and their functions. Front Biol 6:69–75
32. Das S, Ferlito M, Kent OA et al (2012) Nuclear miRNA regulates the mitochondrial genome in the heart. Circ Res 110:1596–1603
33. Wang X, Zhang J, Li F et al (2005) MicroRNA identification based on sequence and structure alignment. Bioinformatics 21:3610–3614
34. Zhang Y, Yang Y, Zhang H et al (2011) Prediction of novel Pre-microRNAs with high accuracy through boosting and SVM. Bioinformatics 27:1436–1437
35. Huang TH, Fan B, Rothschild MF et al (2007) MiRFinder: an improved approach and software implementation for genome-wide fast microRNA precursor scans. BMC Bioinformatics 8:341
36. Brameier M, Wiuf C (2007) Ab initio identification of human microRNAs based on structure motifs. BMC Bioinformatics 8:478
37. Teune J, Steger G (2010) NOVOMIR: de novo prediction of MicroRNA-coding regions in a single plant-genome. J Nucleic Acids 2010:495904
38. Zhao D, Wang Y, Luo D et al (2010) PMirP: a pre-microRNA prediction method based on structure-sequence hybrid features. Artif Intell Med 49:127–132
39. Han D, Tang G, Zhang J (2012) MicroRNAfold: premicroRNA secondary structure prediction based on Modified NCM model with thermodynamics-based scoring strategy. Int J Min Bioinform 6(3):272–291
40. Ahmed F, Ansari HR, Raghava GPS (2009) Prediction of guide strand of microRNAs from its sequence and secondary structure. BMC Bioinformatics 10:105
41. Dezulian T, Remmert M, Palatnik JF et al (2006) Identification of plant microRNA homologs. Bioinformatics 22:359–360
42. http://miracle.igib.res.in/miracle/.
43. Han D, Tang G, Zhang J (2013) A parallel algorithm for predicting the secondary structure of polycistronic MicroRNAs. Machine Learning and Applications (ICMLA), Ninth International Conference, 509–514
44. Xiao J, Tang X, Li Y et al (2011) Identification of microRNA precursors based on random forest with network-level representation method of stem-loop structure. BMC Bioinformatics 12:165
45. Boutz PL, Chawla G, Stoilov P et al (2007) MicroRNAs regulate the expression of the alternative splicing factor nPTB during muscle development. Genes Dev 21:71–84
46. Makeyev EV, Zhang J, Carrasco MA et al (2007) The MicroRNA miR-124 promotes neuronal differentiation by triggering brain-specific alternative pre-mRNA splicing. Mol Cell 27:435–448
47. Smith P, Hashimi AA, Girard J et al (2011) In vivo regulation of amyloid precursor protein

neuronal splicing by microRNAs. J Neurochem 116:240–247
48. Zeng Y, Cullen BR (2004) Structural requirements for pre-microRNA binding and nuclear export by Exportin 5. Nucleic Acids Res 32:4776–4785
49. Kozlowski P, Starega-Roslan J, Legacz M, Magnus M, Krzyzosiak WJ (2008) Structures of microRNA precursors. In Current perspectives in microRNAs (miRNA) (pp. 1–16). Springer Netherlands
50. Ramachandran V, Chen X (2008) Degradation of microRNAs by a family of exoribonucleases in Arabidopsis. Science 321:1490–1492

Chapter 12

Computational Methods for MicroRNA Target Prediction

Hamid Hamzeiy, Jens Allmer, and Malik Yousef

Abstract

MicroRNAs (miRNAs) are important players in gene regulation. The final and maybe the most important step in their regulatory pathway is the targeting. Targeting is the binding of the miRNA to the mature RNA via the RNA-induced silencing complex. Expression patterns of miRNAs are highly specific in respect to external stimuli, developmental stage, or tissue. This is used to diagnose diseases such as cancer in which the expression levels of miRNAs are known to change considerably. Newly identified miRNAs are increasing in number with every new release of miRBase which is the main online database providing miRNA sequences and annotation. Many of these newly identified miRNAs do not yet have identified targets. This is especially the case in animals where the miRNA does not bind to its target as perfectly as it does in plants. Valid targets need to be identified for miRNAs in order to properly understand their role in cellular pathways. Experimental methods for target validations are difficult, expensive, and time consuming. Having considered all these facts it is of crucial importance to have accurate computational miRNA target predictions. There are many proposed methods and algorithms available for predicting targets for miRNAs, but only a few have been developed to become available as independent tools and software. There are also databases which collect and store information regarding predicted miRNA targets. Current approaches to miRNA target prediction produce a huge amount of false positive and an unknown amount of false negative results, and thus the need for better approaches is evermore evident. This chapter aims to give some detail about the current tools and approaches used for miRNA target prediction, provides some grounds for their comparison, and outlines a possible future.

Key words Bioinformatics, Computational biology, miRNA, MicroRNA, Target prediction, Machine learning

1 Introduction

Initially identified two decades ago, microRNAs (miRNAs) are now considered to have a central role in the RNA revolution. This has focused the scientific community's attention to these small RNAs, and vigorous research efforts have resulted in the accumulation of a significant body of data related to miRNA biogenesis and function. This can be seen quite clearly in the super linear increase of miRBase [1] entries. Most of the 17,000 miRNA sequences currently available in the miRBase database are yet to

Malik Yousef and Jens Allmer (eds.), *miRNomics: MicroRNA Biology and Computational Analysis*, Methods in Molecular Biology, vol. 1107, DOI 10.1007/978-1-62703-748-8_12, © Springer Science+Business Media New York 2014

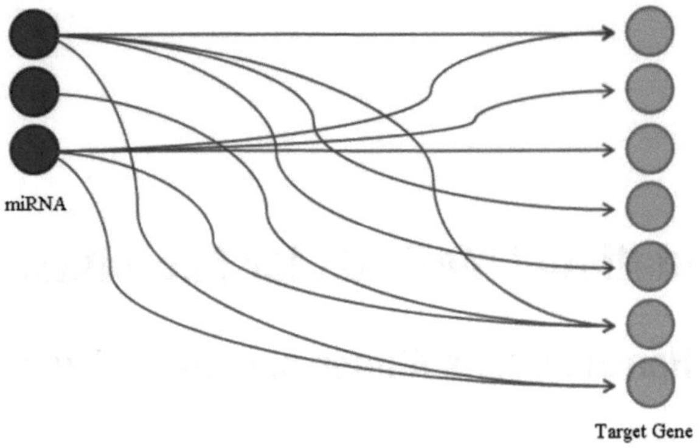

Fig. 1 A schematic representation of the interactions between miRNAs and their target genes

have validated targets, and thus there is a clear need for evermore precise and accurate miRNA target prediction.

A single miRNA has the potential to regulate hundreds of target mature RNAs (mRNAs), and multiple miRNAs may compete for the regulation of the same mRNA [2–4]. Having considered this fact it is not surprising to have more target genes than miRNAs (Fig. 1). TarBase 6.0 [5] currently has more than 65,000 experimentally validated miRNA targets. It is estimated that as much as 90 % of all human genes are somewhat regulated by miRNAs [6]. On average a single miRNA family is thought to have around 300 conserved targets which would mean that a large number of mammalian genes are miRNA regulated [7]. Self-regulatory pathways for miRNA biogenesis such as the inhibition of the synthesis of the Dicer protein which has an essential role in the miRNA biosynthetic pathway have also been identified [8–10]. This autoregulatory pathway leads to the establishment of a negative feedback system which could be exploited to control miRNA expression and thus miRNA-mediated regulatory pathways.

Before miRNA target prediction tools were available, possible miRNA target sites were determined manually. These target sites were later confirmed by laborious and inefficient techniques such as site-directed mutagenesis and other experimental methods (*see* Chapter 14). The identification of the first targets for the let-7 and lin-4 miRNAs led to the idea that miRNAs have a pattern in targeting genes which could be used to develop target prediction algorithms [11].

Gene targeting by miRNAs is generally believed to be the result of their binding to the 3′UTR of the target mRNA. Other studies [12–16] have also confirmed gene regulation as a result of the binding of the miRNA to the coding region (commonly seen in plants [17]) as well as the 5′UTR. Computational evidence suggests that regulation via the binding of the miRNA to the coding region differs in comparison to the binding pattern seen at the

3′UTR [15]. It is suggested that miRNAs target the coding regions of mRNAs with short 3′UTRs [16]. 3′UTRs are prone to change under different conditions which might result in the elimination of the target site [18]. This phenomenon presents an opportunity for the cell to regulate the function of the miRNA (*see* Chapter 18 for more details on miRNA regulation). Binding in the coding region on the other hand may present an evolutionary advantage for the cell as it could help in the preservation of the miRNA-binding site [19]. Regulation of the miRNA function on this level may also be controlled by the inclusion or the exclusion of the binding site as a result of alternative splicing [13, 20].

2 MicroRNA Target Prediction

Targeting patterns are different between plants and animals. Plants show a near-perfect complement between their miRNA and their target mRNA, and similar to the action of siRNAs, this could cause the cleavage of the double-stranded RNA (dsRNA) [21, 22]. This makes target prediction easier in plants, in comparison to animals, reducing the targeting problem to using computational methods for sequence similarity search [23]. On the other hand animal miRNAs bind their targets with only partial complementarity (Figs. 2, 3). A region of about six to eight nucleotides in length within the structure of the miRNA which is called the seed region is of crucial importance in the targeting. This seed sequence binds

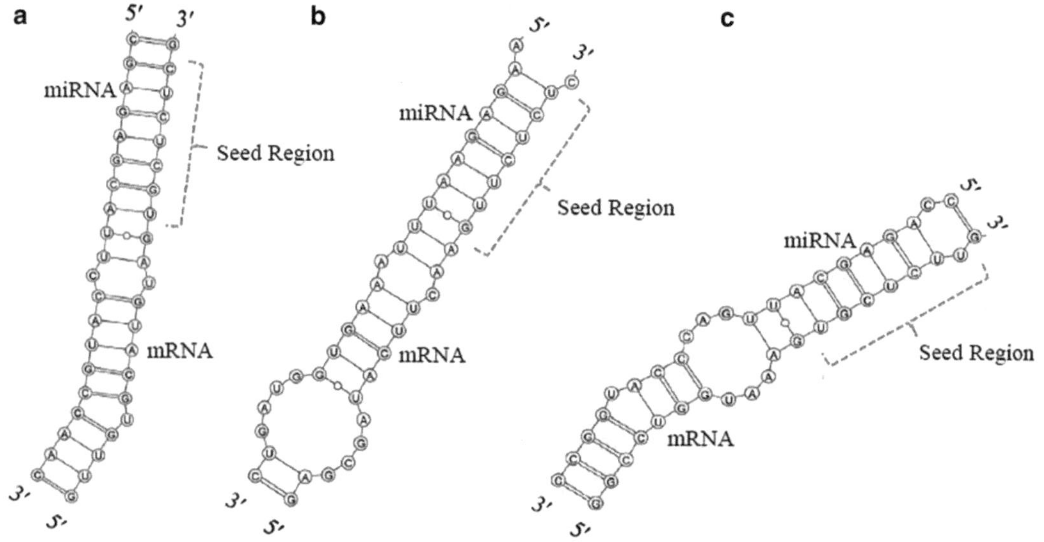

Fig. 2 The hypothetical secondary structures of the main types of miRNA:mRNA duplexes in animals drawn using VARNA which is a tool for drawing and visualization of RNA secondary structure [78]. **a**) Perfect complementarity at the 5′ end of the miRNA (seed region) with a bulge and a mismatch towards the 3′ end. **b**) The seed region contains a mismatch and a G-U wobble and the 3′ end has two bulges. **c**) The seed region contains a bulge and the 3′ end has a bulge and a mismatch

to the target mRNA leading to the regulation of the gene in question [3, 24]. Other than the seed region, two other regions, namely, the extended seed region and the delta seed region, are also deemed important [25, 26]. Binding at the 3′UTR is usually preferred over binding in the coding region or the 5′UTR, but the reasons are yet to be unraveled and contradictory studies have made it difficult to reach a conclusion [13, 19, 20]. Binding in the coding region is known to be effective in plants [17, 20], but in animals it is proposed that coding region binding is only effective where there is a high degree of complementarity (similar to plants). This may lead to the disruption of the interaction of the transcript and the ribosome and thus to the inhibition of translation [20].

2.1 Target Prediction Methodologies

Several different methods and approaches are currently in use for the prediction of miRNA targets [27, 28]. The seed region is one of the most commonly used miRNA traits for miRNA target prediction, and many studies [3, 29–31] have pointed out the importance of binding between the seed region, located at the 5′ end of the miRNA, and its target mRNA. Other characteristics of the miRNA targeting pathway which are currently used for target prediction include the binding pattern of the seed region, the minimum free energy of the binding between the miRNA and its target mRNA, and the accessibility of the target site [32]. Other studies [33, 34] have also looked at base pairing between the miRNA and its target outside of the seed region. They suggest that binding beyond the seed region will compensate for weak binding of the seed region. Conserved sequences around the seed region (adenines for animals in particular [3]) may also play a role in finding targets for miRNAs in different species. Even though this approach helps to eliminate a significant amount of false positive results, it may also result in loosing targets which are less conserved. Furthermore a study [35] suggested that at least 30 % of the experimentally validated target sites are non-conserved suggesting that the conservation of the miRNA target site alone is not enough.

2.1.1 Sequence-Based Methods

The first thing that comes to mind when talking about miRNA targeting is the complementarity between the miRNA and its target. The small size of the miRNA transcript in respect to the genome rules out the possibility to rely solely on sequence complementarity for target predictions. This is because such approaches produce a huge number of potentially false-positive hits. Even though complementarity is very important and useful in target prediction, other properties of this interaction such as bulges and mismatches complicate matters. The seed region is the main focus when sequence-based methods are considered [3, 24]. Most tools look at the 3′UTR of the target gene when searching for complementarity, but others have suggested looking at the 5′UTRs and the coding regions, too. Maybe the most important step in this

```
target : ENSG00000152661 : ENST00000282561
length : 1730
miRNA : miRNA
length : 21

mfe : -20.0 kcal/mol
p-value : 0.999883

position  1594
target 5'  U   UAAU UG   G             U 3'
            UAC     U  UUU   ACAUUCCA
            AUG     A  AAG   UGUAAGGU
miRNA  3'     U     UG   AAA             5'
```

Fig. 3 The typical output of RNAhybrid. The first line gives the name of the FASTA file of the target, the second line is its length, and the third line in the name of the FASTA file of the miRNA followed by its length. The mfe and the pvalue are then given along with a semi graphic representation of the hybridization

method is the information regarding the sequence of the genome. The 3′UTRs for many mammal genomes are not well characterized. This complicates matters when searching for miRNA targets [36] within their bounds. When the boundaries of the 3′UTR are not properly defined they can be estimated by taking the downstream flanking sequence from the stop codon with an average corresponding to the 3′UTR length. Although this may partially solve the problem of undefined 3′UTRs, it is far from the precision needed for accurate predictions.

2.1.2 Structure-Based Methods

Structure-based methods focus mostly on the thermodynamic stability of the miRNA:mRNA duplex. Several different programs are available for the prediction and analysis of the secondary structure and hybridization of miRNAs including Mfold [37] and the Vienna RNA Package [38]. Some target prediction algorithms [2, 24, 37, 39] use these tools to check for the thermodynamic stability of the predicted duplex using sequence complementarity. Other algorithms [40, 41] on the other hand rely on thermodynamics as the initial factor in target prediction.

2.1.3 Homology-Based Methods

As mentioned before looking at conserved targets within different species helps to reduce the number of false positive results [24, 39, 42], but this may also causes an increase in the number of false negatively identified targets. Homology-based methods usually focus on the seed region (Fig. 4a). The choice of genomes to look for conservation in this approach is very important, and genomes which are very similar to each other should be avoided (Fig. 4b). This is because at least 99 % of the transcript will be conserved and maybe it would be better if the genomes were analyzed with larger evolutionary distance in mind.

2.2 Available Tools: Overview

Currently there are more than a dozen algorithms (Table 1) which claim to predict miRNA targets by applying some of the features mentioned above. Among these are tools which combine experimental and computational methods hoping to achieve better predictions. An example for this approach would be the Diana-microT [40] which claims to be able to reproduce all known *C. elegans* miRNA targets. On the other hand programs like miRanda [43] rely on dynamic programming to find the most optimal complementation between a given miRNA and its target mRNA, and RNA secondary structure prediction algorithms like Mfold work by finding complementary regions. PicTar [2] was developed by performing multiple sequence alignments of the 3′UTR of eight vertebrates. PicTar uses a statistical approach and emphasizes the importance of the conservation of the miRNA target site. A different approach based only on sequence information was applied by TargetBoost [44] which is essentially a machine learning algorithm. This approach set a trend towards applying machine learning algorithms to miRNA target predictions, and other studies [34, 45, 46] later used this method. MicroTar [47] is another program which does not rely on the conservation of the miRNA target; instead, it predicts miRNA targets by considering RNA duplex energies. Finally RNA22 [6] aims to find miRNA targets by searching for patterns in the 3′UTR. In the following, TargetScanS [24] and RNAhybrid [41, 48] are discussed in more detail.

2.2.1 TargetScanS

TargetScanS is introduced as an extension to the TargetScan algorithm with some new features including the addition of two more species to the three which were originally in TargetScan. It predicts miRNA targets by looking at conserved target sequences between human, mouse, dog, rat, and chicken. This helps to reduce the number of false positive results, and when tested, it was able to successfully identify targets for 5,300 human genes which were known to be targeted by miRNAs. The algorithm requires perfect binding in the seed region and then looks at binding beyond the seed region. The developers came to notice that the eighth nucleotide of the target is usually an adenosine and that the eighth nucleotide often formed a Watson–Crick pair in the duplex. TargetScan tested the binding sites for their thermodynamic stability using RNAfold from the Vienna RNA Package but TargetScanS does not. The absence of the thermodynamic stability measure and the requirement for several hits in the 3′UTR for each miRNA helped to reduce the runtime for TargetScanS. TargetScanS results are available via their web server (http://genes.mit.edu/tscan/targetscanS2005.html).

2.2.2 RNAhybrid

RNAhybrid aims to predict potential targets for miRNAs by looking at the most energetically favorable hybridization sites between two separate RNA sequences and does not allow base

Table 1
A non-comprehensive list of miRNA targeting programs is given below

Name	Summary	Clade	Link
TargetScanS [3]	Modeling of adenosines flanking the seed region. Similar to TargetScan	Vertebrate	http://genes.mit.edu/tscan/targetscanS2005.html
TargetScan [24]	5′ seed sequence-, homology-, and thermodynamics-based modeling	Mammal, worm, fly	http://www.targetscan.org/
PicTar [2]	Stringent seed pairing for at least one target, target clustering, and duplex stability	Vertebrate, fly, nematode	http://pictar.mdc-berlin.de/
miRanda [43]	Position-specific complementarity, optimization, and interspecies conservation	Vertebrate	www.microrna.org
EMBL [49]	Finds anti-targets in the 3′UTR and miRNA-binding sites	Animal	N/A
DIANA-microT [40]	Experimental rule generation and duplex binding energy	Human and mouse	http://diana.pcbi.upenn.edu/cgi-bin/micro_t.cgi
RNA22 [6]	Identifies clustered targets from patterns and finds corresponding miRNAs	Animal, worm, fly	http://cbcsrv.watson.ibm.com/rna22.html
PITA Top [50]	Target site's sterical accessibility energy model	Animal, fly, worm	http://genie.weizmann.ac.il/pubs/mir07/
miRU [23]	Sequence similarity with adjustable mismatch settings	Plant	http://bioinfo3.noble.org/miRNA/miRU.htm
EIMMo [51]	Homology-based Bayesian prediction and term enrichment	Mammal, fly, worm, fish	http://www.mirz.unibas.ch/ElMMo3/
RNAhybrid [41]	Hybridization energy, no bifurcations, and no fixed seed region	Animal	http://bibiserv.techfak.uni-bielefeld.de/rnahybrid/
TargetBoost [44]	Determines position-specific sequence motives using machine learning	N/A	https://demo1.interagon.com/targetboost/
mirWIP [52]	Structural accessibility, free energy of hybridization, and topology of seed pairing	Worm	http://146.189.76.171/query.php

(continued)

Table 1
(continued)

Name	Summary	Clade	Link
miRGator [53]	Integrates miRanda, PicTar, and TargetScanS results with additional information	Vertebrate	http://genome.ewha.ac.kr/miRGator/
SigTerms [54]	MS Excel-based tool to simplify results of miRanda, PicTar, and TargetScan results	Vertebrate	http://sigterms.sourceforge.net/
MiRTif [55]	Support vector machine (SVM)-based filtering of predictions from other tools	N/A	http://mirtif.bii.a-star.edu.sg/
TopKCEMC [56]	Integrates a number of other tools and evaluates the results statistically	N/A	http://www.stat.osu.edu/~statgen/SOFTWARE/TopKCEMC/
N/A [57]	Gene expression profiles, SVM, and duplex base pairing	Arabidopsis	http://www.biomedcentral.com/content/supplementary/1471-2105-10-S1-S34-S1.xls
GenMIR++ [58]	Gene expression profiles, Bayesian inference, and uses TargetScanS predictions	Vertebrate	http://www.psi.toronto.edu/genmir/
MIR [59]	Gene expression profiles, target enrichment, and binding energy	N/A	http://homes.gersteinlab.org/people/cc59/InferMiRNA/infermir.html
psRNA Target [60]	Extension and incorporation of new rules for miRU	Plant	http://plantgrn.noble.org/psRNATarget/
NBmiRTar [46]	MiRanda score, folding energy and Naive Bayes score	Vertebrate	http://wotan.wistar.upenn.edu/NBmiRTar
miRecords [61]	Integrates the predictions of other tools	Animal	http://mirecords.umn.edu/miRecords/
N/A [42]	Complementarity	Drosophila	http://www.russell.embl.de/miRNAs
miRWalk [62]	Complementarity and integration of 8 other prediction tools	Human, mouse, rat	http://www.umm.uni-heidelberg.de/apps/zmf/mirwalk/index.html
miTarget [63]	SVM	Human	http://cbit.snu.ac.kr/~miTarget
miRDB [64]	SVM	Human, mouse, rat, dog, chicken	http://mirdb.org

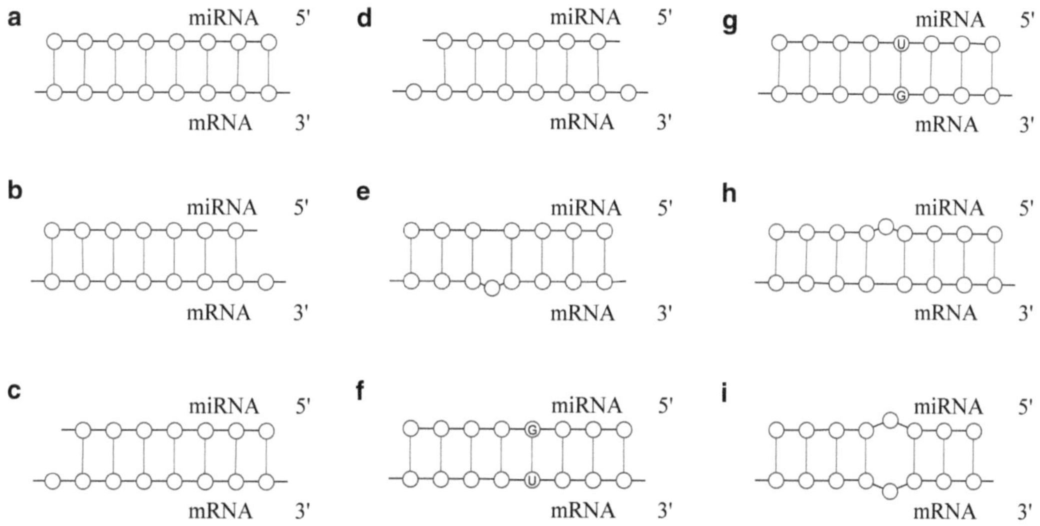

Fig. 4 a) Here are some of the possible different seed region types with a, b, c, and d representing different kinds of perfect complementation and e, h, and i showing different possible binding patterns with 1 mismatch in the middle and a G-U wobble can be seen in f and g. Analyses for the conservation of these seed regions in human, fly, worm, and zebra fish have suggested that perfect matches are more conserved than the G-U pair containing seed regions which are more conserved than the regions with mismatches [53]

pairings between the nucleotides of either of the two molecules. This feature sets it apart from tools such as Mfold and the Vienna RNA Package as they are only able to fold a single sequence. This means that when Mfold or the Vienna RNA Package are used for target prediction a linker sequence would have to be introduced in between the miRNA and the target mRNA sequence which could easily lead to errors in folding and thus target prediction. Another feature of RNAhybrid which sets it apart from other methods is its robust statistical modeling. RNAhybrid claims to be able to predict multiple miRNA-binding sites in larger RNAs and to be easy, fast, and flexible for the prediction of miRNA targets. For target prediction in humans RNAhybrid only looks at the 3′UTRs. Figure 3 shows a typical output of the program. Several different versions of the program are available for different platforms and are available for download from the Bielefeld Bioinformatics Server (http://bibiserv.techfak.uni-bielefeld.de/rnahybrid/). The application is simple and comes with adequate documentation.

3 Methods for Filtering of Predicted Targets

As outlined before, gene regulation by miRNAs is often achieved by their targeting of the 3′UTR region of an mRNA (Fig. 5). Recently, it has been shown that, at least for cyclin D1, only 7 of 45 predicted targets could be experimentally confirmed [65].

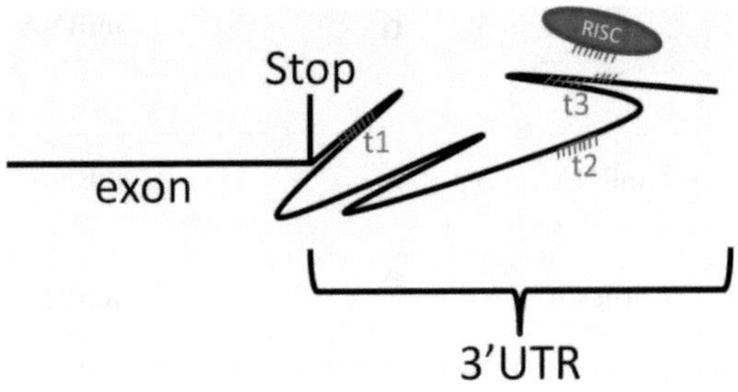

Fig. 5 A highly simplified view of targeting. A RISC-bound mature miRNA is displayed abstractly with available bonds symbolized by *sticks*. Three targets are displayed by bonds represented as *sticks*. Target 1 (t1) is close to the stop of the translation and inaccessible, t2 is freely accessible, and t3 is partially accessible

From this it can be gathered that many of the assumptions that the miRNA target prediction algorithms are based on could be improved. This is even more supported by another more recent experimental study [4]. Therefore, it is advisable to use these tools for guidance rather than accepting their results as ground truths.

There are several challenges regarding miRNA target prediction among which is the fact that a gene can be targeted by multiple miRNAs. However challenging this may be that it actually provides further criteria for discriminating true and false target predictions. For instance, if several miRNA target sites are found in a 3′UTR they would confirm each other and the resulting confidence would be raised. The location of the miRNA target site within the 3′UTR can also be used for better target prediction. The target site should not be too close to the stop codon, and it should also not be in the middle of the 3′UTR due to structural reasons. Figure 5 shows a target site (t1) which is close to the translation stop and may thus not be a good target. It further is within a secondary structure and can therefore not easily be accessed by the RNA-induced silencing complex (RISC) complex bound mature miRNA. Figure 5 also has two other target sites, one of which is fully accessible and is therefore a valid target while the last one (t3) is only partially accessible. In this case it would be important to calculate the minimum free energy (mfe) of the miRNA:mRNA duplex and compare it with the free energy of the present structure. The target is considered valid only if the mfe of the fold is higher than the mfe of the 3′UTR's structure. Below is a list of features which can potentially be used for discriminating true miRNA targets from false positive ones:

- Strong seed region pairing with minimal mismatches.
- The miRNA:mRNA duplex free energy should be minimal.
- Conserved adenosines around the seed region for animals [3].
- Multiplicity control for a gene increases significance [39].
- Proximity among target sites [26, 66].

- Target site secondary structure should be accessible [32].
- Gene expression profiles can validate regulation [57].
- Capping and polyadenylation can be useful [67].

There is also growing evidence that targeting outside the 3′UTR is more common than expected, and in the future target prediction algorithms need to take this into account [19, 68]. It may be beneficial to combine the output of several target prediction programs [67], but since they are largely built on the same assumptions important targets may be missed nonetheless [69]. Since many of the tools in target prediction are based on machine learning algorithms which learn by example, it is clear that only results similar to known examples can be found.

Many target prediction algorithms have been described and implemented, many of which are listed in Table 1. Each of these algorithms uses one or several of the criteria listed above in order to find putative target sites and then to score the significance of the predictions. Many algorithms for predicting folding of RNA sequences have been written, but the tools in Table 1 mostly use Mfold [70], RNAHybrid [48], or the Vienna RNA package [71].

4 MicroRNA Target Databases

Predicted and identified targets and other miRNA-related information need to be stored in a safe and easy-to-access environment for future use. Relevant databases have emerged by manual gathering of data from large numbers of experimentally validated miRNA targets and from high-throughput techniques. As such databases grow important issues such as the need for advanced searching and result filtering capabilities in order to accurately retrieve miRNAs or genes of specific interest become evident. Metadata and further enhancement of the currently available databases with added information from external sources will enable efficient data mining of available experimentally validated results. This is important as it will give way to producing useful novel observations [5]. Currently miRTarBase [72] provides a collection of miRNA–target interactions with experimental support. It has accumulated more than 3,500 miRNA targets by manually surveying the relevant literature. This is done after a systematic data mining step to filter research articles related to functional studies of miRNAs. Maybe the most comprehensive miRNA-related database is miRBase which houses information on both miRNA and target sequences along with predicted targets (for more information on databases pertaining to small RNAs please refer to Chapter 5). TarBase on the other hand houses manually curated targets for different species with information on the target site and the miRNA:mRNA duplex.

It also gives information about the type of experiment used for targeting and validation along with references to relevant publications. Argonaute [73] contains information on mammalian miRNAs including their gene of origin and regulated target genes which are collected from literature and other databases. Animal miRNA targets and predictions from 11 different miRNA target prediction tools are stored in miRecords [61].

5 Conclusion

The number of computational methods for miRNA target prediction is increasing, and new methods promise to deliver better results. Whether or not these methods are successful in keeping up with their promises is a subject for debate. One can expect to see better methods come by as our understanding of the miRNA regulatory pathway increases. The most important factor in developing such new algorithms and tools will be the accurate and precise computational modeling of the new scientific knowledge. This can range from better sequencing data and better classification of the 3′UTRs and splice sites to the biosynthetic pathway of miRNAs and its regulators. This calls for extensive databases which can collect, store, and provide fast and efficient recalls of such scientific data. Whether existing databases are revised or updated, or new databases are designed, this may be one of the most important factors in the development of new and effective methods for miRNA–target predictions.

6 Outlook

The current speed of advancements in miRNA-related studies is staggering. In less than 20 years miRNAs have had a huge impact in biological sciences. If the advancements in target predictions keep up with the current pace one can predict that miRNA–target predictions will be an important player in many applications such as the development of new therapeutics. While predicting targets is possible on a per miRNA basis, genome-wide studies are suffering from a large pool of possibilities, and therefore we will see a trend towards incorporating all filtering mechanisms for miRNA–target prediction, introduced in this work, and potentially further ones to increase the number of true positive identifications. Since no ground truth data is available, more and more small datasets (e.g., microarray data) providing a part of the truth will be incorporated in future studies.

Acknowledgements

H.H. would like to thank Associate Professor Dr. Jens Allmer for his kind guidance, encouragement, and advice and would also like to thank his parents for their ever-increasing support. JA would like to thank his wife Açalya for the tolerance towards late hours spent on this and other work in this volume and his son Lukas Aren for providing fun distractions during the process.

References

1. Kozomara A, Griffiths-Jones S (2011) miRBase: integrating microRNA annotation and deep-sequencing data. Nucleic Acids Res 39: D152–D157
2. Krek A, Grün D, Poy MN et al (2005) Combinatorial microRNA target predictions. Nat Genet 37:495–500
3. Lewis BP, Burge CB, Bartel DP (2005) Conserved seed pairing, often flanked by adenosines, indicates that thousands of human genes are microRNA targets. Cell 120:15–20
4. Wu S, Huang S, Ding J et al (2010) Multiple microRNAs modulate p21Cip1/Waf1 expression by directly targeting its 3′ untranslated region. Oncogene 29:2302–2308
5. Vergoulis T, Vlachos IS, Alexiou P et al (2012) TarBase 6.0: capturing the exponential growth of miRNA targets with experimental support. Nucleic Acids Res 40:D222–D229
6. Miranda KC, Huynh T, Tay Y et al (2006) A pattern-based method for the identification of microRNA binding sites and their corresponding heteroduplexes. Cell 126:1203–1217
7. Friedman RC, Farh KK-H, Burge CB et al (2009) Most mammalian mRNAs are conserved targets of microRNAs. Genome Res 19:92–105
8. Xie Z, Kasschau KD, Carrington JC (2003) Negative feedback regulation of Dicer-Like1 in Arabidopsis by microRNA-guided mRNA degradation. Curr Biol 13:784–789
9. Johnson CD, Esquela-Kerscher A, Stefani G et al (2007) The let-7 microRNA represses cell proliferation pathways in human cells. Cancer Res 67:7713–7722
10. Tokumaru S, Suzuki M, Yamada H et al (2008) let-7 regulates Dicer expression and constitutes a negative feedback loop. Carcinogenesis 29: 2073–2077
11. Mazière P, Enright AJ (2007) Prediction of microRNA targets. Drug Discov Today 12: 452–458
12. Place RF, Li L-C, Pookot D et al (2008) MicroRNA-373 induces expression of genes with complementary promoter sequences. Proc Natl Acad Sci U S A 105:1608–1613
13. Tay Y, Zhang J, Thomson AM et al (2008) MicroRNAs to Nanog, Oct4 and Sox2 coding regions modulate embryonic stem cell differentiation. Nature 455:1124–1128
14. Ørom UA, Nielsen FC, Lund AH (2008) MicroRNA-10a binds the 5′UTR of ribosomal protein mRNAs and enhances their translation. Mol Cell 30:460–471
15. Forman JJ, Legesse-Miller A, Coller HA (2008) A search for conserved sequences in coding regions reveals that the let-7 microRNA targets Dicer within its coding sequence. Proc Natl Acad Sci U S A 105:14879–14884
16. Reczko M, Maragkakis M, Alexiou P et al (2012) Functional microRNA targets in protein coding sequences. Bioinformatics 28:771–776
17. Jones-Rhoades MW, Bartel DP (2004) Computational identification of plant microRNAs and their targets, including a stress-induced miRNA. Mol Cell 14:787–799
18. Selbach M, Schwanhäusser B, Thierfelder N et al (2008) Widespread changes in protein synthesis induced by microRNAs. Nature 455: 58–63
19. Lytle JR, Yario TA, Steitz JA (2007) Target mRNAs are repressed as efficiently by microRNA-binding sites in the 5′ UTR as in the 3′ UTR. Proc Natl Acad Sci U S A 104: 9667–9672
20. Gu S, Jin L, Zhang F et al (2009) Biological basis for restriction of microRNA targets to the 3′ untranslated region in mammalian mRNAs. Nat Struct Mol Biol 16:144–150
21. Vaucheret H (2006) Post-transcriptional small RNA pathways in plants: mechanisms and regulations. Genes Dev 20:759–771
22. Rhoades MW, Reinhart BJ, Lim LP et al (2002) Prediction of plant microRNA targets. Cell 110:513–520
23. Zhang Y (2005) miRU: an automated plant miRNA target prediction server. Nucleic Acids Res 33:W701–W704
24. Lewis BP, Shih I, Jones-Rhoades MW et al (2003) Prediction of mammalian microRNA targets. Cell 115:787–798

25. Liu J (2008) Control of protein synthesis and mRNA degradation by microRNAs. Curr Opin Cell Biol 20:214–221
26. Grimson A, Farh KK-H, Johnston WK et al (2007) MicroRNA targeting specificity in mammals: determinants beyond seed pairing. Mol Cell 27:91–105
27. Sethupathy P, Megraw M, Hatzigeorgiou AG (2006) A guide through present computational approaches for the identification of mammalian microRNA targets. Nat Methods 3:881–886
28. Rajewsky N (2006) microRNA target predictions in animals. Nat Genet 38(Suppl):S8–S13
29. Doench JG, Sharp PA (2004) Specificity of microRNA target selection in translational repression. Genes Dev 18:504–511
30. Lai EC (2002) Micro RNAs are complementary to 3′ UTR sequence motifs that mediate negative post-transcriptional regulation. Nat Genet 30:363–364
31. Rajewsky N, Socci ND (2004) Computational identification of microRNA targets. Dev Biol 267:529–535
32. Du T, Zamore PD (2005) microPrimer: the biogenesis and function of microRNA. Development 132:4645–4652
33. Brennecke J, Stark A, Russell RB et al (2005) Principles of microRNA-target recognition. PLoS Biol 3:e85
34. Yan X, Chao T, Tu K et al (2007) Improving the prediction of human microRNA target genes by using ensemble algorithm. FEBS Lett 581:1587–1593
35. Sethupathy P, Corda B, Hatzigeorgiou AG (2006) TarBase: a comprehensive database of experimentally supported animal microRNA targets. RNA 12:192–197
36. Hubbard T (2002) The Ensembl genome database project. Nucleic Acids Res 30:38–41
37. Zuker M (2003) Mfold web server for nucleic acid folding and hybridization prediction. Nucleic Acids Res 31:3406–3415
38. Hofacker IL (2003) Vienna RNA secondary structure server. Nucleic Acids Res 31: 3429–3431
39. Enright AJ, John B, Gaul U et al (2003) MicroRNA targets in Drosophila. Genome Biol 5:Rl
40. Kiriakidou M, Nelson PT, Kouranov A et al (2004) A combined computational-experimental approach predicts human microRNA targets. Genes Dev 18:1165–1178
41. Krüger J, Rehmsmeier M (2006) RNAhybrid: microRNA target prediction easy, fast and flexible. Nucleic Acids Res 34:W451–W454
42. Stark A, Brennecke J, Russell RB et al (2003) Identification of Drosophila microRNA targets. PLoS Biol 1:E60
43. John B, Enright AJ, Aravin A et al (2004) Human microRNA targets. PLoS Biol 2:e363
44. Saetrom O, Snøve O, Saetrom P (2005) Weighted sequence motifs as an improved seeding step in microRNA target prediction algorithms. RNA 11:995–1003
45. Kim S-K, Nam J-W, Lee W-J et al (2005) A Kernel method for microRNA target prediction using sensible data and position-based features. 2005 IEEE symposium on computational intelligence in bioinformatics and computational biology, IEEE, pp 1–7
46. Yousef M, Jung S, Kossenkov AV et al (2007) Naïve Bayes for microRNA target predictions—machine learning for microRNA targets. Bioinformatics 23:2987–2992
47. Thadani R, Tammi MT (2006) MicroTar: predicting microRNA targets from RNA duplexes. BMC Bioinformatics 7(Suppl 5):S20
48. Rehmsmeier M, Steffen P, Hochsmann M et al (2004) Fast and effective prediction of micro RNA/target duplexes. RNA 10:1507–1517
49. Stark A, Brennecke J, Bushati N et al (2005) Animal microRNAs confer robustness to gene expression and have a significant impact on 3′UTR evolution. Cell 123:1133–1146
50. Kertesz M, Iovino N, Unnerstall U et al (2007) The role of site accessibility in microRNA target recognition. Nat Genet 39:1278–1284
51. Gaidatzis D, van Nimwegen E, Hausser J et al (2007) Inference of miRNA targets using evolutionary conservation and pathway analysis. BMC Bioinformatics 8:69
52. Hammell M, Long D, Zhang L et al (2008) mirWIP: microRNA target prediction based on micro RNA-containing ribonucleoprotein-enriched transcripts. Nat Methods 5:813–819
53. Nam S, Kim B, Shin S et al (2008) miRGator: an integrated system for functional annotation of microRNAs. Nucleic Acids Res 36: D159–D164
54. Creighton CJ, Nagaraja AK, Hanash SM et al (2008) A bioinformatics tool for linking gene expression profiling results with public databases of microRNA target predictions. RNA 14:2290–2296
55. Yang Y, Wang Y-P, Li K-B (2008) MiRTif: a support vector machine-based microRNA target interaction filter. BMC Bioinformatics 9(Suppl 12):S4
56. Lin S, Ding J (2009) Integration of ranked lists via cross entropy Monte Carlo with applications to mRNA and microRNA Studies. Biometrics 65:9–18
57. Joung J-G, Fei Z (2009) Computational identification of condition-specific miRNA targets based on gene expression profiles and sequence information. BMC Bioinformatics 10(Suppl 1):S34
58. Huang JC, Morris QD, Frey BJ (2007) Bayesian inference of MicroRNA targets from

sequence and expression data. J Comput Biol 14:550–563
59. Cheng C, Li LM (2008) Inferring microRNA activities by combining gene expression with microRNA target prediction. PLoS One 3:e1989
60. Dai X, Zhao PX (2011) psRNATarget: a plant small RNA target analysis server. Nucleic Acids Res 39:W155–W159
61. Xiao F, Zuo Z, Cai G et al (2009) miRecords: an integrated resource for microRNA-target interactions. Nucleic Acids Res 37:D105–D110
62. Dweep H, Sticht C, Pandey P et al (2011) miRWalk-database: prediction of possible miRNA binding sites by "walking" the genes of three genomes. J Biomed Inform 44: 839–847
63. Kim S-K, Nam J-W, Rhee J-K et al (2006) miTarget: microRNA target gene prediction using a support vector machine. BMC Bioinformatics 7:411
64. Wang X (2008) miRDB: a microRNA target prediction and functional annotation database with a wiki interface. RNA 14:1012–1017
65. Jiang Q, Feng M-G, Mo Y-Y (2009) Systematic validation of predicted microRNAs for cyclin D1. BMC Cancer 9:194
66. Saetrom P, Heale BSE, Snøve O et al (2007) Distance constraints between microRNA target sites dictate efficacy and cooperativity. Nucleic Acids Res 35:2333–2342
67. Barbato C, Arisi I, Frizzo ME et al (2009) Computational challenges in miRNA target predictions: to be or not to be a true target? J Biomed Biotechnol 2009:803069
68. Kloosterman WP, Wienholds E, Ketting RF et al (2004) Substrate requirements for let-7 function in the developing zebrafish embryo. Nucleic Acids Res 32:6284–6291
69. Peter ME (2010) Targeting of mRNAs by multiple miRNAs: the next step. Oncogene 29: 2161–2164
70. Mathews DH, Sabina J, Zuker M et al (1999) Expanded sequence dependence of thermodynamic parameters improves prediction of RNA secondary structure. J Mol Biol 288:911–940
71. Hofacker I, Fontana W, Stadler P et al (1994) Fast folding and comparison of RNA secondary structures. Monatsh Chem 125:167–188
72. Hsu S-D, Lin F-M, Wu W-Y et al (2011) miRTarBase: a database curates experimentally validated microRNA-target interactions. Nucleic Acids Res 39:D163–D169
73. Shahi P, Loukianiouk S, Bohne-Lang A et al (2006) Argonaute—a database for gene regulation by mammalian microRNAs. Nucleic Acids Res 34:D115–D118
74. Darty K, Denise A, Ponty Y (2009) VARNA: interactive drawing and editing of the RNA secondary structure. Bioinformatics 25:1974–1975

Chapter 13

MicroRNA Target and Gene Validation in Viruses and Bacteria

Debora Baroni and Patrizio Arrigo

Abstract

Noncoding RNAs (ncRNAs) constitute an evolutionary conserved system involved in the regulation of biological functions at posttranscriptional level. The capability to rapidly adapt their metabolism is essential for the survival of organisms. NcRNAs are a valuable means used by cells to rapidly transfer and internalize an external signal. NcRNAs are capable not only to influence the translational phase but also to affect epigenetic processes. They have been identified in almost all kingdoms of life (from archaea to human and plants). In this chapter we outline the currently available resources that could be used for the screening of viral and bacterial ncRNAs.

Key words Virus, Prokaryotes, Bioinformatics, ncRNAs

1 Introduction

Noncoding RNAs (ncRNAs) are detectable in almost all kingdoms of life. They have been discovered in eukaryotes [1], bacteria [2], and viruses [3]. More recently small RNAs have also been found in archaea, in particular in extremophiles [4]. Beside their involvement in the regulation of gene expression at posttranscriptional level, ncRNAs are involved in the control of a wide range of cellular processes. In fact, a rapid modularly cell response is essential for the adaptation of cells to environmental changes, to stress factors (biotic and abiotic), and to protect against infection. The rapid response to each of these factors requires a highly coordinated system that involves both ncRNAs and protein regulators.

Prokaryotes use two basic ways to transfer genetic information (Mobile Genetic Elements): the endosymbiotic gene transfer (EGT) and the horizontal gene transfer (HGT) [5]. Although both processes have been mainly associated with the transfer of DNA elements, some information is now available about the role played by small RNAs in these mechanisms. Endosymbiosis is a process that

Malik Yousef and Jens Allmer (eds.), *miRNomics: MicroRNA Biology and Computational Analysis*, Methods in Molecular Biology, vol. 1107, DOI 10.1007/978-1-62703-748-8_13, © Springer Science+Business Media New York 2014

allows free-living prokaryotic cells to reside within another cell that not necessarily belongs to the same phylogenetic kingdom. The transfer of genetic material from an endosymbiotic cell to the recipient one takes advantages of the host's cellular system. Conversely, HGT implies the exchange of genetic material between two different and separated biological systems that could be evolutionary distant [6]. The interest in HGT between bacteria and higher vertebrates is demonstrated by the activation of the Human Microbiome Project (http://commonfund.nih.gov/hmp/) [7, 8], aimed at unraveling the mechanisms that facilitates the HGT between bacteria and human cells.

A human being is exposed not only to bacterial HGT but also to the viral genetic transfer. We could also add to the notion of HGT all the genetic process modifications that allow a virus to control the metabolism of its host. The existence of viral ncRNAs is well established. Viruses are capable of using their ncRNAs to remodulate the host's metabolism [9]. For instance it is well known that the transition between the lysogenic and the lytic cycle of the Epstein-Barr virus is controlled by several viral miRNAs. It is worth to note that some studies have demonstrated that viral miRNAs show some functional similarity with human ones [10].

Actually, only a relatively limited number of investigations have been carried out aimed at unraveling the potential interference of exogenous (bacterial, viral, and plant) ncRNAs on the host posttranscriptional gene silencing process. From an environmental health perspective, the deep investigation of this kind of interference could be helpful for the comprehension of the onset of complex (multifactorial) diseases. In this regard, the discovery of circulating ncRNAs [11], in particular miRNAs, has pointed out the necessity to reconsider the mechanisms of cell–cell communication and, as a consequence, has underlined, once more, the possibility that exogenous transposable genetic elements could have a deep influence on the metabolism or on the processing of the genetic information of the recipient. From a biotechnological and bioinformatics perspective, the capability to discriminate between endogenous and exogenous ncRNAs is a significant challenge, because this knowledge will permit to predict the role of ncRNAs in the HGT process, in the emergence of drug resistance, or unapparent (latent) infectious diseases. In this chapter we outline the available computational resources, for viral and bacterial ncRNAs, that could support the investigation of the interplay between exogenous and endogenous ncRNA on the host's posttranscriptional process regulation.

2 Viral Noncoding RNA Resources

Similar to the human microbiome, it is possible to define a human virome as the viral community that resides, without apparent pathological signs, in the human body [12]. A large number of papers have demonstrated the critical role played by ncRNAs in the viral escape from host natural immunity [13]. Viruses possess the capability to bypass host defense by expressing proteins capable of blocking this process in infected cells [14]. The Baltimore viral classification [15] organizes different groups of viruses on the basis of their nucleic acid composition. Among the different classes of viruses, we underline some, such as the Rheovirus, that uses RNAs (both ssRNA and dsRNA) for their replication. Another very important class of viruses is that of Retrovirus [16] (HIV, Epstein-Barr virus (EBV), Herpes simplex virus, etc.) that is nowadays widely accepted being able to move to induce a stable posttranscriptional gene silencing in cultured cells. The genome of a Retrovirus is an RNA capable of producing DNA depending on a reverse transcriptase. It is out of the scope of this chapter to describe in detail the mechanism of viral infection, but it is important to underline that these infective agents are able to produce ncRNAs. The retroviral ncRNAs, and in particular the retroviral miRNAs, have been extensively studied. MicroRNAs are an efficient system, for the virus, to infect the host and to control the switch from the latent phase to virulent phase.

In general, viral microRNAs play a multitude of functions inside the host's cell. They can not only trigger the rise of viral pathologies but also of more complex ones such as cancer. Taking only human cells into account, a viral microRNA can, for example, mimic the function of oncogenic microRNAs [17]. Thus it is possible to state that viral ncRNAs could interfere with host functionality at different levels. We can schematically hypothesize two different possible routes of action for viral miRNAs: (1) they could compete with the host's miRNAs for the same mRNA target or (2) alternatively they could have a synergistic effect on the host's target mRNA. In order to underline differences and similarities between viral and human microRNA precursors we show, in Fig. 1, two miRNA precursors (viral and human) which closely resemble each other structurally.

The computational screening of viral ncRNAs, in particular of miRNAs, has attracted the interest of many groups. Bioinformatics plays a central role in this search [18]. In order to study viral microRNAs several resources have become available. Table 1 summarizes some of them. The miRBase resource [19] is the more general one. Its organization is well known and it contains also viral microRNAs. Additional and more specific data repositories are available. Vir-mir [20] allows the prediction of potential miRNA coding regions in viral genomes. Another complementary resource is the ViTa database [21]. The authors have developed and implemented a system to predict the targets of viral miRNAs. A recently developed resource is the vHoT database [22] which also contains information to predict the interaction between viral miRNAs

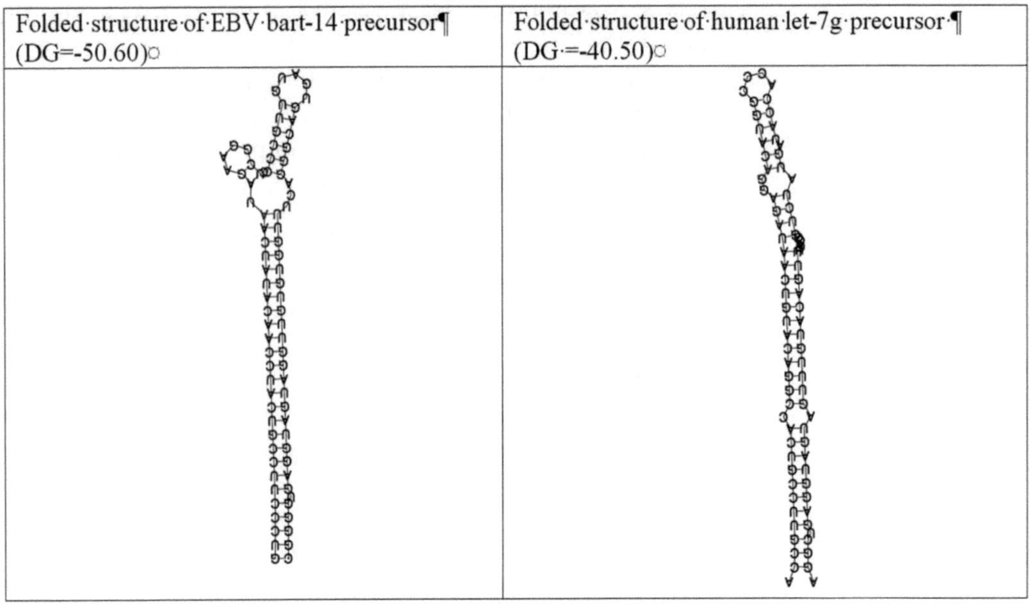

Fig. 1 Comparison between Viral and human miRNA precursors (Model obtained by MFOLD [36]). (**a**) Folded structure of EBV bart-14 precursor; (**b**) Folded structure of human let-7g precursor

Table 1
Viral noncoding RNAs resources

Database name	URL	Data accessibility
miRBase	http://www.mirbase.org/	Web Service data, download
Vir-Mir	http://alk.ibms.sinica.edu.tw	Web Service
vHoT	http://best.snu.ac.kr/vhot/	Web Service
RepTar	http://bioinformatics.ekmd.huji.ac.il/reptar/	Web Service data, download
VirSiRNA	http://crdd.osdd.net/servers/virsirnadb/index.php	Web Service
ViTa	http://vita.mbc.nctu.edu.tw/	Web Service, data download

and the host genome. A more comprehensive database is the VIRsiRNAdb [23] (http://crdd.osdd.net/servers/virsirnadb). This repository contains data about siRNA/shRNA that are able to target viral genomes. One challenge, in the investigation of viral function lead by ncRNAs is, without any doubt, the identification of the host's potential targets and to establish if a specific viral ncRNA could be secreted by infected cells. The suprementoned databases constitute a good bioformetics staiting point to solve this task.

3 Bacterial Noncoding RNAs

Contrary to viruses, less information is available about the potential interference of bacterial ncRNAs with the host cells. Regulatory ncRNAs adjust bacterial physiology in response to environmental cues. ncRNAs can base pair to mRNAs and change their translation efficiency and/or their stability, or they can bind to proteins and modulate their activity. ncRNAs have been discovered in several species throughout the bacterial kingdom [24]. The recent discovery and characterization of bacterial small regulatory RNAs has attracted the interest of many scientists. The acronym sRNA has been defined in the 1960s, to identify soluble RNAs. Nowadays, the abbreviation indicates a large number of relatively small ribonucleic acids that originate from bacteria to generally activate a fast response against environmental stresses. The availability of new high-throughput RNA deep sequencing methods has enhanced the number of screened sRNAs. This discovery has disclosed a new perspective to study the mechanism of gene expression in bacteria [25]. Bacterial sRNAs are longer than eukaryote small ncRNAs. Their length is between 50 and 250 nucleotides and they are highly structured. They do not require a perfect complementarity such as, for instance, in the case of plant miRNAs. The sRNAs seem to be capable of mimicking the functionality of other nucleic acids. The sRNAs are roughly classified into *trans*-encoded and *cis*-encoded base-paired RNAs. The *cis-encoded* sRNAs are transcribed in antisense orientation in respect to the transcription of their target gene. The *trans*-encoded sRNAs are instead transcribed in the intergenic regions and they are able to target multiple genes. Current knowledge suggests that *trans-encoded* base-paired RNAs require, for their functionality, the interaction with chaperon proteins. In respect to this, the most extensively studied protein is Hfq, an *Escherichia coli* host factor essential for replication of the bacteriophage Qβ [26], that is present in Gram-negative bacteria. The regulatory capability of bacteria seems to be associated with some conformational properties of their mRNA, but a relevant role is also played by small RNAs. Research activities are continuously adding new knowledge about the bacterial small RNAs. Figure 2 exemplifies the 2D conformation of a bacterial sRNA.

The identification of bacterial and, more generally of prokaryotes and viral, ncRNAs follows the following main guidelines: (1) Identification of structural homology and (2) target prediction.

The structural perspective has the objective to identify the conformational determinants conserved during the evolution [27]. This kind of approach embeds two different aspects:

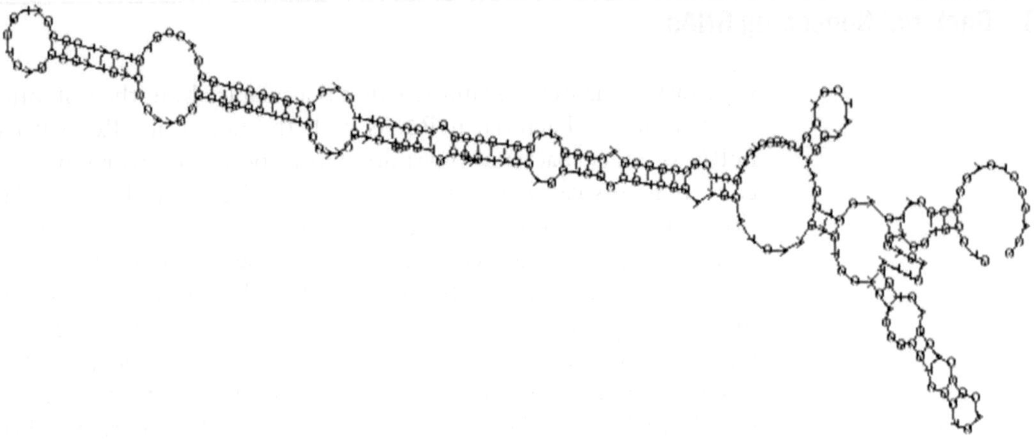

Fig. 2 An example of *Mycobacterium tuberculosis cis*-encoded sRNA (BSRD code:smtu1953.1)

1. Identification of structural homologies. This approach aims to determine a consensus structure taking multiple alignments into account. It is important to find the conservation of base pairing of the nucleotide involved in the pairing.
2. Identification of ncRNA genes. In this case genome alignment is required. The alignment allows to predict those gene that have conserved RNA structure.

The prediction of bacterial (prokaryotes) ncRNA targets is the most complicated task to be solved. The search for potential targets requires, analogously to eukaryotes, different operational steps. The pivotal phases are briefly summarized below:

1. Identification of complementary regions. The aim of this approach is to identify those nucleotide regions that are mainly involved in ncRNA:mRNA interaction. Bacterial ncRNA:mRNA matching has a hyphen flexibility and thus complexity respect to the eukaryotic miRNA:mRNA interaction.
2. Estimation of duplex characteristics. It is important to estimate the degree of folding conservation in the selected regions. This analysis allows to estimate, even if with many limitations, all the possible RNA–RNA interactions [28]. This analysis allows to determine those conformations that have the highest probability to be functional.
3. Estimation of the best folded local structure in order to obtain a high-ordered co-folded structure.
4. Estimation of site accessibility.

It is worth to note that structural bioinformatics tools play a pivotal role for the prediction of targets for bacterial sRNAs. These structural approaches are, in any case, dependent on the availability of annotated resources. We briefly describe some of the more relevant resources for bacterial ncRNA screening and analysis. Among the accessible resources that specifically contain information about bacterial ncRNAs we can consider the Bacterial Small Regulatory Database (BSRD) [29]. The BSR database contains the largest number of experimentally validated sRNAs and their known targets. The BSR database also contains information about the combinatorial structure of transcriptional regulatory networks of sRNAs.

RNAspace (http://rnaspace.sourceforge.net) is an open source platform that allows to predict the conformation of ncRNAs not only for bacteria but also for eukaryotes and archaea [30]. The sRNAMap is another integrative resource to find information about bacterial sRNAs (http://srnamap.mbc.nctu.edu.tw/) [31]. An interesting tool called nocoRNAc has been developed by Herbig and Nieselt to predict ncRNA gene in bacteria [32]. The sRNAdb [33] is a suite that allows the user to analyze, by alignment, nCRNAs of gram-positive bacteria and to search their potential targets. The Rho element is an important feature of bacterial transcription that could be modulated by some bacterial ncRNAs. RNIE is a suite to predict rho-independent terminators. This information could be valuable to investigate the interaction between ncRNA and transcriptional regulation. Bacterial target identification could be also carried out by sRNA-Target [34]. This is a software suite specifically developed to investigate this task. A comprehensive system that contains information about ncRNAs and their function is the BSR database [29]. It contains not only information about nucleotide sequences but also about targets and proteins. A genomic-based system for bacterial target prediction is the program called RNApredator [35]. This system permits to screen the putative sRNAs in bacterial genomes and plasmids. The search can be performed only on a selected specific strain or plasmid. These resources are summarized in Table 2.

We underline that we did not include general repositories (Sanger Center, TIGR, or NCBI) because our aim was to highlight the repositories that are more specialized to the screening of bacterial sRNAs. It is important to underline that viral ncRNAs are included in miRBase, but bacterial and archaea ncRNA are developed in an independent manner. Taking our initial considerations about inter-kingdom genetic transfer into account we hope that these different sources will be integrated and become more useful.

Table 2
Bacterial noncoding RNA data repositories

Database name	URL	Data accessibility
sRNAMap	http://srnamap.mbc.nctu.edu.tw/	Web Service, Data download
nocoRNAc	http://www-ps.informatik.uni-tuebingen.de	Data download (software)
sRNAdb	http://bioinfo.mikrobio.med.unigiessen.de/sRNAdb/Home	Web Service, data download
RNIE	http://github.com/ppgardne/RNIE	
sRNATarget	http://ccb.bmi.ac.cn/srnatarget/index.php	Web Service
BSRD	http://bac-srna.org/BSRD/index.jsp#	Web Service, data download
RNApredator	http://rna.tbi.univie.ac.at/RNApredator	Web Service

4 Conclusion

The aim of this chapter is to convey a survey about some of currently available data sources for viral and bacterial ncRNAs. One of the major challenges for ncRNA bioinformatics is the capability to discriminate between endogenous and exogenous ncRNAs in biological samples. This knowledge will allow to improve the design of new diagnostic tools capable, for instance, to detect circulating miRNA associated with diseases with the molecular detect exogenous ncRNAs that can interfere with the molecular diagnostic screening.

References

1. Szymanski M, Barciszewski J (2006) RNA regulation in mammals. Ann N Y Acad Sci 1067: 461–468
2. Repoila F, Darfeuille F (2009) Small regulatory non-coding RNAs in bacteria: physiology and mechanistic aspects. Biol Cell 101:117–131
3. Grundhoff A, Sullivan CS (2011) Virus-encoded microRNAs. Virology 411:325–343
4. Marchfelder A, Fischer S, Brendel J et al (2012) Small RNAs for defence and regulation in archaea. Extremophiles 16:685–696
5. Brown JR (2003) Ancient horizontal gene transfer. Nat Rev Genet 4:121–132
6. Syvanen M (2012) Evolutionary implications of horizontal gene transfer. Annu Rev Genet 46:341–358
7. Liu L, Chen X, Skogerbø G et al (2012) The human microbiome: a hot spot of microbial horizontal gene transfer. Genomics 100:265–270
8. Gevers D, Knight R, Petrosino JF et al (2012) The human microbiome project: a community resource for the healthy human microbiome. PLoS Biol 10:e1001377
9. Zhou R, Rana TM (2013) RNA-based mechanisms regulating host-virus interactions. Immunol Rev 253:97–111
10. Yu G, He Q-Y (2011) Functional similarity analysis of human virus-encoded miRNAs. J Clin Bioinforma 1:15
11. Fu Y, Yi Z, Wu X et al (2011) Circulating microRNAs in patients with active pulmonary tuberculosis. J Clin Microbiol 49:4246–4251

12. Wylie KM, Weinstock GM, Storch GA (2012) Emerging view of the human virome. Transl Res 160:283–290
13. Cullen BR (2013) MicroRNAs as mediators of viral evasion of the immune system. Nat Immunol 14:205–210
14. Cazalla D, Yario T, Steitz JA et al (2010) Down-regulation of a host microRNA by a Herpesvirus saimiri noncoding RNA. Science 328:1563–1566
15. Baltimore D (1971) Expression of animal virus genomes. Bacteriol Rev 35:235–241
16. Kurth R, Bannert N (2010) Beneficial and detrimental effects of human endogenous retroviruses. Int J Cancer 126:306–314
17. Kincaid RP, Burke JM, Sullivan CS (2012) RNA virus microRNA that mimics a B-cell oncomiR. Proc Natl Acad Sci U S A 109: 3077–3082
18. Grundhoff A (2011) Computational prediction of viral miRNAs. Methods Mol Biol 721: 143–152
19. Griffiths-Jones S, Saini HK, van Dongen S et al (2008) miRBase: tools for microRNA genomics. Nucleic Acids Res 36:D154–D158
20. Li S-C, Shiau C-K, Lin W-C (2008) Vir-Mir db: prediction of viral microRNA candidate hairpins. Nucleic Acids Res 36:D184–D189
21. Hsu PW-C, Lin L-Z, Hsu S-D et al (2007) ViTa: prediction of host microRNAs targets on viruses. Nucleic Acids Res 35:D381–D385
22. Kim H, Park S, Min H et al (2012) vHoT: a database for predicting interspecies interactions between viral microRNA and host genomes. Arch Virol 157:497–501
23. Thakur N, Qureshi A, Kumar M (2012) VIRsiRNAdb: a curated database of experimentally validated viral siRNA/shRNA. Nucleic Acids Res 40:D230–D236
24. Liu JM, Camilli A (2010) A broadening world of bacterial small RNAs. Curr Opin Microbiol 13:18–23
25. Storz G, Vogel J, Wassarman KM (2011) Regulation by small RNAs in bacteria: expanding frontiers. Mol Cell 43:880–891
26. Brennan RG, Link TM (2007) Hfq structure, function and ligand binding. Curr Opin Microbiol 10:125–133
27. Backofen R (2012) Bioinformatics of bacterial sRNAs and their targets. In: Hess WR, Marchfelder A (eds) Regulatory RNAs in prokaryotes. Springer, Vienna, pp 221–239
28. Salari R, Backofen R, Sahinalp SC (2010) Fast prediction of RNA-RNA interaction. Algorithms Mol Biol 5:5
29. Li L, Huang D, Cheung MK et al (2013) BSRD: a repository for bacterial small regulatory RNA. Nucleic Acids Res 41:D233–D238
30. Cros M-J, de Monte A, Mariette J et al (2011) RNAspace.org: an integrated environment for the prediction, annotation, and analysis of ncRNA. RNA 17:1947–1956
31. Huang H-Y, Chang H-Y, Chou C-H et al (2009) sRNAMap: genomic maps for small non-coding RNAs, their regulators and their targets in microbial genomes. Nucleic Acids Res 37:D150–D154
32. Herbig A, Nieselt K (2011) nocoRNAc: characterization of non-coding RNAs in prokaryotes. BMC Bioinformatics 12:40
33. Gardner PP, Barquist L, Bateman A et al (2011) RNIE: genome-wide prediction of bacterial intrinsic terminators. Nucleic Acids Res 39: 5845–5852
34. Cao Y, Zhao Y, Cha L et al (2009) sRNATarget: a web server for prediction of bacterial sRNA targets. Bioinformation 3:364–366
35. Eggenhofer F, Tafer H, Stadler PF et al (2011) RNApredator: fast accessibility-based prediction of sRNA targets. Nucleic Acids Res 39: W149–W154
36. Zuker M (2003) Mfold web server for nucleic acid folding and hybridization prediction. Nucleic Acids Res 31:3406–3415

Chapter 14

Gene Reporter Assay to Validate MicroRNA Targets in Drosophila S2 Cells

Bünyamin Akgül and Çağdaş Göktaş

Abstract

Bioinformatics programs have helped tremendously in identifying the targets of microRNAs, which are small noncoding RNAs that regulate gene expression posttranscriptionally. However, the partial complementarity between miRNAs and their targets hinders the accuracy of target prediction, necessitating the use of experimental validation procedures. Here, we describe a gene reporter assay typically used in our lab to validate putative miRNA–mRNA interactions in Drosophila S2 cells.

Key words Reporter assay, Target validation, Luciferase, S2, Drosophila

1 Introduction

MicroRNAs (miRNAs) are small noncoding RNAs of 17–25 nucleotides in length that posttranscriptionally regulate gene expression by perfectly or imperfectly base pairing with their target mRNAs in plants and animals, respectively [1, 2]. MicroRNAs are involved in a number of fundamental cellular processes ranging from metabolism to cell growth and differentiation to apoptosis [3–5]. The current estimate is that over half of the mammalian protein-coding genes are controlled posttranscriptionally through various mechanisms by miRNAs [6]. A single miRNA can regulate the expression of multiple target genes, but at the same time, one target gene can be independently or cooperatively regulated by multiple miRNAs.

Advances in sequencing technology have led to a great progress in the identification of novel miRNAs in a wide range of species [7–9]. This has been further supplemented with sophisticated target prediction algorithms to effectively predict miRNA–mRNA interactions [10–13]. Since the base pairing between miRNAs and their targets is imperfect in animals, the interactions predicted by bioinformatics tools have to be validated experimentally to avoid potential false-positive target predictions.

Malik Yousef and Jens Allmer (eds.), *miRNomics: MicroRNA Biology and Computational Analysis*, Methods in Molecular Biology, vol. 1107, DOI 10.1007/978-1-62703-748-8_14, © Springer Science+Business Media New York 2014

A number of different approaches have been used in the experimental identification of putative miRNA–mRNA interactions [14, 15]. Broadly, these approaches have been categorized under transcriptome analyses, biochemical approaches, and proteome analyses. The choice primarily depends upon the intended number of targets. For example, if the aim is to identify multiple miRNA targets at once, a genomics approach such as transcriptome or proteome analyses, following miRNA knockdown or overexpression, would be highly desirable. The handicap of this approach is that it would be cumbersome to distinguish between direct or indirect effects of miRNAs on target expression. To avoid indirect miRNA effects in a genomics target screen, AGO or RISC immunoprecipitation can be combined with microarray or deep-sequencing analysis of total RNA isolated from the immunoprecipitate in the presence and absence of an miRNA of interest ([16, 17], Chapter 6). By doing so, the changes in the mRNA contents, thus miRNA targets, of RISC complexes can be directly identified when miRNAs are over-expressed or knocked-down.

Direct demonstration of miRNA targets usually involves experimental identification of target mRNAs individually. This approach nicely eliminates indirect off-target effects since the function of the reporter mRNA is determined exclusively by a physical interaction between the miRNA of interest and the reporter construct. One disadvantage of this approach is that it is quite labor-intensive. The results should still be interpreted carefully especially when miRNAs are over-expressed transiently. Any superficial increase in the intracellular miRNA concentration beyond its physiological concentration could potentially generate false-positive results by interfering with the interaction between other miRNAs and RISC complexes [18].

In this chapter, a gene reporter assay is described in which the 3′UTR of a target mRNA is cloned into an expression vector bearing a reporter gene (e.g., luciferase). When co-expressed with the miRNA, the miRNA-mediated suppression of the reporter function, miRNA:target interaction, can easily be measured. Cells without miRNA are used as negative controls. Additionally, constructs containing 3′UTRs with mutated target sites serve as additional negative controls. This approach can be further solidified by using miRNA inhibitors in the experimental design.

2 Materials

2.1 Molecular Cloning Components

- PCR Cloning Kit (Fermentase InsTAclone™ PCR cloning kit #K1213 which contains the TA cloning vector pTZ57R/T, 5× ligation buffer, T4 DNA ligase and nuclease-free water).
- Fragment isolation kit (Invitrogen PureLink gel extraction kit, K2100-12).

- Plasmid purification Kits (Fermentas GeneJET™ plasmid miniprep kit K0503; Invitrogen PureLink™ HiPure plasmid filter midiprep kit K2100-15).
- Restriction enzymes.
- Taq DNA polymerase (Fermentas).
- Squishing buffer (10 mM Tris–Cl, pH 8.2, 1 mM EDTA, 25 mM NaCl, and 200 μg/ml proteinase K).
- SOC medium (0.5 % yeast extract, 2 % Tryptone, 10 mM NaCl, 2.5 mM KCl, 10 mM $MgCl_2$, 10 mM $MgSO_4$, 20 mM glucose).

2.2 S2 Cell Maintenance and Transfection Components

- Schneider's Drosophila medium (Invitrogen 11720-034).
- Heat inactivated fetal bovine serum (Invitrogen, GIBCO 10500).
- Penicillin G (Biochrom AG, A321-44), Streptomycin (Biochrom AG, A331-44).
- Calcium-phosphate transfection kit (Invitrogen K2780-01), (2× HEPES-buffered saline (HBS), 2 M $CaCl_2$, tissue culture sterile water).

Drosophila Schneider 2 embryonic stem cells are maintained in Drosophila Schneider medium (Invitrogen) supplemented with L-Glutamine, 10 % FBS (GIBCO) and 2 % Penicillin–Streptomycin (Biochrom AG) at 25 °C without CO_2. Passages of cells are performed twice a week.

2.3 Luciferase Assay Components

- Luciferase assay kit (Promega, E1960).
- Passive cell lysis buffer (glycerol (50–75 %), CDTA (1.0–5.0 %), *N*,*N*-Bis(3-D-glukonamidopropyl) cholamide (<1.00 %)).
- 1× PBS [(1) 8 g NaCl, (2) 0.2 g KCl, (3) 1.44 g Na_2HPO_4, (4) 0.24 g KH_2PO_4, (5) in 800 ml of distilled H_2O, (6) adjust the pH to 7.4 with HCl. Add H_2O to 1 l].

3 Methods

Three main constructs should be prepared to perform a complete luciferase dual reporter assay. The first construct contains the miRNA precursor, which is expressed under the control of a potent promoter such as actin. The second one is a chimeric construct carrying the open reading frame of the luciferase firefly gene fused to the 3′UTR of the target mRNA of interest. The third construct carries the luciferase renilla which is used to normalize the transfection efficiency.

3.1 MicroRNA Overexpression Constructs

There are at least three different ways of over expressing miRNAs in cell lines such as Drosophila S2 cells. Several commercial companies provide experimentally validated synthetic mature miRNAs or their precursors (pre-miRNA). Transient transfection of these molecules is easy and quick but requires dose kinetics to determine expression levels comparable to their endogenous concentrations in the cell line used. The third approach involves the expression of miRNAs on an expression vector carrying a strong promoter. The use of an inducible promoter would be desirable, if possible, to modulate the expression levels of miRNA transcripts. In this section, the third approach will be described.

The cloning of an miRNA gene into an expression vector can be carried out in two ways. In the first option, the PCR-amplified miRNA sequence is directly cloned into the expression vector. This approach saves time by eliminating a second cloning step. However, the preparation of the fragments and the vector carrying the matching restriction sites can be difficult especially for beginners. In this situation, it may be easier to clone the miRNA sequence into a TA cloning vector such as pTZ57R/T (Fermentase). The fragment can then be easily transferred from this sub-cloning vector into the expression vector. Unless specified by the product provider, the protocols described by Sambrook and Russell are usually followed [19].

3.1.1 PCR Amplification of an MicroRNA Gene from Genomic DNA and Cloning into Expression Vectors

Although the isolation procedure may vary depending upon the cell/tissue, singly fly genomic DNA isolation procedure [20] was used to prepare the genomic DNA for PCR amplification of the miRNA precursor. The same protocol was applied to isolate genomic DNA from S2 cells. The following procedure has given excellent results especially with Drosophila embryos or adult flies.

(a) A single fly is mashed five to ten times by a pipet tip in 100 μl squishing buffer.

(b) The mixture is incubated at 37 °C for 30 min following the addition of proteinase K (200 μg/ml). The powder proteinase K should be dissolved in pure water, divided in aliquots, and stored at −20 °C until use to minimize the loss of activity.

(c) Proteinase K is then inactivated by heating the sample at 95 °C for 2 min. As an option, RNAse can be added to the mixture followed by phenol–chloroform extraction to eliminate contaminating RNAs. Usually, RNA should not interfere with DNA amplification, though. The genomic DNA can be stored at −20 °C until use.

(d) Forward and reverse primers are designed to amplify candidate miRNA precursors by PCR. An example is presented in Fig. 1. Both forward and reverse primers can be about 20–25 nucleotides in length. A restriction enzyme recognition site should be placed on the 5′-ends of primers to facilitate cloning in the subsequent steps. Having the same restriction site on both primers makes it easy to prepare the fragments for cloning.

5'-CAGCACACAGGTCACCATTCCAAAAGAGGTGGGCGCATATATTTTCATGATTT
ATAGAATTTAACTAATATAGTGTTCGTTGTGTTTCAGCTGGCTGGTGCCATCATTG
GTAAGGGAGGTGGCCGCATCCGTCGCATCCGCAACGAGTCCAGTGCGTACATCAC
CTCGACGAGCCCCTGCCAAACTCGAACGATCGTAT*CATCACCATCTCGGGCACGC*C
GAAGCAAATACAAATGGCCCAGTATCTGCTGCAACAGAGGTTGGTGTCGCAATCT
GAATAAGAAGTATGCACATCCTATCATTTGTCTAACCACCCATCCCCACAATAAC
ACTCTAACACAAAAAAAAAAAAAAAACGCGTGATTAATTTGGAAGGAAAGGTGTC
TGCTGTGCGTCCCGGTCCCGAGTGTTAAAATATGTGCTTGATCGTAACTCCATCCA
AACTCGATATTAACTAACCGATTGGTCCTCCTGGGAGTGCATTCCGTATGGAAGA
CTAGTGATTTTGTTGTTTGGTCTTTGGTAATAACAATAAATCCCTTGTCTTCTTACG
GCGTGCATTTGTGCTCTTCATTCTATCGATGGTTAACCAATAAAACTAAACACGG
CATTGGAAACTACCTAACTAAACGTGTCACAATTCATCCTTGTCCCAGCTCGCAA
AAAAAAAAAAAAATAATAATCATCCATCGTGTATAACTATTTCACCCAAAAACGA
ACTCCCCAGTACCCTCCCATCCCACTACATCCACATCCACATTCATTTAGCTGTAT
CTGCTGTTCATTAAAAAGTTCATTAATTATGTTTTCCTTTGGCTTGCTGCCATCCTG
CCGCCAACTCTTTCCATTGCAGCGTACA*CGAGAATGGCAGGCGAAACA*TTTAAGTG
GGCAAGGACTTGAACAGCTACAACAGCAATAGCAACACCATTCGACAAAACAAC
AAATTCAACAACTACAATAACCATCGAGAATTGCTGCGTTTCACGTTTAAATTTA
AATGT

Fig. 1 Primer design to PCR-amplify the *dme*-miR7 gene. The miRNA sequence was obtained from flybase. Placing the mature miRNA sequence in the middle (*gray-shadowed* and *underlined*), approximately 150–200 bp from each region is amplified. Sequences complementary to the primer sequences are *italicized* and *underlined*. One of the restriction sites that does not cut the miRNA gene but is present in the multiple cloning site of the expression vector is then placed on each primer with a few extra nucleotide sequences (depending on the restriction site chosen). Such a design would result in a forward primer 5′ GG**GGATCC**CATCACCATCTCGGGCACGC 3′ and a reverse primer 5′ GG**GAGCTC**TGTTTCGCCTGCCATTCTG 3′. The restriction sites are *bolded* in the primer sequences. The two additional 5′ G residues are used to facilitate more efficient restriction digestion

However, the fragment can be cloned in either orientation, which requires verification of the correct orientation. To avoid this problem, different restriction sites should be placed on each primer. The selected restriction sites should be present only in the multiple cloning site of the expression vector and absent in the other regions of the vector or the candidate miRNA gene. Such non-cuter sites can be detected by programs freely available on the web (e.g., RESTRICTION MAPPER: http://www.restrictionmapper.org/). Primers are designed to amplify at least a region 150–200 bp flanking from each side of the mature miRNA sequence.

(e) PCR amplification may require optimization of the annealing temperature and the template amount but the setup described below may be a good start. Mix the following in a clean Eppendorf tube: 2.5 μl 10× Taq buffer (Fermentase), 1.5 μl 25 mM $MgCl_2$, 0.5 μl 10 mM dNTP mix (Fermentase), forward primer (5–10 mM), reverse primer (5–10 mM), 0.75–1 unit Taq polymerase (Fermentase), template genomic DNA (50–500 ng), and dH_2O to 25 μl.

(f) Amplify the miRNA gene in a thermocycler by using the following program: initial denaturation at 94 °C for 5 min followed by 25 cycles for initiation at 94 °C for 1 min, annealing at 50 °C for 1 min, and elongation at 72 °C for 1 min.

The program is finished with a final extension at 72 °C for 10 min. The PCR products are run on 1 % agarose gel to check the size of the fragment against a marker. Note that the annealing temperature depends upon the melting temperature of the primers, which will vary depending on the primer sequence.

(g) The PCR-amplified fragment should be purified from agarose gel to eliminate contaminating genomic DNA and incomplete PCR products. The purelink quick gel extraction kit (Invitrogen) yielded consistent results in our lab. Since the company provides detailed instructions, this particular procedure will not be explained here. It is important to keep in mind that elution of the fragment into distilled water increases the efficiency of downstream enzymatic reactions.

(h) The PCR-amplified and gel-purified products of miRNA precursor includes adenine residues at their 3′ ends due to extension by Taq polymerase. By taking advantage of this property, the fragments are easily cloned into TA cloning vectors using the instructions of the manufacturer. We commonly use pGEM T easy (Promega) or pTZ57R/T (Fermentase) for this purpose. 50 ng vector and 4.5–5 ng extracted PCR product of approximately 300–500 bp is sufficient to set up an efficient ligation reaction. Note that the ligation reaction can be completed in 1 h or left overnight if desired.

3.1.2 Transformation of Ligation Products to DH5α Competent Cells

Since blue-white colony screening is performed to select the correct transformant carrying the miRNA precursor, 40 μl X-Gal, and 7 μl IPTG should be spread onto an agar plate before starting the transformation procedure. If commercial competent cells are used, we advise that the manufacturer's instructions are followed. The following procedure originally described by Sambrook and Russell [19] works well for home-made competent cells. Remember to prepare a water-bath at 42 °C before you start the transformation.

(a) Immediately place 50 μl competent cells into an ice-cold Eppendorf tube and let it thaw on ice.

(b) Gently add 5 μl ligation mixture into the competent cells and incubate on ice for 20 min after mixing the tube content gently.

(c) Heat shock the cells at 42 °C for 45 s and immediately place the tube on ice for 2 min.

(d) Add 950 μl SOC medium into the tube and shake 1 h in a shaker (100 RPM at 37 °C).

(e) Spread 100 μl onto an agar plate containing the appropriate selectable antibiotic. Incubate the plate in an incubator at 37 °C overnight (16–24 h).

(f) A single white colony is then streaked onto an agar plate again to obtain a pure single colony. One of the colonies is inoculated into 8 ml LB medium and the plasmid is purified via a plasmid purification kit (e.g., Fermentase GeneJET™ plasmid miniprep kit K0503) according to the manufacturer's instructions.

3.1.3 Transfer of the MicroRNA Precursor from the Cloning Vector into an Expression Vector

The fragment containing the miRNA precursor sequence is released from the sub-cloning vector by digesting it with the appropriate restriction enzymes (e.g., 5′ *Bam*HI and 3′ *Sal*I). The expression vector is also digested by the same enzymes to generate sticky ends for the ligation of the miRNA precursor released from the sub-cloning vector. The last step involves ligating the miRNA-carrying fragment into the expression vector with the sticky ends. The basic molecular biology protocols described by Sambrook and Russell [19] can be used to complete this step. Two control reactions should be included to minimize the number of false-positive transformants. Control ligation I is basically the same as the test reaction but does not contain any ligase enzyme. Control ligation II should not contain any insert. When transformed into *E. coli*, no transformants should be observed from these two control ligation reactions if the expression vector is prepared properly. Following the transformation, the cloning efficiency can be checked as in Subheading 3.1.2.

3.2 Construction of a Chimeric Gene Carrying the Open Reading Frame of the Luciferase Firefly Gene Fused to the 3′ UTR of a Target mRNA

The chimeric gene includes firefly luciferase gene whose 3′UTR is replaced with that of the miRNA target which contains a single or multiple binding sites for the miRNA of interest. This chimera is cloned into an expression vector (e.g., pAct-5c for expression in Drosophila S2 cells) to examine the effect of miRNA:target 3′UTR interaction on the expression of the reporter luciferase activity.

(a) Forward and reverse primers are designed as in Subheading 3.1.1 to amplify firefly luciferase without its 3′UTR. The selected restriction sites should be present in the correct orientation in the expression vector. The pGL4.12[luc2CP] construct can be used as a template to amplify firefly luciferase as in Subheading 3.1.1. The only difference is that the template DNA is pGL4.12 for this reaction.

(b) Double-digest the expression vector pAct-5c and the PCR products with the same restriction enzymes included in the primers to amplify the firefly luciferase and extract from 1 % agarose gel by a fragment isolation kit.

(c) Ligation, transformation, and verification of the insert can be performed as in Subheading 3.1.

(d) **Steps 3.2a–c** are repeated but with a pair of primers complementary to the 3′UTR of target mRNA to insert the 3′UTR of the miRNA target downstream from the firefly luciferase.

3.3 Construction of the Normalization Vector Containing Renilla Luciferase

It is important to normalize the transfection efficiency in all transfection reactions. This is accomplished by transfecting the cells with a different reporter gene such as luciferase renilla. The cloning rationale is almost the same as in Subheading 3.1 except for the template and the pair of primers. The commercial plasmid pGL4.74(hRluc/TK, Promega) is used as a template instead.

3.4 S2 Cell Transfection

(a) 24 h before transfection, Drosophila S2 cells (5×10^5 per well) are seeded in a 6-well plate (Jet Biofil, TCP011006) in complete medium containing 10 % FBS and 2 % Penicillin–Streptomycin (At least four wells should be seeded to measure one miRNA-target dual reporter assay—referred to as well 1, 2, 3, and 4). Well 1 contains the control S2 cells that are not transfected with any plasmids. It is used to assess the effect of the transfection reagent on S2 cells. Well 2 is another control that is used for transfection with the empty expression vector (pAct-5c) to measure the effect of plasmids, if any, to the cells. Well 3 contains S2 cells transfected with the chimeric reporter gene (firefly luciferase and target mRNA 3′UTR). Well 4 contains the S2 cells transfected with the miRNA expression plasmid (e.g., pAct-5c-miRX) and the chimeric reporter gene. The cells in the wells 2, 3, and 4 are also transfected with equal amounts of renilla luciferase to normalize the transfection efficiency.

(b) For transfection, the calcium-phosphate transfection kit (Invitrogen) offers high transfection rate, and 10 μg plasmid DNA is sufficient. It is very important to use a Pasteur pipette to slowly add the solutions A2, A3, and A4 (dropwise) to the solutions B2, B3, and B4, respectively, while bubbling air through by another Pasteur pipette. This is a slow process executed over 1 or 2 min.

3.5 Measurement of Luciferase Reporter Function

The manufacturer company's instructions were followed to measure the luciferase activity. The measurement consists of three main steps: (1) cell lysis, (2) measurements of luciferase firefly, and (3) luciferase renilla.

(a) 48 h after transfection, the cell medium in the wells is removed and cells are rinsed with 1× PBS at least twice. The rinsing solution should be removed completely.

(b) 500 μl PLB is added into each well and the cell culture plate is placed on an orbital shaker (~50 RPM) for 15 min at room temperature. If desired, cells are scraped in PLB directly and rinsed with 1× PBS in microfuge tubes. 500 μl PLB is added and cells are lysed by pipetting.

(c) Luciferase reporter function can be conveniently measured in a 96-well plate using a luminometer (VarioScan, Thermo). The luminometer is programmed to read a 2-s premeasurement delay followed by a 10-s measurement period for each reporter assay. 20 μl cell lysate is usually sufficient for each assay.

(d) Cells may possess very little amount of luciferase. Thus, the background enzymatic activity should be subtracted from the total luminescence. For this purpose, non-transfected control cells are lysed in PLB and used as a negative control. Any luminescence obtained from this sample is subtracted from all other readings.

4 Conclusion

Advances in sequencing technology have nearly exhausted the identification of potential miRNAs in mammals. Nowadays, it has become more important to identify potential targets of miRNAs and to understand the significance of miRNA:mRNA interactions in biological systems. Although sophisticated target prediction algorithms effectively predict miRNA–mRNA interactions, experimental validation is essential in animals due to imperfect pairing between miRNAs and their targets.

Typically, transcriptome analyses, biochemical approaches, and proteome analyses are used to identify miRNA:mRNA interactions. Gene reporter assays are preferred over other approaches as they nicely avoid off-target effects associated with transcriptome or proteome analyses. Direct demonstration of a single miRNA:mRNA interaction increases specificity. Reporter assays are also attractive due to their simplicity and cost-effectiveness.

Acknowledgements

This work was supported by the Scientific and Technical Research Council of Turkey (104T144 to BA). We also thank the IZTECH Center for Biotechnology for their help.

References

1. Ghildiyal M, Zamore PD (2009) Small silencing RNAs: an expanding universe. Nat Rev Genet 10:94–108
2. Kim VN, Han J, Siomi MC (2009) Biogenesis of small RNAs in animals. Nat Rev Mol Cell Biol 10:126–139
3. Bartel DP, Chen CZ (2004) Micromanagers of gene expression: the potentially widespread influence of metazoan microRNAs. Nat Rev Genet 5:396–400
4. Pauli A, Rinn JL, Schier AF (2011) Non-coding RNAs as regulators of embryogenesis. Nat Rev Genet 12:136–149
5. Yekta S, Tabin CJ, Bartel DP (2008) MicroRNAs in the Hox network: an apparent link to posterior prevalence. Nat Rev Genet 9:789–796
6. Huntzinger E, Izaurralde E (2011) Gene silencing by microRNAs: contributions of translational repression and mRNA decay. Nat Rev Genet 12:99–110
7. Berezikov E, Guryev V, van de Belt J et al (2005) Phylogenetic shadowing and computational identification of human microRNA genes. Cell 120:21–24
8. Lai EC, Tomancak P, Williams RW et al (2003) Computational identification of Drosophila microRNA genes. Genome Biol 4:R42

9. Lim LP, Glasner ME, Yekta S et al (2003) Vertebrate microRNA genes. Science 299:1540
10. Bartel DP (2009) microRNAs: target recognition and regulatory functions. Cell 136:215–233
11. John B, Enright AJ, Aracin A et al (2004) Human microRNA targets. PLoS Biol 2:e363
12. Krek A et al (2005) Combinatorial microRNA target predictions. Nat Genet 37:495–500
13. Lewis BP, Burge CB, Bartel DP (2005) Conserved seed pairing often flanked by adenosines, indicates that thousands of human genes are microRNA targets. Cell 120:15–20
14. Orom UA, Lund AH (2009) Experimental identification of microRNA targets. Gene 451:1–5
15. Thomson DW, Bracken CP, Goodall GJ (2011) Experimental strategies for microRNA target identification. Nucleic Acids Res 39:6845–6853
16. Karginov FV, Conaco C, Xuan Z et al (2007) A biochemical approach to identifying microRNA targets. Proc Natl Acad Sci U S A 104:19291–19296
17. Chi SW, Zang JB, Mele A et al (2009) Argonaute HITS-CLIP decodes microRNA-mRNA interaction maps. Nature 460: 479–486
18. Bracken CP, Gregory PA, Kolesnikoff N et al (2008) A double-negative feedback loop between ZEB1-SIP1 and the microRNA-200 family regulates epithelial-mesenchymal transition. Cancer Res 68:7846–7854
19. Sambrook J, Russell D (2001) Molecular cloning: a laboratory manual. Cold Spring Harbor Laboratory, New York
20. Gloor GB, Preston CR, Johnson-Schlitz DM et al (1993) Type I repressors of P element mobility. Genetics 135:81–95

Chapter 15

Computational Prediction of MicroRNA Function and Activity

Hasan Oğul

Abstract

Inferring microRNA (miRNA) functions and activities has been extremely important to understand their system-level roles and the mechanisms behind the cellular behaviors of their target genes. This chapter first details methodologies necessary for prediction of function and activity. It then introduces the computational methods available for investigation of sequence and experimental data and for analysis of the information flow mediated through miRNAs.

Key words Regulatory networks, Transcriptional modules, Biclustering, Bipartite graphs, Multiway analysis

1 Introduction

Understanding the mechanisms of gene regulation has been a central problem in functional genomics. Since the first paper on genome-wide cluster analysis of gene expression data [1], there has been a tremendous research effort in better elucidation of individual and joint regulation of genes. The majority of this research has focused on either grouping genes with similar behaviors into clusters [2] or finding pairwise regulatory relationships between upstream factors and their targets [3]. The discovery of miRNAs at the end of 1990s has shifted the direction of research in this field to more integrative techniques due to the changes in understanding of regulatory networks. It has already been shown that these tiny regulatory molecules are abundantly present in many organisms, and they can play pivotal roles in several processes such as cell growth, proliferation, differentiation, development, and apoptosis [4]. Their consequent effects in many other pathways and the development of several diseases have motivated the use of intelligent data analysis techniques to facilitate better comprehension of facts regarding miRNAs and the production of new testable hypotheses [5].

Malik Yousef and Jens Allmer (eds.), *miRNomics: MicroRNA Biology and Computational Analysis*, Methods in Molecular Biology, vol. 1107, DOI 10.1007/978-1-62703-748-8_15, © Springer Science+Business Media New York 2014

Inferring individual targets for given miRNAs using either computational or experimental methods can provide valuable information for understanding of the basic mechanisms behind regulatory interactions of miRNAs with other genes. Since miRNA target selection is coordinated in a sequence-specific manner [6], target prediction methods are usually based on the information derived from the potency of binding between miRNA and putative target. Though useful, this approach is limited to explain miRNA behavior in practice. The limitation is imposed by four major reasons. First, the reliability of target predictions is quite low [7]. Since the mechanism behind target selection and binding is still elusive, especially for animals, the sequence-based models developed so far have been built upon hypothetical rules such as sequence complementarity and thermodynamic stability. Consequently, it is usually difficult to observe a consensus among the prediction results of different tools [7]. Second, potential binding between an miRNA and an mRNA does not necessarily imply an active regulatory relationship between them [8]. Third, miRNAs may act together to regulate one or more genes [9]. Target prediction results can only explain pairwise interactions but not many-to-many networks. And lastly, other factors may simultaneously contribute to the regulation of genes [10]. Therefore, we need integrative methods which could explain the active behaviors of miRNAs in a cooperative manner. The model should also allow researchers to link these behaviors to other functional consequences such as diseases or metabolic pathways.

This chapter aims to introduce the computational methods used to analyze microRNA activities and functions from a systems biology perspective. Here, computational miRNA function analysis means the prediction and analysis of the information flow originated, redirected, inhibited, or terminated by single or multiple miRNAs. Following sections will introduce the previous works and basic terminologies regarding gene regulatory networks (GRNs), bipartite graphs, and biclustering techniques, which are frequently used for multiway analysis of miRNA functions and activities.

2 Regulatory Networks: New Actors and Changing Paradigms

Systems biology aims to model living organisms using the interaction of all entities involved in the process under consideration. To this end, several network-based methods have been developed to explain the directed relationships between genes, proteins, metabolites, or other biomolecular entities. The most famous approach in that sense is the concept of GRN, which attempts to model regulatory relationships between genes and other potential regulators, such as transcription factors. A GRN is composed of nodes, representing the entities involved, and of edges, representing the molecular interactions between the entities. Edges are usually directed to show the course of the regulatory effect. The inference of GRNs is usually based on measurements at the transcript level. Since this reconstruction is based on available experimental data, the whole process of GRN inference can be considered reverse engineering. GRNs are able to discern a comprehensive view of all interactions either on a genome-wide level or as a local process. The major challenge with the reconstruction of GRNs is the fact that, due to the combinatorial nature of the problem, the available data is often insufficient for accurate and efficient inference of optimal model parameters. However, recent advances in high-throughput technologies have largely promoted the use of GRNs to enable the researchers to analyze large-scale interactions among thousands of transcripts.

Various network architecture models have been used to infer GRNs, which can be broadly classified into four categories:

1. Information theoretical models.
2. Differential equation models.
3. Boolean networks.
4. Bayesian networks.

In *information theoretical models*, the interaction between two genes is simply represented by the correlation coefficient of their expressions. In a simple undirected graph where each node represents a gene, an edge is present between two nodes if the correlation coefficient of two corresponding genes is above a predefined threshold. Other similarity or distance measures, such as Euclidean distance or mutual information, can be used instead of correlation coefficient to model the pairwise interaction.

Information theory models parameterize only the correlations between genes. A *differential equation model*, on the other hand, uses a system of equations which describe the temporal value of gene expression as a function of the expression of other genes and potentially other factors. This enables the model to capture the dynamic behavior of all network components in a more quantitative manner. The model is expected to learn the differential equations and their parameters.

Boolean networks are the models of discrete equations over binary variables which define the temporal states of the genes in the network. Therefore, continuous expression values need to be transformed to binary data. Inference of a *Boolean network* is equivalent to finding Boolean functions which may explain observed data.

Bayesian networks are the most frequently used techniques to model GRNs. A *Bayesian network* is a probabilistic graphical model which represents the knowledge in an uncertain domain but for which prior knowledge may exist. It is simply modeled by a graph, where each node in the graph represents a random variable, e.g., expression of a gene or a transcription factor, and the edges between the nodes represent probabilistic dependencies between corresponding random variables. Reverse engineering of a Bayesian network is defined as the estimation of these conditional dependencies in the graph using known statistical methods.

For a formal definition of Bayesian networks (also refer to Chapter 7), let us first define what a graph is. A graph is a symbolic representation of a set of entities and their relationships. A simple graph $G(V, E)$ consists of a nonempty set representing vertices, V, and a set of unordered pairs of elements of representing edges, E, where each edge can connect two vertices. Edges can be weighted by any value to represent the relation between two vertices depending on the problem modeled by the graph. A Bayesian network B can be defined as an acyclic graph that represents a joint probability density over a set of random variables V. The network B is defined by a 2-tuple (G, Θ), where G is the directed acyclic graph whose nodes $X_1, X_2, \ldots, X_n$ represents random variables, and whose edges represent the direct dependencies between these variables. Each variable X_i is assumed to be independent of its non-descendents given its parents in G. Θ denotes the parameters set, which consists of the parameters defined over conditional probabilities, given by $\theta x_i|\pi_i = P_B(x_i\ |\pi_i)$, x_i is a realization of X_i conditioned on π_i, and π_i denotes parents of X_i. A unique joint probability density is then given in Eq. 1.

$$P_B(X_1, X_2, \ldots, X_n) = \prod_{i=1}^{n} P_B(X_i \mid \pi_i) = \prod_{i=1}^{n} \theta_{X_i|\pi_i} \tag{1}$$

Given a joint probability density function, inference of a Bayesian network's parameters is usually achieved through a set of greedy steps [11].

The desire for analyzing groups of co-regulated genes together with their pairwise interactions with regulators has introduced the new paradigm of ***transcriptional modules.*** A transcriptional module is defined as a collection of genes under control of the same regulation factors which bind to these genes in a similar manner. Genes in a transcriptional module are expected to have similar regulatory elements in their promoter regions and should undergo similar changes in mRNA expression in response to changes in environmental conditions or cellular state. Therefore, the identification of transcriptional modules is based on either the regulatory elements obtained from the sequences of genes or their expression profiles over different conditions. Segal et al. combined the ideas of Bayesian networks and transcriptional modules and introduced a new technique to analyze gene regulation, called module networks [12]. A module network is an extension of a Bayesian network where the nodes in the tree are gathered into so-called modules. All node values in the same module follow the same distribution, that is, all nodes in a module have the same parent nodes. Assignment of nodes to particular modules is considered as a part of network optimization. A Bayesian score is defined by the logarithm of the posterior probability of the dependency structure S and assignment A of the nodes in the modules given the observed values D of the nodes (Eq. 2)

$$score(A,S \mid D) = \log p(D \mid A,S) + \log p(A,S) \quad (2)$$

This score is used for network optimization for which a two-stage greedy approach is used. The first stage is to search for the best structure. A Bayesian score is computed for every possible local modification, i.e., all additions and deletions of single edges in the current network. The modification which leads to the highest increase in score is chosen for the next iteration step. The second stage is for module assignment. An iterative procedure is applied to find better assignments by trying to move one node at a time to other modules.

GRNs have been intensively used in the analysis of gene expression data [12]. However, we have seen only a few studies which have explored the functions of miRNAs in the context of GRNs [13–17]. The main reason that complicates the reconstruction of GRNs comprising of miRNAs is the fact that several types of regulatory circuits which could create inconvenient structures may appear due to distinct type of regulatory information flow among miRNAs and other regulatory elements such as transcription factors. For example, in a Bayesian network topology, due to its probabilistic nature, loops are not allowed. However, miRNAs and TFs might have bidirectional regulatory relationships among them which represent a loop and thus create a problem for using a Bayesian network model. For instance, a TF may repress the transcription of a gene while activating the transcription of an miRNA which inhibits the translation of the same target. Several such configurations might be observed in a complete network of interactions. Tsang et al. classified these network motifs into two types: type-I and type-II [13]. Type-I circuits include miRNAs and target genes which are both positively or negatively co-regulated by an upstream regulator. On the other hand, the transcription rates of miRNAs and target genes are differentially regulated by the upstream factor in type-II circuits (Fig. 1). They discovered that both type-I and type-II motifs are abundant in mammals.

To simplify the model, miRNAs can be considered independently from TFs to analyze their functional effects in certain development processes. An example work is presented by Liu and Olson [14]. Through the reconstruction of miRNA regulatory networks, they investigated the roles of miRNAs as regulators in cardiovascular development and showed that the resulting network may offer

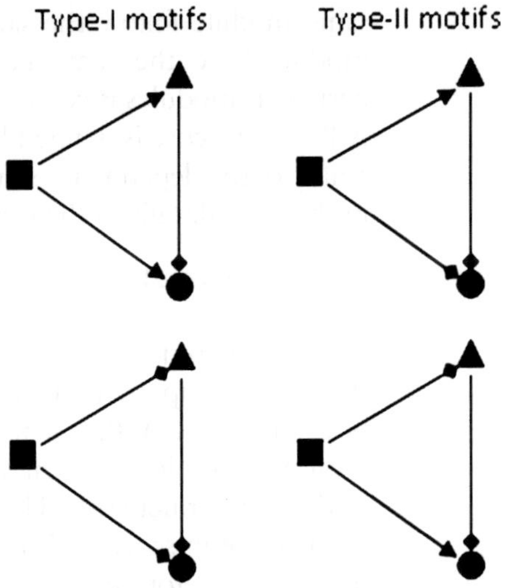

Fig. 1 Possible types of regulatory network motifs comprising TFs (*squares*), miRNAs (*triangles*), and target genes (*circles*). A connection from one entity to another denotes an up-regulation if it ends in an *arrow head*, a down-regulation otherwise

opportunities for therapeutically modulating cardiac function through the manipulation of pathogenic and protective miRNAs. A similar approach was used to analyze miRNA function in the myeloid lineage development and differentiation [15], hyperthermia [16], and human colorectal cancer [17]. A specialized Bayesian network was proposed by Liu et al. [18] to model the miRNA-mediated gene regulation. It was recently found that the miRNA regulation might be condition-specific, that is, an miRNA can perform up-regulation in some cases while it can repress translation in other cases [19]. They took this information into consideration to develop a new framework which can model more complex miRNA–target interactions.

An application of module networks in miRNA function analysis is presented by Bonnet et al. [20]. They select a set of miRNAs as regulators and build a module network of genes, where both genes and miRNAs are grouped into modules constructing a regulatory network among the modules instead of pairwise edges between individual genes. They identified important modules related to prostate cancer in a data of expression profiles obtained from prostate cancer and control samples.

Two popular module network applications are freely available: Genomica [21] and Lemone [22]. Genomica is a Windows-based application where data can be browsed, analyzed, and visualized. It was developed as general-purpose analytical tool to integrate several types of data such as gene expression, sequence, and annotations. Module network application is provided as a Genomica module for the application of necessary statistics to extract transcriptional modules and their relationships with predefined regulators. Lemone is a specialized tool for module network analysis. It does not provide a graphical user interface, rather it allows the user to call necessary commands for step-by-step analysis of expression data. Final modules can be visualized as separate files with enriched GO annotations for each.

3 Bipartite Graphs: Modeling Combinatorial Interactions

The activity of an miRNA can be best described by the behaviors of its targets. Therefore, target identification is crucially important in understanding miRNA function. The algorithms and tools for identifying or predicting of potential targets for a given miRNA are introduced in previous chapters (Chapters 12–14). Although current systems biology can significantly benefit from available target prediction tools, their explanatory power is limited. While the target prediction process is built upon a pairwise relationship between miRNA and its putative target, the case is more complicated. An miRNA can regulate several targets and an mRNA can be simultaneously regulated by several miRNAs. Target prediction tools fail to answer the questions regarding how multiple miRNAs can work in cooperation to regulate a group of genes or a disease-related pathway.

In 2005, Yoon and De Micheli introduced the term miRNA regulatory modules (MRMs) to define the coordinated activity of miRNAs with their targets [23]. They used a special data structure, called weighted bipartite graph or relation graph, to model the pairwise interactions simultaneously with the clusters based on their regulatory behaviors. A bipartite graph is an undirected graph whose vertices can be divided into two disjoint sets U and V such that no two graph vertices within the same set are adjacent and every edge connects a vertex in U to one in V (*see* Fig. 2).

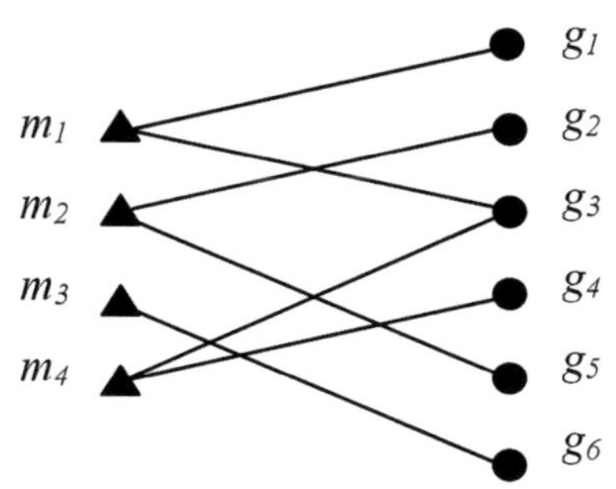

Fig. 2 An example weighted bipartite graph which models the interactions between a set of miRNAs and their target genes. Here, miRNA m_2 has two targets; g_2 and g_5, and g_3 is regulated by two miRNAs; m_1 and m_4 cooperatively

A weighted graph is an appropriate model for representing many-to-many relationships between miRNAs and their targets since one disjoint set corresponds to miRNA nodes, while the other set can represent their targets. In this setup, miRNAs and targets can make up clusters among themselves. The overall method has three major steps: (1) From a pool of single miRNAs and mRNA transcripts, select a set of miRNA–mRNA pairs using target prediction tools. (2) Build a weighted bipartite graph, where the nodes in first disjoint set are miRNAs, the nodes in second disjoint set are mRNAs, and the edges between the nodes correspond to the pairs found in first step. Each pair is weighted using a function which can be determined by principal component analysis on a feature space obtained from miRNA–target duplex formation. The feature space is built over local sequence alignment scores and free energy of putative duplex. (3) Iteratively update the graph and collect a set of regulatory modules, starting from an initial set of miRNAs that have similar binding patterns in the first graph.

The result after creating the bipartite graph model is a set of miRNA–mRNA pairs, each of which corresponds to a distinct MRM with refined duplex scores for each pair in the module. This approach has been extended later by several research groups to improve pairwise selection and module selection. Joung et al. [24] integrated gene expression data to capture instant behavior of miRNAs and target genes. They defined a fitness function based on the weighted sum of an expression coherence score and sequence-based binding scores for predicted miRNA–mRNA duplexes in present modules. Using an evolutionary learning algorithm, the fitness function is optimized with a prior setting of weight parameters from a random initial configuration of modules. The algorithm releases new configurations in each generation and is finalized with a solution which attains best fitness value.

Classical regulatory networks can only reveal the interactions among the entities in a generic way, that is, they cannot explain the dynamic behavior due to varying conditions. However, incorporation of condition specificity in regulatory networks may provide better insights in several cases such as understanding local networks due to specific stress conditions or identifying disease-related genes or miRNAs. Differential expression of several miRNAs in distinct cancer types has been recently reported [25]. To deal with this phenomenon, Liu et al. [26] introduced a condition-specific approach to identify miRNA groups with their targets for normal and cancer samples. Instead of unsupervised construction of unknown number of modules, they attempted to cluster miRNAs into two groups to understand the correlation between miRNA activities and corresponding disease. Their algorithm has two independent steps. First, a set of putative networks, i.e., group of miRNA–target pairs, are identified using sequence-based target predictions. Second, a set of condition-specific modules, which they called functional miRNA regulatory modules (FMRMs), are built using miRNA and gene expression profiles. The model represents miRNA–target relationships via a bipartite graph and employs gene expression data to get an updated graph by pruning the initial one using association rule mining techniques.

Peng et al. [27] considered inverse expression relationships while inferring miRNA–mRNA regulatory modules associated with hepatitis C virus. Similar to Yoon and De Micheli's work, they pruned an initial bipartite graph and use a number of post-processing steps to extract multiple relationships between the miRNAs and their targets. They additionally proposed some integrative evaluation steps before constructing the relation graph. First, an miRNA–mRNA correlation matrix is calculated using the similarities in expressions across samples. Then, a binary relation matrix is deduced based on a series of thresholds which minimize false detection rates. Final the matrix is combined with the predicted binding matrix to infer an initial relation graph.

4 Biclustering: A Multiway Look

Due to their cooperative effects, main focus of the research in functional analysis of miRNAs has shifted to the inference of mechanisms associated with functional groups as opposed to individual miRNAs that exhibit similar expression patterns across a set of conditions or cellular states. One standard way of inferring these groups is to cluster them with respect to their similarity derived from their expression profiles over all given conditions (*see* Chapter 7 for detailed description of clustering). In general, clustering is a useful tool to elucidate the function of miRNAs by collecting the functionally related entities and analyzing their behavior over the enriched patterns on their groups. Since miRNAs involved in the same pathway usually respond to specific stimuli that could appear only in certain conditions, one-way clustering approach may fail to unveil the coordinated activity of miRNAs which act cooperatively in only a subset of given conditions. Biclustering can address this challenge by identifying the clusters of entities that may exhibit similar patterns for only certain conditions but not over all. Since the entities are both grouped and associated with feature or condition subsets, biclustering provides a natural framework with increased interpretability in comparison with traditional one-way clustering.

Biclustering is defined as simultaneous grouping of both row and column sets in a data matrix. More formally, given a set of entities $M=\{m_1, m_2, \ldots, m_n\}$ and a set of conditions $C=\{c_1, c_2, \ldots, c_k\}$, a bicluster B is a subset of (M, C), which contains the entities that exhibit coherent patterns in a subset of conditions. In other words, given the data as a $(n \times k)$ matrix of real values, where each entry $e_{i,j}$

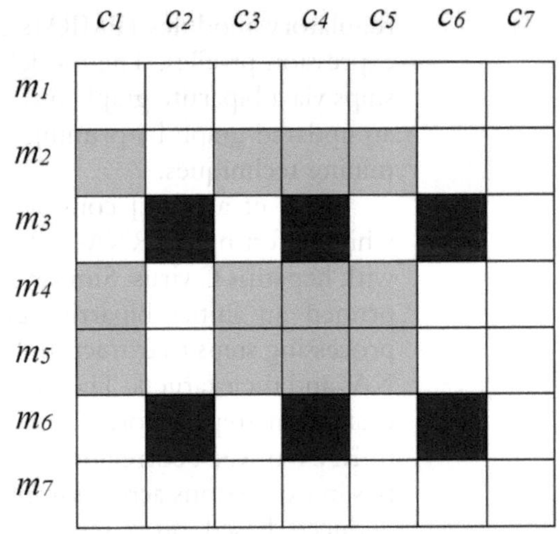

Fig. 3 An example of an miRNA biclustering over a set of conditions. Let each m_i be an miRNA and each c_j be a condition where miRNA expression is measured and let each cell store the corresponding expression values. Two biclusters are shown: In the first one, the miRNAs denoted by m_2, m_4, and m_5 make a bicluster under conditions c_1, c_3, c_4, and c_5. The second bicluster covers m_3 and m_6 under conditions c_2, c_4, and c_6

corresponds to related measurement, e.g., an expression value of the entity m_i, e.g., a gene, under the condition c_j, e.g., a diseased cell line, a bicluster is a submatrix (I, J) that exhibits correlation in its rows and columns. Here, I denotes the set of indices of the rows in a row cluster and J is the set of indices of the columns in a column cluster. The resulting submatrix (I, J) is the bicluster of the entities and conditions, where $|I| \leq n$ and $|J| \leq k$. An example of bicluster is schematically shown in Fig. 3.

Biclustering has been often used for analysis of gene expression data [28]. Early approaches applied biclustering to obtain one bicluster at a time. This can be iterated several times to obtain additional biclusters [29]. Other approaches can identify a set of biclusters simultaneously [30, 31]. In general, finding biclusters can be considered to be an optimization problem, where the best submatrix is chosen which attains the highest level of coherence. Three major approaches exist to identify biclusters. The first and most straightforward method is to use one of the traditional clustering algorithms to obtain clusters in rows and columns, separately, and then to integrate the results by an iterative procedure. This scheme has been used by several biclustering algorithms [30, 32]. The second approach employs the divide-and-conquer paradigm of algorithm development, that is, the problem is divided into a set of similar subproblems, conquered recursively to obtain the solutions of smaller problems, and combined to obtain the global result [33]. The greedy search is the third main approach used for biclustering.

In this scheme, a locally optimal step is taken in the hope that later steps will reach the optimal solution. In the earliest paper on biclustering applied on gene expression data, Cheng and Church [29] used a greedy iterative search algorithm to identify biclusters. They define the goal of the algorithm as to minimize a mean squared residue score H (Eq. 3).

$$H_{I,J} = \frac{1}{|I||J|} \sum_{i \in I, j \in J} h_{i,j}^2 \quad (3)$$

Where $h_{i,j}$ is defined over the elements of the bicluster matrix indexed by I and J, by Eq. 4.

$$h_{i,j} = e_{i,j} - e_{i,J} - e_{I,j} + e_{I,J} \quad (4)$$

Here, $e_{i,J}$ is the mean of elements in row i whose column indices are in J, $e_{I,j}$ is the mean of elements in column j whose row indices are in I, and $e_{I,J}$ is the mean of all entries in the bicluster. A submatrix (I, J) is defined as a δ-bicluster of $H < \delta$ for some δ. To find δ-biclusters, they proposed several greedy steps such as the removal or addition of rows or columns. A number of iterations make it possible to identify biclusters with optimized H and required number of biclusters.

Caldas and Kaski introduced a generative model to identify hierarchical biclusters using a non-parametric Bayesian formulation and applied their method to miRNA expression analysis [34]. Their model can jointly group the samples in a hierarchical manner and assign genes to the nodes in that hierarchy. With this representation scheme, it is possible to view clusters in a tree structure and to explicitly show the features in the data which are most relevant to final groupings. When applied to a set of miRNA expression profiles over healthy tissues, tumors, and cell lines, they could identify differentially expressed genes between different tumor types, in addition to a comprehensive view of connected nodes which consist of miRNAs responsible for distinct mechanisms.

An R implementation, called *biclust*, is available at no cost as a CRAN package to run various biclustering algorithms. It provides methods for preprocessing, visualization, and validation of biclusters. Caldas and Kaski also make their hierarchical biclustering tool available as C++ package [34].

5 Other Tools for miRNA Activity and Function Analysis

Several software tools have been released for functional annotation of miRNAs. MMIA is web-based tool for integrative analysis of miRNA function [35]. It is designed to provide information about miRNA-associated phenotypes and biological functions.

In the first step of MMIA analysis, miRNA expression is used to identify significant differential regulations. The second step uses traditional target prediction tools to find targets for differentially regulated miRNAs. The third step identifies down-regulated mRNAs from mRNA expression data. In the final step gene set analysis is applied to find the intersection between predicted and down-regulated mRNAs. Further analysis can provide output for miRNA-associated diseases, relevant pathways, and gene ontology annotations.

FAME (functional assignment of miRNAs via enrichment) is another framework based on weighted graphs [36] to identify miRNA activities with two alternative applications: (1) Using targeted gene clusters with common annotations to relate miRNA functions and (2) Matching miRNA and mRNA expression profiles to predict miRNA-based regulation. The framework uses target predictions to construct an initial graph, in which the weights of miRNA–mRNA edges are assigned using context scores of miRNA–target sites in prediction results. It then measures the significance of the overlap between predicted targets of miRNAs (or genomic miRNA clusters) and designated target sets on generated permutations of the original graph. Significance tests can unravel some relationships between miRNAs and functional annotations or regulatory effects.

MAGIA is a novel online tool which can provide a combination of several traditional statistical inference methods to analyze whole expression profiles together with target predictions by other tools [37]. It can offer a visual analysis of combinatorial miRNA activity by a regulatory network. Like MMIA, they provide a four-step analysis framework; predicting targets from sequence, analyzing gene expression profiles, building posttranscriptional regulatory networks, functional annotation, and enrichment analysis.

CORNA is a method which considers cooperative effects of miRNAs on gene sets to dissect miRNA functions [38]. It takes a list of genes and an miRNA as input and examines whether there is a regulatory association between the miRNA and the gene set. Having predicted targets in miRbase database, CORNA employs three standard statistical procedures (HyperGeometric test, Fisher's exact test, and Chi-square test) for enrichment analysis and can report a list of associated gene sets for a given miRNA, or alternatively a list of miRNAs which potentially regulate a given gene set. It is implemented as an R package and made freely available in sourceforge.net.

6 Conclusion and Future Directions

Due to the high cost of experimental procedures, computational approaches have been serving as good alternatives for miRNA function analysis like several other bioinformatics applications. In this domain, two major problems have come into prominence: interaction modeling and functional clustering. This chapter introduced fundamental models used in this context. Bipartite graphs are essential tools to model interactions, whereas biclustering approaches are useful in identifying functionally similar or cooperative miRNA groups. Regulatory networks can serve as a means of both interaction prediction and clustering at the same time with properly chosen parameter sets.

It is anticipated that the integration of data from different sources will be a focal point in the analysis of miRNA functions and activities. The need for data integration may serve several challenges from both biological and computational views. Which data to integrate is the main question. We have already witnessed that using expression data with sequence information may enhance the exploratory performance of current models. We expect the contribution of metabolic or other behavioral data in upcoming models.

Analysis of the upstream region of miRNAs will help to understand their implicit regulatory activities and enlighten more complex interactions among regulating and regulated entities. As a result of such distinct actors and interactions in regulation processes, the experimental data obtained may appear in multiple views. Therefore, the increase in the available data will motivate the generalization of biclustering in several ways. Multiway, multisource, and multitask paradigms of computational inference [39] will become more of an issue in future computational miRNA research.

Acknowledgement

This study was supported by the Scientific and Technological Research Council of Turkey (TUBITAK) under the Project 110E160.

References

1. Eisen MB, Spellman PT, Brown PO, Botstein D (1998) Cluster analysis and display of genome-wide expression patterns. Proc Natl Acad Sci USA 95:14863–14868
2. Ben-Dor A, Shamir R, Yakhini Z (1999) Clustering gene expression patterns. J Comput Biol 6:281–297
3. Tompa M, Li N, Bailey TL et al (2005) Assessing computational tools for the discovery of transcription factor binding sites. Nat Biotechnol 23:137–144
4. Bartel DP (2004) MicroRNAs, genomics, biogenesis, mechanism, and function. Cell 116: 281–297
5. Mendes ND, Freitas AT, Sagot MF (2009) Current tools for the identification of miRNA genes and their targets. Nucleic Acids Res 37:2419–2433
6. Saito T, Saetrom P (2010) MicroRNAs—targeting and target prediction. N Biotechnol 27:243–249
7. Alexiou P, Maragkakis M, Papadopoulos GL et al (2009) Lost in translation, an assessment and perspective for computational microRNA target identification. Bioinformatics 25:3049–3055
8. Barbato C, Arisi I, Frizzo ME et al (2009) Computational challenges in miRNATarget predictions, to be or not to be a true target? J Biomed Biotechnol 2009:803069
9. Krek A, Grun D, Poy MN et al (2005) Combinatorial microRNA target predictions. Nat Genet 37:495–500

10. Wang J, Lu M, Qiu C et al (2010) TransmiR, a transcription factor–microRNA regulation database. Nucleic Acids Res 38:D119–D122
11. Heckerman D (1998) Tutorial on learning with Bayesian networks. In: Jordan M (ed) Learning in graphical models. Adaptive computation and machine learning. MIT Press, Massachusetts, pp 301–354
12. Segal E, Shapira M, Regev A et al (2005) Learning module networks. J Mach Learn Res 6:557–588
13. Tsang J, Zhu J, van Oudenaarden A (2007) MicroRNA-mediated feedback and feedforward loops are recurrent network motifs in mammals. Mol Cell 26:753–767
14. Liu N, Olson EN (2010) MicroRNA regulatory networks in cardiovascular development. Dev Cell 18:510–525
15. El Gazzar M, McCall CE (2011) MicroRNAs regulatory networks in myeloid lineage development and differentiation, regulators of the regulators. Immunol Cell Biol. doi:10.1038/icb.2011.74
16. Stingo FC, Chen YA, Vannucci M et al (2010) A Bayesian graphical modeling approach to microRNA regulatory network inference. Ann Appl Stat 4:2024–2048
17. Tang J, Fang J (2009) MicroRNA regulatory network in human colorectal cancer. Mini Rev Med Chem 9:921–926
18. Liu B, Li J, Tsykin A et al (2009) Exploring complex miRNA-mRNA interactions with Bayesian networks by splitting-averaging strategy. BMC Bioinformatics 10:408
19. Vasudevan S, Tong Y, Steitz JA (2007) Switching from repression to activation: MicroRNAs can up-regulate translation. Science 318:1931–1934
20. Bonnet E, Michoel T, Van de Peer Y (2010) Prediction of a gene regulatory network linked to prostate cancer from gene expression, microRNA and clinical data. Bioinformatics 26:i638–i644
21. Segal E, Shapira M, Regev A et al (2003) Module networks, discovering regulatory modules and their condition specific regulators from gene expression data. Nat Genet 34:166–176
22. Michoel T, Maere S, Bonnet E et al (2007) Validating module networks learning algorithms using simulated data. BMC Bioinformatics 8:S5
23. Yoon S, Micheli G (2005) Prediction of regulatory modules comprising microRNAs and target genes. Bioinformatics 21:i93–i100
24. Joung JG, Hwang KB, Nam JW et al (2007) Discovery of microRNA-mRNA modules via population-based probabilistic learning. Bioinformatics 23:1141–1147
25. Lu J, Getz G, Miska EA et al (2005) MicroRNA expression profiles classify human cancers. Nature 435:834–838
26. Liu B, Li J, Tsykin A (2009) Discovery of functional miRNA-mRNA regulatory modules with computational methods. J Biomed Inform 42:685–691
27. Peng X, Li Y, Walters KA et al (2009) Computational identification of hepatitis c virus associated microRNA-mRNA regulatory modules in human livers. BMC Genomics 10:373
28. Madeira SC, Oliveira AL (2004) Biclustering algorithms for biological data analysis, a survey. IEEE/ACM Trans Comput Biol Bioinform 1:24–45
29. Cheng Y, Church GM (2000) Biclustering of expression data. Proc 8th int conf intel syst mol biol, pp 93–103
30. Getz G, Levine E, Domany E (2000) Coupled two-way clustering analysis of gene microarray data. Proc Natl Acad Sci U S A 97: 12079–12084
31. Yang J, Wang W, Wang H (2003) Enhanced biclustering on expression data. Proc 3rd IEEE conf bioinform bioeng, pp 321–327
32. Tang C, Zhang L, Zhang I et al (2001) Interrelated two-way clustering, an unsupervised approach for gene expression data analysis. Proc 2nd IEEE int sym bioinform bioeng, pp 41–48, 2001
33. Hartigan JA (1972) Direct clustering of a data matrix. J Am Stat Assoc 67:123–129
34. Caldas J, Kaski S (2011) Hierarchical generative biclustering for microRNA expression analysis. J Comput Biol 18:251–261
35. Nam S, Li M, Choi K et al (2009) MicroRNA and mRNA integrated analysis (MMIA), a web tool for examining biological functions of microRNA expression. Nucleic Acids Res 37: W356–W362
36. Ulitsky I, Laurent LC, Shamir R (2010) Towards computational prediction of microRNA function and activity. Nucleic Acids Res. doi:10.1093/nar/gkq570
37. Sales G, Coppe A, Bisognin A et al (2010) MAGIA, a web-based tool for miRNA and genes integrated analysis. Nucleic Acids Res 38:W352–W359
38. Wu X, Watson M (2009) CORNA, testing gene lists for regulation by microRNAs. Bioinformatics 25:832–833
39. Huopaniemi I, Suvitaival T, Nikkilä J et al (2010) Multivariate multi-way analysis of multi-source data. Bioinformatics 26:i391–i398

Chapter 16

Analysis of MicroRNA Expression Using Machine Learning

Henry Wirth, Mehmet Volkan Çakir, Lydia Hopp, and Hans Binder

Abstract

The systematic analysis of miRNA expression and its potential mRNA targets constitutes a basal objective in miRNA research in addition to miRNA gene detection and miRNA target prediction. In this chapter we address methodical issues of miRNA expression analysis using self-organizing maps (SOM), a neural network machine learning algorithm with strong visualization and second-level analysis capabilities widely used to categorize large-scale, high-dimensional data. We shortly review selected experimental and theoretical aspects of miRNA expression analysis. Then, the protocol of our SOM method is outlined with special emphasis on miRNA/mRNA coexpression. The method allows extracting differentially expressed RNA transcripts, their functional context, and also characterization of global properties of expression states and profiles. In addition to the separate study of miRNA and mRNA expression landscapes, we propose the combined analysis of both entities using a covariance SOM.

Key words Microarrays, Self-organizing maps, Feature selection, mRNA–miRNA coexpression, Gene set enrichment analysis

1 Introduction

A major goal of using machine learning in research is to recognize complex patterns and to make intelligent decisions based on the data and the particular objective of interest with the help of computers. Usually two research problems involving miRNAs are tackled with computational methods, namely, detecting miRNA genes and predicting miRNA targets. The use of machine learning methods has been shown to improve the outcome of both miRNA gene detection and target prediction (see previous chapters in this volume, ref. 1 for a review, and refs. 2–4 presenting primary works). Here, machine learning approaches typically make use of sequence information (e.g., of short six- to eight-nucleotide miRNA-binding motifs), secondary structure (e.g., of stem–loops using thermodynamic modeling), and phylogenetic conservation to classify possible

Henry Wirth and Mehmet Volkan Çakir have contributed equally to this chapter.

Malik Yousef and Jens Allmer (eds.), *miRNomics: MicroRNA Biology and Computational Analysis*, Methods in Molecular Biology, vol. 1107, DOI 10.1007/978-1-62703-748-8_16, © Springer Science+Business Media New York 2014

candidates according to their relevance as miRNA genes or targets using methods such as hidden Markov models, random forest classifiers, or support vector machines (*see* Chapters 7 and 10).

Recently another machine learning technique, namely, self-organizing maps (SOM), was applied for miRNA target prediction based on clustering of short 3′-untranslated regions [5]. SOM machine learning was developed by Kohonen about 30 years ago [6]. It projects data from high dimensional space to reference vectors of lower dimension. First studies applying SOM to microarray gene expression data were published by Tamayo et al. [7] and Törönen et al. [8]. These, and later applications of the SOM method to expression data, emphasized either a gene-centered perspective to cluster genes or a sample-centered mode to map individual samples onto the SOM grid enabling the classification of samples into a small number of diagnostic or prognostic groups [9, 10]. The SOM method can also be configured in such a way that it combines sample- and gene-centered perspectives [9]. In such application SOM analysis seems not only well suited to accomplish downstream analysis tasks specified above, but it also offers several advantages that make this method superior in many aspects compared to other ones such as correlation clustering or nonnegative matrix factorization [10].

In this contribution we address methodical issues of miRNA expression analysis with the special focus on this combined gene and sample-centered SOM technique. To the best of our knowledge this method was applied here for the first time to the analysis of miRNA expression data. The work is divided into two parts: First, we shortly review selected experimental and theoretical aspects of miRNA expression analysis. Thereafter, our SOM method is presented in more detail with special emphasis on miRNA/mRNA coexpression. In this part the details of the method are exemplified using a combined dataset of miRNA and mRNA expression in healthy and tumor tissue samples taken from Lu et al. [11].

2 Methods of MicroRNA Expression Analysis

2.1 MicroRNA Detection

Overviews over different methods of miRNA detection including Northern blot, in situ detection, bead techniques, and quantum dots are given in refs. 12, 13. Currently, three methods are most commonly applied to measure mRNA and miRNA expression:

1. Real-time quantitative PCR (qPCR).
2. Microarray hybridization.
3. Massively parallel next-generation sequencing (NGS) (*see* Chapter 6, [14–16], and references cited therein).

Application of these methods to miRNAs faces basically two challenges compared to their use in mRNA analysis: the short length of mature miRNA sequences (~18–24 nt) and the nearly identical sequences of miRNAs of the same family whose members can differ by as little as one nucleotide and nevertheless can exhibit differential expression. The specificity required to differentiate between such closely related short RNA fragments surpasses requirements for conventional mRNA detection and raises problems due to the constrained probe design (microarrays) and sequencing errors (NGS). In contrast to mRNA profiling technologies miRNA profiling must also take into account the difference between mature miRNAs and their precursors which also can produce detection signals (which usually address only one transcript, if one neglects alternative splicing).

The technical merits and drawbacks of qPCR, microarrays, and sequencing of miRNAs are similar to their application in mRNA or genomic DNA quantitation. The clear advantage of high-throughput sequencing compared to microarrays is that it is not hindered by the variability of probe affinities and cross hybridization of nearly identical miRNA family members. Moreover, analysis of read patterns allows to identify novel miRNAs [17]. On the other hand, RNA ligation and PCR amplification steps (see below) and also library preparation bear inherent biases paralleled, e.g., by systematic preferential representation of the miRNA complement [18]. Moreover, NGS of miRNAs can be influenced by sequencing errors and often requires search and removal of adapter sequences before the miRNA sequence itself can be elucidated.

As in mRNA analysis, microarrays are still a good choice for a standardized genome-wide assay that is amenable to high-throughput applications. The differences between available platforms (e.g., Agilent; Exiqon, Illumina, Ambion, Combimatrix, Invitrogen, Affymetrix) range from surface chemistry and printing technology, through probe design and labeling techniques to the required amount of material for hybridization and costs (*see* Chapter 6 and [12, 15, 16]). Several attempts have been made in surface chemistry (e.g., with probes containing locked nucleic acid (LNA) bases) and probe structure (e.g., with "stem–loop" probes) to improve the array sensitivity for discriminating mispaired bases in a better way and to equalize the probe affinities for miRNA target binding, however, sometimes with questionable success [15]. MicroRNA microarrays have high intra-platform reproducibility and comparability as opposed to qPCR. However, the current lineup of commercially available miRNA microarray systems fails to show good inter-platform concordance, probably because of severe divergence in stringency of detection call criteria between different platforms [16]. Unlike for mRNA gene expression [19], only few attempts have been made so far to establish rigorous parameters for the evaluation of miRNA microarray platforms using standardized quality measures [16].

Several studies report low degree of overlap in differentially expressed miRNAs between different detection methods, which are not easily attributable to the strength or the weakness of the different platforms [15, 20]. Quantitative PCR, often considered a "gold standard" in the detection and quantification of gene expression, can be used better as a validation rather than as discovery tool because of relatively large number of miRNAs presently known. Even qPCR seems to fail as a validation method of miRNA microarray data [15, 20]. Hence, despite its descriptive name and the fact that qPCR has been repeatedly used as a validation technique of choice, it is not necessarily appropriate to use qPCR data as an absolute "gold standard." The question of a basal standard in miRNA expression awaits further advances in both technology (e.g., deep sequencing) and computation (normalization and downstream analysis algorithms).

2.2 Data Preprocessing Tasks: Calibration and Normalization

Quantitative PCR, microarray hybridization, as well as NGS methods for miRNA detection face the problem of significant technical and experimental bias. Preprocessing aims at minimizing such systematic errors and thus has significant impact on downstream analysis and particularly on the detection of differentially expressed miRNAs.

1. Calibration, the first subtask of preprocessing, aims at removing systematic biases from raw data to get expression estimates which linearly correlate with the "true" RNA transcript abundance separately in each of the samples. This includes method- and platform-specific steps, such as baseline correction and threshold setting for qPCR analyses, background and affinity correction for microarray technology, or filtering for small RNA-sequence data (NGS). We refer the reader to special literature addressing such method-specific calibration issues [19, 21, 22].
2. The second task, normalization, aims at ensuring comparability of the transcript abundance estimates between the different specimen by adjusting the data to batch effects such as different total RNA concentrations, total read counts (NGS), or residual background levels (microarrays). Normalization is crucial since signal levels may be modulated by the RNA extraction yields and inverse transcription and PCR amplification reaction efficiencies in a sample specific way. In general, there is no best-performing normalization method for any of the three miRNA profiling approaches [23]. Several normalization techniques are currently applied, some of which are similar to mRNA profiling normalization methods, while others consider the specifics of miRNA data.

Normalization, using endogenous control probes, represents a simple but powerful strategy. It is based on the selection of reference miRNAs or other small noncoding (nc) RNAs (e.g., small nucleolar RNA) as predefined invariant endogenous controls [24]. The expression levels of the transcripts measured in each of the samples are then simply scaled by the average expression levels of the controls (preferentially in log-scale) in the respective samples by assuming that their variations are solely caused by technical and experimental factors. Noncoding RNAs in contrast to miRNAs might be problematic because they do not mirror the physicochemical properties of mRNAs and because ncRNA abundance might not reflect the overall activity of the miRNA processing machinery. It especially raises problems if the total miRNA level alters as in comparisons of multiple tissues or cell lines [15]. Selection of invariant "housekeeping" miRNAs identified by different algorithms is superior over small ncRNA-based normalization (*see* ref. 23 and references cited therein).

Another normalization strategy uses global RNA expression measures as intrinsic control. It assumes that the overall transcription level is constant and one can use the median or the average expression of all transcripts measured in each sample as a reference. Other, more sophisticated methods such as quantile normalization (leveling the expression frequency distributions) [25] or LOESS [26, 27] (local regression; leveling the local mean) scale the frequency distribution of expression values of all samples or their local, expression-dependent mean, respectively.

Global miRNA expression patterns (and potentially also the expression levels of endogenous controls), however, are thought to change dramatically in response to Drosha and histone deacetylase levels, cell division status, neoplastic transformation, developmental stage(s), circadian rhythms, cellular stress, and other factors. Hence, the assumptions—common to many mRNA expression profiling experiments—that overall RNA transcription is constant and that a low percentage of individual transcripts are changed under different test conditions are mostly not applicable to miRNA studies. The nature of miRNA profiling data which mirrors the distinct biogenesis and physicochemical nature of miRNAs can challenge conventional normalization methods originally developed for mRNA expression data. Innovative approaches are required in order to reconcile miRNA profiling data analyses with the specifics of miRNA biology. This can make use of combinations of qPCR, microarray, and NGS data for mutual validation, possibly circumventing the need for external references.

2.3 Downstream Analysis Tasks

Downstream analysis follows preprocessing. Preprocessing options which were discussed in the previous section do not lead to immediately useful results, and they need to be complemented with appropriate data analysis in order to extract meaning from the acquired data. In the following we discuss these downstream analysis tasks.

1. It includes tasks such as differential analysis, also known as marker selection which is the search for genes that are differentially expressed in distinct phenotypes, treatment conditions, developmental stages, etc. Differential expression can be accessed using different scores such as simple fold change measures or *t*-test statistics (see below).
2. Another task, supervised learning and class prediction, is the search for a gene expression signature that predicts class (phenotype) membership.
3. Class discovery (unsupervised learning) is the search for biologically relevant but unknown groups of samples identified by a gene expression signature or a biologically relevant set of co-expressed genes. The basic methodology for class discovery is clustering.
4. Finally, functional context analysis is the search for sets of genes differentially expressed in distinct phenotypes using enrichment techniques. We address these tasks below in the context of SOM machine learning.

2.4 Data

We selected a dataset consisting of healthy and tumor tissues to illustrate different aspects of expression analysis using SOM machine learning in the form of a case study. This so-called *LU-cancer* dataset contains miRNA and mRNA measurements from the same samples of seven healthy and tumor tissues (colon, kidney, bladder, prostate, uterus, lung, breast) [11]. MicroRNAs were measured using a bead-based profiling method which allowed estimating the abundance of 217 miRNAs. Messenger RNA expression was determined using microarrays.

3 Discovering MicroRNA Expression Phenotypes Using SOM

3.1 Input Data

In general, we analyze the expression levels, E_{nmi}, of $n = 1, \ldots, N$ genes measured under $m = 1, \ldots, M$ different conditions such as different sample types (e.g., tissues or cell lines), time points (e.g., in a time series after perturbation), treatments (e.g., using different chemicals), or patients (e.g., from a cohort study). Each condition defines a different molecular expression phenotype which can be measured in $i = 1, \ldots, R_m$ replicates. Replicates might be technical (e.g., by analyzing the RNA extracts several times) or biological (e.g., by extracting RNA from different equally treated specimen). In addition, we define phenotype classes as groups of samples of a common functional context such as tissue categories (e.g., nervous or muscle tissues and cancer or healthy samples). The choice of classes depends on the study. For miRNA and mRNA expression studies the number of different transcripts is typically about

N=200–1,000 and 10,000–40,000, respectively. Combined datasets thus contain about 2×10^6–4×10^7 pairwise combinations of miRNA/mRNA features. Below we use the termini "transcript" and "gene" as synonyms.

3.2 Preprocessing

In the first step, raw expression data of each of the $\sum_{m=1}^{M} R_m$ measurements are calibrated and normalized. For mRNA GeneChip expression data we used hook calibration of the raw probe intensities combined with quantile normalization of the expression values as described previously [10]. MicroRNA expression data were taken from the original publication and then quantile normalized. The replicated expression values are optionally log-averaged over all replicates i for each condition, $e_{nm} \equiv \log_{10} E_{nm} = \frac{1}{R_m}\sum_{i=1}^{R_m}\log_{10} E_{nmi}$. Alternatively each replicate can be processed individually to characterize the specifics of its expression. In this case the sample index m also runs over the replicates. Finally, the log-expression values of each transcript were centered with respect to their mean expression, $e_{n\bullet}$, averaged over all conditions studied, $\Delta e_{nm} \equiv e_{nm} - e_{n\bullet}$.

3.3 Training the SOM

SOM machine learning was applied to all preprocessed expression data. The expression data are considered as N vectors of dimensionality M defining the expression profiles of the genes over all phenotypes studied, $\Delta\vec{e}_n \equiv \{\Delta e_{n1},\ldots,\Delta e_{nm},\ldots,\Delta e_{nM}\}$ (n=1, …, N). The algorithm initializes K so-called metagene expression profiles also representing vectors of length M, $\Delta\vec{e}_k(t=0) \equiv \{\Delta e_{k1},\ldots,\Delta e_{kM}\}$ (k=1, …, K, we use the same symbol Δe as above but substitute the gene index n by the metagene index k; the argument t=0, …, T denotes the iteration step). The metagenes are arranged in a two-dimensional grid of rectangular topology $K=K_x{\cdot}K_y$ with K_x=10–60 and $K_y \approx K_x$ tiles per x- and y-dimension, respectively. Then a single gene n' is picked from the list, and its profile vector $\Delta\vec{e}_{n'}$ is compared with all metagene profiles using the Euclidean distance $\|\ldots\|$ as similarity measure. The gene picked is associated with the metagene profile of closest similarity, $\min\{\Delta\vec{e}_{n'} - \Delta\vec{e}_k \|\}$ for $k' = k$. This "winner" metagene profile k' is then modified, such that it a bit more closely resembles the expression profile of the selected gene. In addition, the neighboring metagene vectors in the two-dimensional grid adjacent to this winning metagene are also modified, so that they also resemble the expression vector a little more closely. The update rule for the metagene vectors at step t+1 can be written as $\Delta\vec{e}_k(t+1) = \Delta\vec{e}_k(t) + \eta(t)\cdot h_{k'k}\cdot(\Delta\vec{e}_{n'} - \Delta\vec{e}_k)$ where $0<\eta(t)<1$ is the learning rate decaying with progressive iteration. $h_{k'k}$ denotes the neighborhood kernel around the winner metagene. Different neighborhood kernels such as bubble (only nearest neighbors are considered) or Gaussian (i.e., a smooth decaying neighborhood) can be chosen. This process is applied to all genes and repeated a few hundred thousand times.

The radius of considered neighbors decreases with progressing iterations. As a consequence, less metagene vectors are affected by smaller amounts of change. The metagene vectors therefore asymptotically settle down. The resulting map becomes organized because the similarity of neighboring metagenes decreases with increasing distance in the map. The algorithm ensures that all "single" genes are assigned to "their" metagene vector of closest similarity. The training of the SOM also ensures that the obtained metagene profiles cover the manifold of different single-gene profiles seen in the experiment.

3.4 Staining the SOM: Individual Phenotype Portraits

In our application the method sorts the individual genes into K metagene clusters. Each cluster is characterized by one metagene profile which is used for visualizing the expression pattern of each phenotype. To create an image that portrays the individual expression state of all genes in one of the phenotypes we first normalize the metagene expression data to values between −1 and +1 according to $\Delta e_{km}^{norm} = 2\left(\Delta e_{km} - \Delta e_{\bullet m}^{\min}\right) / \left(\Delta e_{\bullet m}^{\max} - \Delta e_{\bullet m}^{\min}\right) - 1$ where $\Delta e_{\bullet m}^{\max/\min}$ denotes the maximum/minimum values of Δe_{km} for a given m. Then, each tile of the grid is stained according to its $\Delta e_{km}{}^{\text{norm}}$ value by applying an appropriate color (or grayscale) code. Our standard "logFC"-scale linearly transforms the normalized logged fold change, $\text{logFC} = \Delta e_{km}{}^{\text{norm}}$, into suitable colors for increasing $\Delta e_{km} \geq 0$ and for decreasing $\Delta e_{km} \leq 0$. Other color codes can be applied to highlight, for example, intermediate and low expression degrees (for details *see* ref. 10, 28).

The SOM algorithm arranges similar metagene profiles together into neighbored tiles of the map, whereas more different ones are located at more distant positions. In consequence, neighbored metagenes tend to be colored similarly owing to their similar expression values. Therefore, the obtained mosaic portraits show typically a smooth blurry texture with spotlike regions referring to clusters of over- and underexpressed metagenes. These blurry images portray the expression landscape of each particular phenotype in terms of a visual image. Metagenes from the same spot are co-expressed in the experimental series, whereas different, well-separated overexpression spots in the same image refer to metagenes commonly overexpressed in the particular phenotype but differently expressed in other phenotypes.

3.5 Clustering the Metagenes: Spots and Blurs

Figure 1 shows the tissue-specific SOM portraits of miRNA and mRNA expression in normal and cancer samples. The spots are clusters of over/underexpressed metagenes in their respective samples. In general, the spot patterns of the individual portraits are relatively heterogeneous. Mean subtype-specific portraits are calculated as average value of each metagene expression over all phenotype portraits of one class, $\langle \Delta e_{km} \rangle_{m \in \text{class}}$. The averaged portraits (see large mosaics in the right part of Fig. 1) reveal an antagonistic pattern of up- and downregulated spots in normal and diseased tissues.

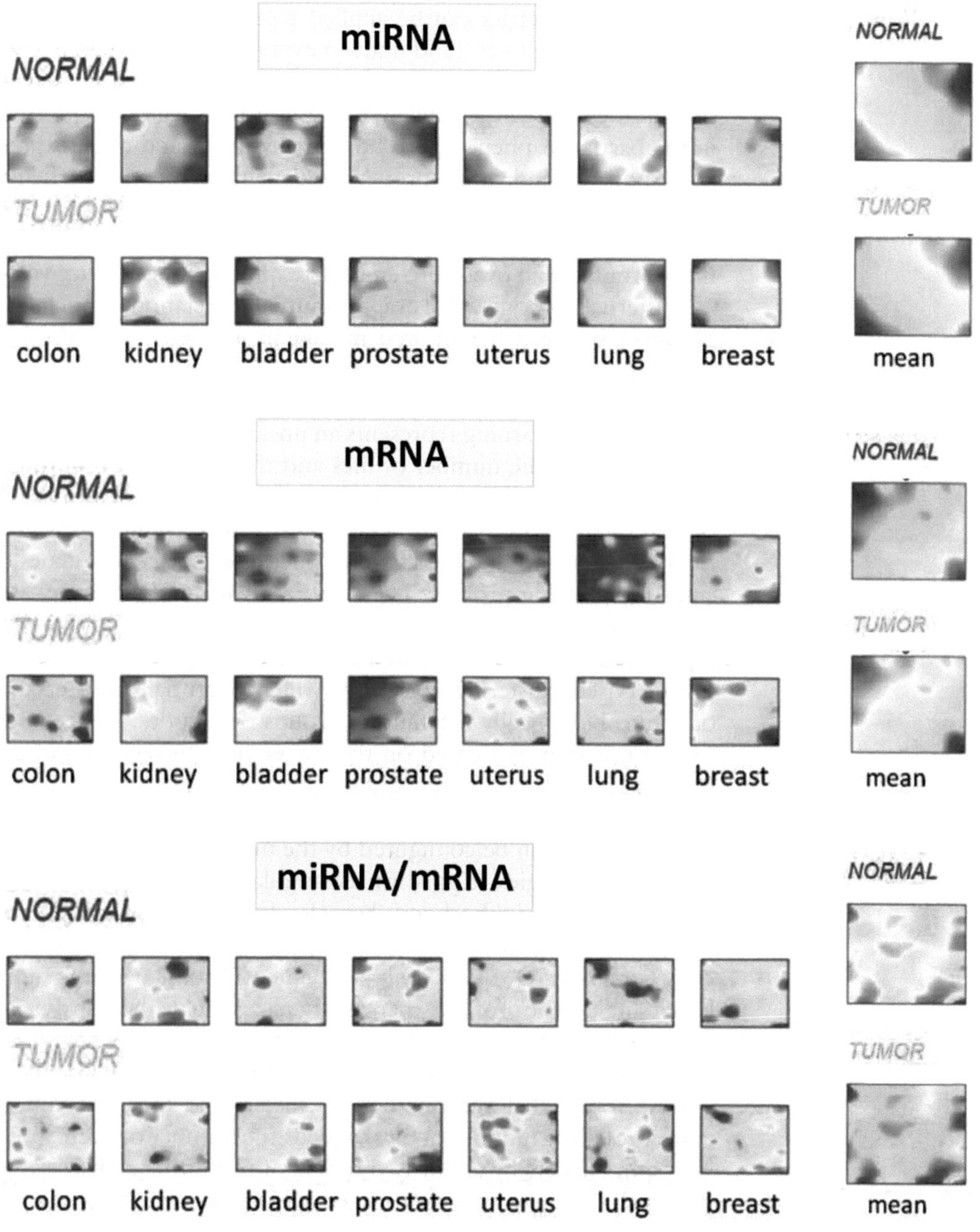

Fig. 1 Gallery of miRNA, mRNA, and combined miRNA/mRNA expression portraits of normal and tumor tissues taken from the LU dataset. The larger portraits on the *right* are mean portraits averaged over all individual normal and cancer portraits, respectively

In the miRNA and mRNA maps only two to three upregulated spots in normal tissues and downregulated in tumor tissues (and vice versa) can be observed. They are located preferentially in opposite corners of the map.

Different metrics can be applied to select metagene clusters. Firstly, we define over- (and under-) expression spots by applying a simple percentile criterion which selects a certain fraction (usually 2 %) of the metagenes showing the largest (or smallest) expression in the particular phenotype. The obtained over- and underexpression spots are individual properties depending on the particular metagene expression in each sample. They can change their size from phenotype to phenotype, and they can even disappear or transform from an over- into an underexpression spot or vice versa.

Alternatively one can also apply mutual correlation or Euclidean distances between neighbored metagene profiles of the SOM as similarity measures for appropriate clustering [10].

3.6 Adjusting the SOM

SOM machine learning represents an unsupervised clustering algorithm whereby the number of tiles and thus the resolution of the map are predefined by the researcher and therefore constitutes the option of supervised adjustment of the results. Neighboring tiles might cluster into one spot together because they collect genes of similar expression profiles. These spots, their number, shape, and size, depend on the intrinsic expression landscape of the phenotypes studied. In this sense SOM spot clustering is a higher order unsupervised clustering algorithm which potentially clusters the data into biologically meaningful groups or "modes." This mode selection is however based on the underlying "pixelation" of the expression landscape which should be chosen such that the SOM algorithm produces a stable and consistent spot pattern.

The SOM can be configured by the number of tiles per image, different topologies (e.g., with rectangular or hexagonal lattices), and different neighborhood kernels describing the range and strength of interactions between the nodes during the training process. For small SOM sizes each metagene will contain a large number of single-gene profiles, whereas large sizes enable the distribution of the genes over a larger number of metagene clusters which more specifically adapt to details of the expression landscape.

We found that the number of clusters and their assignment converges if the number of tiles exceeds the number of overexpression spot clusters by about two orders of magnitude. This asymptotic behavior indicates that larger SOM sizes essentially do not further improve the information content of the map and that the obtained clusters indeed reflect intrinsic properties of the overall expression pattern. Alternative topologies such as the hexagonal ones and different neighborhood kernels only weakly affect the obtained spot textures (see supplementary text in [10]).

3.7 Messenger RNA/miRNA Coexpression

The SOM training described in the previous subsection applies to differential expression of single genes. Hence, mRNA and miRNA data are treated separately providing separate SOMs. One can combine both data if measured under the same series of conditions by substituting Δe_{nm} by all pairwise products of the expression values of the mRNA and miRNA genes, $\text{cov}_{n1n2m} = \Delta e^{mRNA}_{n1m} \cdot \Delta e^{miRNA}_{n2m}$ ($n1 = 1, \ldots, N1$ and $n2 = 1, \ldots, N2$ are the gene indices of mRNA and miRNA expression values, respectively). The size of the data increases from $N1 \sim 10^4$ (mRNA) and $N2 \sim 10^3$ (miRNA) to $N1 \cdot N2 \sim 10^7$ (combined miRNA/mRNA) which usually exceeds the maximum data capacity of our software application ($\sim 5 \times 10^4$) running on desktop PC machines by several orders of magnitude.

One option to handle this problem is to reduce the number of features by appropriate filtering. We previously showed that SOM appropriately compresses the original data [10]. Particularly we study miRNA/mRNA metagene pairings instead of pairs of single genes, i.e., $\text{cov}_{k1k2m} = \Delta e^{mRNA}_{k1m} \cdot \Delta e^{miRNA}_{k2m}$ ($k1 = 1, \ldots, K1$ and $k2 = 1, \ldots, K2$ are the metagene indices of the mRNA and miRNA expression SOM, respectively). Then, cov_{k1k2m} defines the sample-specific covariance term of both datasets which, in turn, is related to the correlation coefficient of the metagene profiles $k1$ and $k2$, $r_{k1k2} = \sum_m \text{cov}_{k1k2m} / \sqrt{\sum_m \left(\Delta e^{mRNA}_{k1m}\right)^2 \cdot \sum_m \left(\Delta e^{miRNA}_{k2m}\right)^2}$. Large positive values of cov_{k1k2m} thus indicate concerted up- or downregulation of mRNA and miRNA expression, whereas negative values refer to antagonistic changes of both RNA species in sample m. The size of the combined data is $K1 \cdot K2 \sim 10^6$, which requires further reduction. We filtered miRNA and mRNA metagene profiles for the largest variance and population with single genes. Particularly, metagenes are selected whose variance and population exceed the respective mean value averaged over all metagenes.

SOM training then provides meta profiles for the combined data, cov_{km}, in analogy to the separate datasets as described in the previous subsection. Each meta-covariance feature describes a characteristic profile of the combined expression data observed in the dataset. The respective microcluster contains combinations of mRNA and miRNA with similar profiles of their combined expression values as the meta feature, i.e., $\text{cov}_{k1k2m} \propto \text{cov}_{km}$.

The third row of images in Fig. 1 shows the combined SOM portraits of miRNA/mRNA coexpression. The individual portraits and especially the averaged map are more heterogeneous showing more spots than the respective miRNA and mRNA maps. The average combined portrait shows more than five spots of preferentially concerted changes in normal tissues and anti-concerted changes in tumor samples. These different trends presumably reflect deregulation of mRNA and miRNA expression in the respective spots. Concerted changes between both species in healthy tissues change into anti-concerted changes in the tumor samples.

All subsequent analyses apply to metagenes of single differential expression, Δe_{km}, and to metagenes of combined differential expression, cov_{km}, as well. Special analyses considering the pairwise character of the combined data are addressed separately below.

Another alternative option for studying miRNA/mRNA coexpression makes use of spot–spot correlation coefficients, $r_{s1s2} = \sum_m \left(\Delta e_{s1m}^{mRNA} \cdot \Delta e_{s2m}^{miRNA}\right) / \sqrt{\sum_m \left(\Delta e_{s1m}^{mRNA}\right)^2 \cdot \sum_m \left(\Delta e_{s2m}^{miRNA}\right)^2}$, where Δe_{s1m}^{mRNA} is the mean expression of spot and $s1$ is averaged over all included metagenes. The cross correlation coefficient is determined for all pairwise combinations of miRNA and mRNA spots taken from the respective overexpression summary maps (see next subsection, $s1$ and $s2$ are the respective spot indices). This analysis reduces the number of combined features to about 10^2 due to the relatively small number of spots identified in the miRNA and mRNA SOM. An example of such analysis is provided in the accompanying contribution [29].

3.8 Spot Overviews

The texture of the SOM visualizes "local" expression properties in terms of spots due to high and low expression levels in the individual phenotypes. For an overview, all overexpression spots observed in minimum one of the phenotypes are selected into one overexpression summary map. The respective underexpression summary map provides an overview over all observed underexpressed spots.

Figure 2 shows the over- and underexpression summary maps of the LU dataset. Note that the spot patterns are more diverse compared with the averaged maps shown in Fig. 1 because alternating over- and underexpression data in the set of samples are not removed by averaging. The single (mRNA and miRNA) data maps illustrate that regions overexpressed in normal tissues typically become underexpressed in tumor samples and vice versa. Spots showing such antagonistic changes are candidates for further in-detail tumor-relevant features. The combined covariance data provides different patterns of the over- and underexpression spots.

The abundance of each spot ($s = A, B, \ldots$) in the individual portraits is calculated as its relative frequency of appearance in the samples of each class ($c = N, T$), $x_{sc} = n_{sc}/N_c$, where the numerator and denominator define the number of sample portraits n_{cs} which show a particular spot and the total number of samples per class N_c, respectively. The spot abundances are represented as stacked bars for each spot. The integral abundance, $X_s = \sum x_{sc}$, can be interpreted as the mean number of classes showing the respective spot. Its maximum possible value equals the number of classes considered.

The respective bar chart in Fig. 2 shows the clear preference of the selected spots to overexpress metagenes in normal tissues (N_UP) and to underexpress these metagenes in tumor tissues (T_DN) and vice versa (i.e., N_DN and T_UP). These class-specific spots tend to accumulate in regions detected as differentially expressed in the averaged portraits (see ellipses in Fig. 2 referring to N_UP and T_UP spots, respectively). On the other hand, T_UP and N_UP mix in the right upper corner of the mRNA-overexpression map

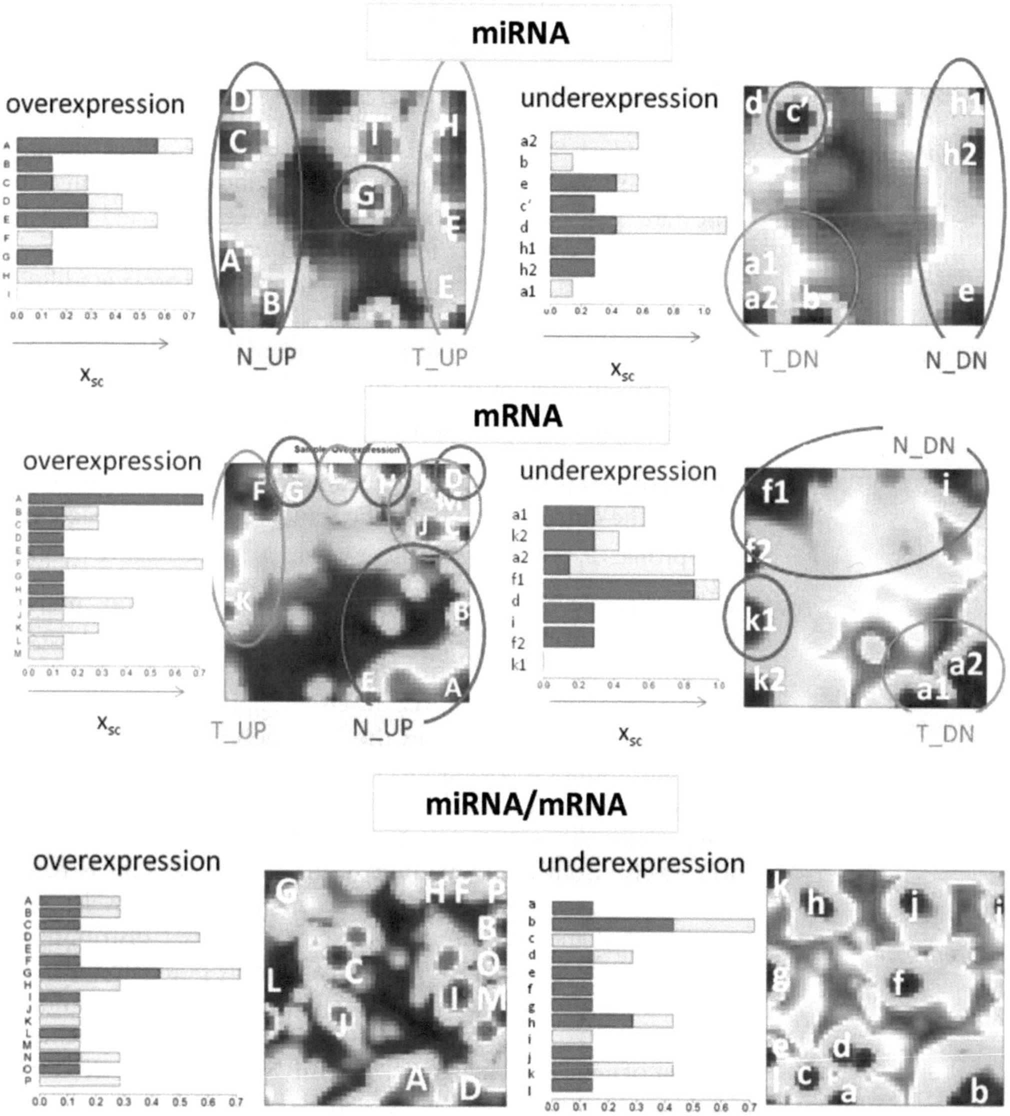

Fig. 2 Over- and underexpression spot summary maps of the miRNA, mRNA, and combined miRNA/mRNA expression portraits shown in Fig. 1. Spots are annotated using capital letters (overexpression) or lower case letters (underexpression) where the latter ones are chosen to agree with respective overexpression spot annotation for spots having a big overlap of metagenes in both summary maps. Regions up- or downregulated in normal (N_up, N_down) and tumor (T_up, T_down) samples are indicated by the *ellipses*. The barplots provide the fractional abundance of each spot in healthy (*dark grey*) and tumor (*light grey*) tissues

reflecting more complex patterns than can be resolved with this study (Fig. 1).

The spots observed in the combined datasets are more or less unique for the individual samples, i.e., only a few of them are found in more than one sample. "Coexpression" spots are more abundant in tumor samples, whereas "antiexpression" spots are more abundant in normal samples. The non-equivalence of "over-" and "underexpression" spots reflects the fact that concerted modes dominate in tumor samples whereas anti-concerted modes dominate in healthy tissues.

3.9 Supporting Maps

The individual SOM portraits partly mask information about the single-gene level. For example, it remains unclear how many single genes are associated with one particular metagene, and thus how importantly it contributes to the overall expression landscape. We therefore defined a series of supporting maps which provide additional information about selected properties of the metagene miniclusters:

The population map plots the number of real genes per metagene in logarithmic scale, log n_k. The variance map illustrates the variability of the metagene profiles, $\text{var}_k = \sum_m (\Delta e_{km} - \Delta e_k)^2 / (M-1)$, where $\Delta e_{k\bullet} = 0$ is the respective mean expression averaged over the profile. The entropy map plots the standard entropy of each metagene profile, $h_k = -\sum_m p_{km} \log_2 p_{km}$, where p_{km} is the relative frequency of the three levels of gene expression: overexpression, underexpression, and non-differential expression of metagene k. Therefore expression values of the metagene profiles of each sample are assigned to one of the three levels by application of a defined threshold (here the 25th and the 75th percentile of all metagene expression values was used). h_m is restricted to values in the interval $[0, \log_2 3]$. An entropy value of 0 represents a perfectly "ordered" state, where all metagenes are assigned to only one of the expression levels. Contrary, maximum value of $\log_2 3 \approx 1.58$ is reached when metagenes uniformly distribute over the three levels.

The 217 miRNAs studied in the LU dataset distribute over $K = 30 \times 30$ metagenes giving rise to a sparsely populated map with a series of empty metagenes (Fig. 3, left part). The mean number of single genes per metagene is $G/M \sim 0.24$, with G being the number of genes per metagene. The mRNA map of size $K = 50 \times 50$ contains 15,500 single genes. It is much denser populated ($G/M \sim 6.2$) with less empty metagenes which, however, accumulate into larger empty areas. These empty regions usually separate different types of profiles defining different modes of regulation. Such modes appear as separate spots of highly variant metagenes in the variance map (Fig. 3, middle part). These maps show that the (Euclidean distance-based) SOM algorithm clusters not only correlated expression profiles in different regions of the SOM but also genes of virtually invariant profiles. These two groups of profiles tend to occupy different regions either along the edges or in the central area of the mosaic image, respectively.

The combined map of size $K = 50 \times 50$ collects 6,100 miRNA/mRNA features ($G/M \sim 2.4$). It is much more fragmented into regulatory modes as indicated by the spot pattern of the variance map. The spots refer to relatively invariant meta profiles forming another kind of separators between different regulatory modes.

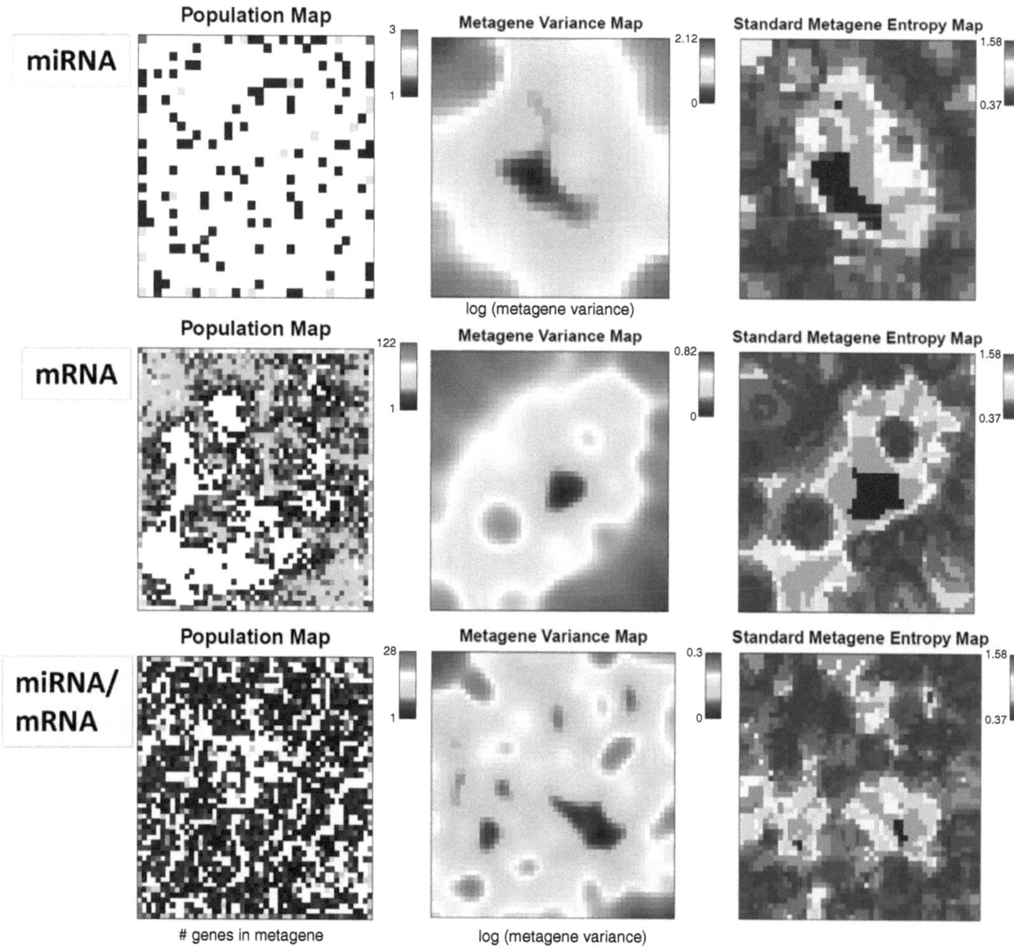

Fig. 3 Supporting maps miRNA, mRNA, and combined miRNA/mRNA expression data of the LU dataset

The entropy maps (Fig. 3, right part) closely resemble the respective variance maps. The color of each pixel visualizes the information content of each metagene profile. Invariant profiles contain no sample-specific information (low entropy), whereas variant profiles are more informative (high entropy). Interestingly, the entropy map is more structured than the variance map.

3.10 Global Portrait Characteristics

In order to estimate global properties of the expression landscape of every phenotype we calculated the variance of metagene expression values in each SOM image, $\mathrm{var}_m = \sum_k (\Delta e_{km} - \Delta e_{m})^2 / (K-1)$, and its entropy, $h_m = -\sum_k p_{km} \log_2 p_{km}$, where p_{km} is the relative frequency of expression as described above for the supporting maps. Here, the relative frequency refers to the expression state m and not to the expression profile k. The global entropy thus characterizes the information content of each portrait. Both, the variance and the entropy assess the expression landscape of phenotype m as seen

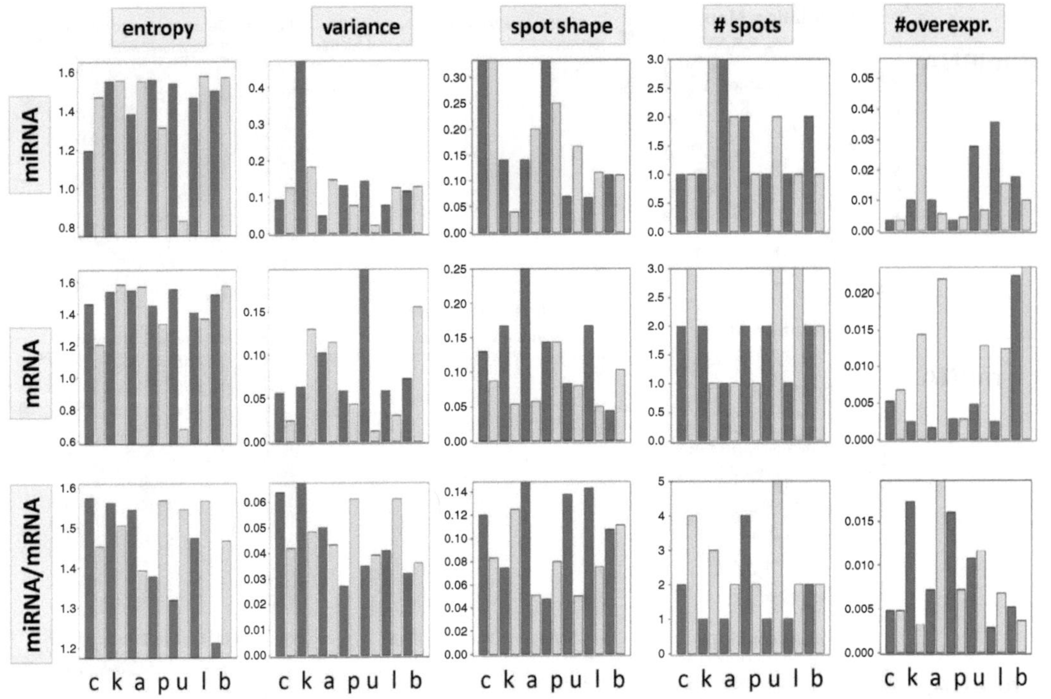

Fig. 4 Sample-related metagene entropy and variance, spot shape, spot number, and area of overexpressed regions of the sample portraits of the LU dataset of healthy (*dark grey*) and tumor (*light grey*) tissues (*c* colon, *k* kidney, *a* bladder, *p* prostate, *u* uterus, *l* lung, *b* breast)

by the SOM portrait. The variance estimates the variability of the metagene expression and the entropy its information content or, in other words, its degree of ordering.

The bar chart in Fig. 4 illustrates that entropy and variance of the expression states mutually correlate. Entropies accentuate highly ordered states (low entropy), whereas variances accentuate strongly variable ones. Entropies and variances of miRNA and mRNA expression states are almost comparable in their order of magnitude reflecting similar complexities of the respective expression landscapes. The combined miRNA/mRNA covariance patterns slightly lose information (i.e., they become more evenly distributed) for part of the cancer tissues (for bladder, prostate, uterus, lung, and breast cancer). This trend might reflect the partial loss of mutual co-regulation between miRNA and mRNA expression in the diseased tissues.

Other global properties of the expression landscapes are the average spot number detected per class, the mean "shape" of the spots, and the fraction of overexpressed metagenes. The shape is defined as $\text{shape}_m = A_m / L_m^2$, where A_m denotes the number of tiles included in all spots observed and L_m is the number of tiles forming the border of the spots with at minimum one adjacent tile outside the spots. It judges the fuzziness of the observed spots in the portraits. One finds that the number of spots per mRNA portrait (and thus the number of distinct regulatory modes) slightly exceeds that of the miRNA portraits. Interestingly, the number of overexpressed miRNA metagenes is smaller in most of the cancer tissues than in the respective healthy tissues. This relation reverses for the number of overexpressed mRNA metagenes: i.e., an increased number of overexpressed metagenes can be observed in cancer. This trend presumably reflects the antagonistic effect of miRNA and mRNA expression expected. Particularly, global downregulation of miRNA expression in cancer (compared with healthy tissues) associates with global upregulation of mRNA expression. The global downregulation of miRNA expression in tumors was reported also in the original paper [11]. It has been hypothesized that global miRNA expression reflects the state of cellular differentiation and that its abrogation is a hallmark of cancer onset.

This trend also corresponds with the loss of information of the covariance landscape discussed above. Note that the number of spots observed in the cancer covariance landscapes in most cases exceeds that observed for normal tissues. This difference can be interpreted in terms of a loss of concerted expression of both miRNA and mRNA. The larger number of spots in the cancer samples is paralleled by a decreased *shape* parameter which simply reflects the increase of fuzziness with increasing spot number.

The global characteristics of the expression landscapes of miRNA, mRNA, and of their combination thus describe distortions of the gene regulation patterns due to disease.

3.11 Phenotype Similarity Analysis

This task aims at establishing the mutual relations between the phenotypes studied to group them into different classes. Particularly, we analyze the hierarchy of similarities and estimate the mutual distances between the expression states. Similarity analysis compares the expression states as seen by the SOM portraits. It consequently uses the metagenes instead of single genes as the basal dataset. Using meta- instead of single genes is advantageous because it improves the representativeness and resolution of the results [10].

We applied second-level SOM analysis as proposed by Guo et al. [30] as a first option to visualize the similarity relations between the individual SOM metagene expression patterns. The method clusters the samples and not the genes as in first-level SOM analysis. In addition, we characterize similarities using the neighbor-joining algorithm based on the Euclidean distances in terms of similarity trees [31].

The separate miRNA and mRNA data well separates healthy and tumor samples in both plots in most cases (Fig. 5). This result reflects the tumor-specific over- and underexpression of selected spots discussed above (*see* Fig. 2). Note that the normal and tumor samples of colon (miRNA) and prostate (mRNA) are found at the

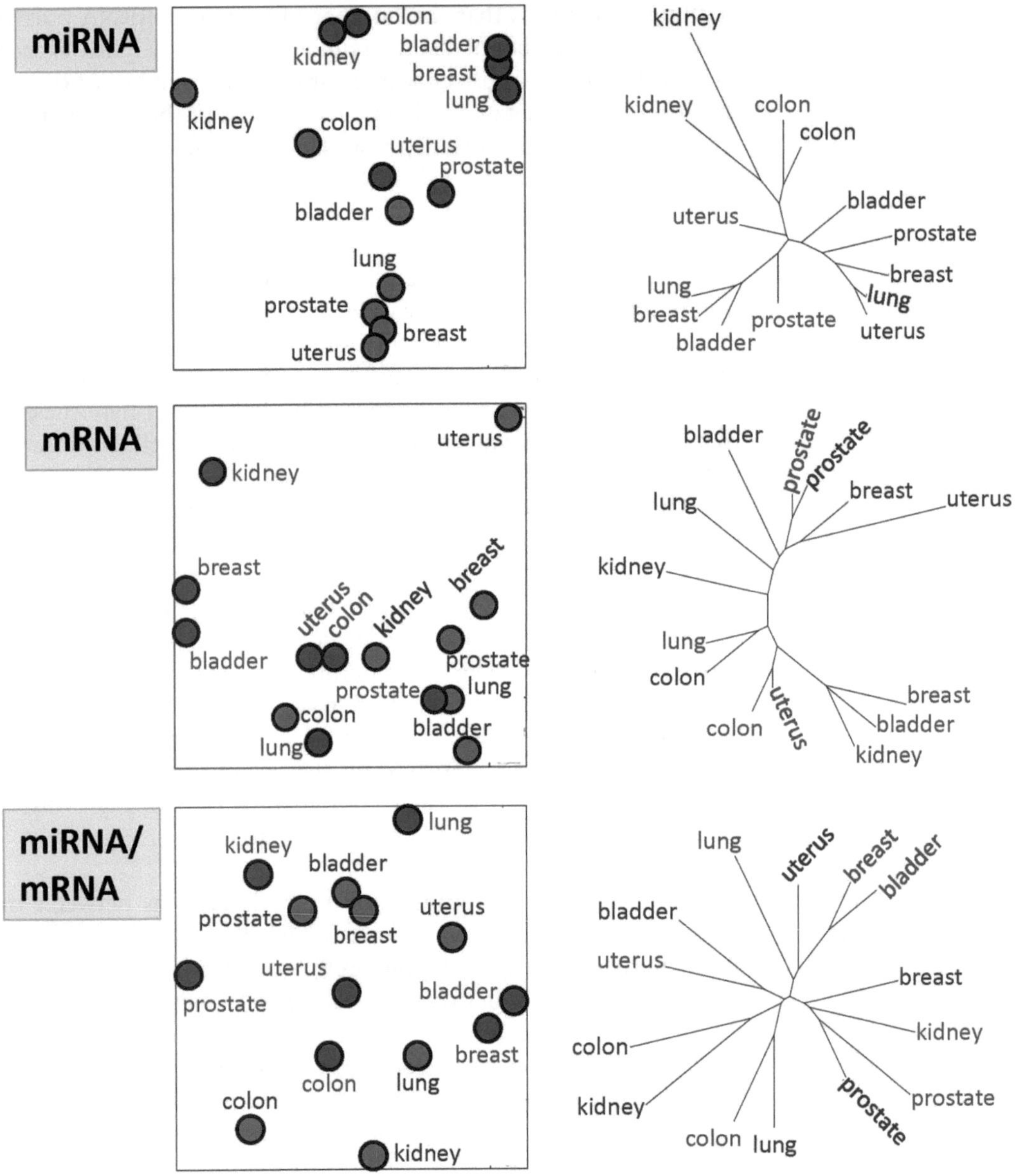

Fig. 5 Similarity analysis using second-level SOM (*left part*) and nearest neighbor joining trees (*right part*) of miRNA, mRNA, and the combined covariance expression landscapes of healthy (*dark grey*) and tumor (*light grey*) samples

same branch of the respective trees reflecting the close similarities of the respective SOM portraits (*see* Fig. 1). The question about the origin of this result is beyond our study. They possibly reflect weak tumor effects or strong contamination of the tumor samples with healthy tissue. The combined miRNA/mRNA data allows virtually no differentiation between tumor and healthy tissues based on a common set of features. Essentially each sample obeys its own specifics. This result reflects the lack of clearly resolved tumor-specific spots in the sample portraits (Fig. 1).

3.12 Differential Expression and Concordance Analysis

The tasks described above characterize the metagene expression landscapes. Usually researchers are interested to select single genes associated with each of the detected spot clusters. Two different analyses are implemented in our method. Concordance analysis estimates the similarity between each metagene profile and the profiles of the associated single genes and ranks them with decreasing agreement using either correlation- or distance-based significance scores for each spot cluster. Concordance analysis thus refers to the whole profiles of all different expression states studied. In contrast, differential expression analysis estimates the most prominently up- and/or downregulated single genes in each spot cluster using scores based on fold change or *t*-statistics.

Correlation-based concordance simply uses the correlation coefficient between each single gene and the respective metagene, $r_{nk} = \sum_m (\Delta e_{nm} \cdot \Delta e_{km}) / \sqrt{\sum_m \Delta e_{nm}^2 \cdot \sum_m \Delta e_{km}^2}$, and then estimates the *p*-value for each gene using the *t*-statistics, $t_{nk} = r_{nk} / \sqrt{(1 - r_{nk}{}^2) / (M - 2)}$. Distance-based concordance uses the sum of squared normalized residual expression, $d_{nk}^2 = \sum_m \left((\Delta e_{nm} - \Delta e_{km})^2 / SD_{nm}{}^2 \right) / (M - 1)$ (SD is the regularized standard deviation of the gene expression in state *m*, see below). Significance is then estimated using χ^2 statistics. The latter distance-based measure allows to identify similarities between virtually invariant profiles, whereas correlation-based measures preferentially select highly variant profiles.

For differential expression analysis a large multitude of various methods are available to assess statistical significance. As the standard method we apply a regularized *t*-score on the single-gene level, $t_{nm} = \Delta e_{nm} / \left(\sqrt{SD_{nm}} / R_m \right)$. The regularized standard deviation, $SD_{nm} \approx \sqrt{\lambda \sigma_{nm}{}^2 + (1 - \lambda) \sigma^{LPE} (e_{nm})^2}$, is calculated as weighted mean of the "individual" standard deviation of the expression of each gene (σ_{nm}) and of a locally pooled error value ($\sigma^{LPE}(e_{nm})$). Both values are combined using the empirical scaling factor $\lambda = 0.5$. Such regularized *t*-scores consistently lead to relatively accurate gene rankings which might outperform simple *t*-statistics or FC scores [32]. The regularized *t*-statistics transforms into *p*-values assuming Student's *t*-distribution. They estimate the significance of differential expression for each gene in a single test. Consideration of the density distribution of the *p*-values of all genes in each phenotype allows to transform the *p*-values into false discovery rates to control the number of false discoveries in the multiple testing problem [33]. Differential features extracted from the LU dataset are given in the accompanying paper [29].

3.13 Gene Set Enrichment Analysis

Gene set analysis aims at evaluating the relevance of selected predefined sets of genes in the expression landscapes of the phenotypes studied. A gene set usually collects genes of common functional context.

This context is given by independent knowledge such as the Gene Ontology (GO) classification (e.g., according to selected GO terms such as "biological process," "molecular function," or "cellular component"), chromosome location, involvement in biochemical pathways, or independent gene expression studies on diseases or toxic effects. For miRNA/mRNA coexpression studies mRNA targets for one selected miRNA provide specific mRNA target sets or vice versa. Similarly, sets of miRNA affecting the same mRNA are collected into miRNA sets.

Basically, gene set analysis estimates the enrichment of genes of the set within a list of genes which is obtained independently. Our SOM method provides a natural choice of gene lists in terms of the genes contained in the spot clusters defined above. Particularly, each gene studied is classified according to two memberships leading to a 2×2 contingency table for further testing: firstly, its membership in the set of functionally related genes of length N_{set} (N_+ "positive" genes in list *and* set and ($N_{\text{set}} - N_+$) "negative" genes in set but *not* in list) and, secondly, its membership in the respective list of length N_{list} (($N_{\text{list}} - N_+$) genes in list but *not* in set ($N - (N_{\text{set}} + N_{\text{list}}) + N_+$) genes neither in list nor in set). The intersection of the set and the list is given by the number of "positive" genes, N_+. For any gene set, right-tailed modified Fisher's exact test was used to determine whether the number of genes with this set is overrepresented in a particular list of genes included in a spot cluster. The hypergeometric distribution then provides a p-value for each set and spot which estimates the cumulative probability to find a stronger overlap between the genes in a spot cluster and the set than expected by chance given a certain total number N of genes studied [28]. This "overrepresentation" analysis assesses the probability to find more members of the set in the list compared with their random appearance.

3.14 Program and Availability

The SOM method is implemented as R program "oposSOM" available on Comprehensive R Archive Network (CRAN, http://cran.r-project.org/) and on our website http://som.izbi.uni-leipzig.de.

4 Notes, Conclusions, and Outlook

SOM machine learning enables analysis of miRNA expression landscapes. The method extracts differentially expressed single features and their functional context and also characterizes global properties of expression states and profiles. Despite the relatively small number of miRNAs, their expression landscapes are of similar heterogeneity as that of the much more numerous mRNAs in the systems studied. To the best of our knowledge, application of SOM portraying to miRNA expression was reported here for the first time. Also the combined analysis of miRNA and mRNA differential expression using covariance terms is novel. It provides a very detailed resolution of the coexpression landscape which however awaits for further discovery and interpretation. SOM machine learning clusters features (miRNA and mRNA expression or miRNA/mRNA covariance) with similar profiles together. Enrichment techniques allow association of functional themes with each of the clusters. The spectrum of available gene sets can be extended in future studies by taking into account sets of mRNA targets of single miRNA, sets of miRNA targeting the same mRNA, or sets of miRNA regulated in the same functional context. The sets might be collected using computational as well as experimental methods. Of special interest in this context are new experimental techniques such as high-throughput sequencing of RNAs isolated by cross-linking immunoprecipitation (HITS-CLIP) that directly decode mRNA–miRNA interactions [34].

Acknowledgements

This publication is supported by LIFE—Center for Civilization Diseases, Universität Leipzig. LIFE is funded by the European ERDF fund and by the Free State of Saxony.

References

1. Mendes ND, Freitas AT, Sagot M-F (2009) Current tools for the identification of miRNA genes and their targets. Nucleic Acids Res 37:2419–2433
2. Kim S-K, Nam J-W, Rhee J-K, Lee W-J, Zhang B-T (2006) miTarget: microRNA target gene prediction using a support vector machine. BMC Bioinformatics 7:411
3. Wang X, El Naqa IM (2008) Prediction of both conserved and nonconserved microRNA targets in animals. Bioinformatics 24:325–332
4. Yousef M, Jung S, Kossenkov AV, Showe LC, Showe MK (2007) Naïve Bayes for microRNA target predictions—machine learning for microRNA targets. Bioinformatics 23:2987–2992
5. Heikkinen L, Kolehmainen M, Wong G (2011) Prediction of microRNA targets in Caenorhabditis elegans using a self-organizing map. Bioinformatics 27(9):1247–1254
6. Kohonen T (1982) Self-organized formation of topologically correct feature maps. Biol Cybern 43:59–69
7. Tamayo P, Slonim D, Mesirov J, Zhu Q, Kitareewan S, Dmitrovsky E, Lander ES, Golub TR (1999) Interpreting patterns of gene expression with self-organizing maps: methods and application to hematopoietic differentiation. Proc Natl Acad Sci U S A 96:2907–2912
8. Törönen P, Kolehmainen M, Wong G, Castrén E (1999) Analysis of gene expression data using self-organizing maps. FEBS Lett 451: 142–146
9. Eichler GS, Huang S, Ingber DE (2003) Gene expression dynamics inspector (GEDI): for integrative analysis of expression profiles. Bioinformatics 19:2321–2322
10. Wirth H, Loeffler M, von Bergen M, Binder H (2011) Expression cartography of human tissues using self organizing maps. BMC Bioinformatics 12:306
11. Lu J et al (2005) MicroRNA expression profiles classify human cancers. Nature 435:834–838
12. Yin JQ, Zhao RC, Morris KV (2008) Profiling microRNA expression with microarrays. Trends Biotechnol 26:70–76
13. Wang Z, Yang B (eds) (2010) MicroRNA expression detection methods. Springer, Heidelberg
14. Kong W, Zhao J-J, He L, Cheng JQ (2009) Strategies for profiling MicroRNA expression. J Cell Physiol 218:22–25
15. Git A, Dvinge H, Salmon-Divon M, Osborne M, Kutter C, Hadfield J, Bertone P, Caldas C (2010) Systematic comparison of microarray profiling, real-time PCR, and next-generation sequencing technologies for measuring differential microRNA expression. RNA 16: 991–1006

16. Sato F, Tsuchiya S, Terasawa K, Tsujimoto G (2009) Intra-platform repeatability and inter-platform comparability of MicroRNA microarray technology. PLoS One 4:e5540
17. Fasold M, Langenberger D, Binder H, Stadler PF, Hoffmann S (2011) DARIO: a ncRNA detection and analysis tool for next-generation sequencing experiments. Nucleic Acids Res 39:W112–W117
18. Linsen SEV et al (2009) Limitations and possibilities of small RNA digital gene expression profiling. Nat Methods 6:474–476
19. Binder, H.; Preibisch, S.; Berger, H. Calibration of microarray gene-expression data. In Methods in Molecular Biology; Grützmann, R.; Pilarski, C., Eds.; Humana Press: New York, 2009; Vol. 575, pp. 376–407
20. Nelson PT, Wang W-X, Wilfred BR, Tang G (2008) Technical variables in high-throughput miRNA expression profiling: much work remains to be done. Biochim Biophys Acta 1779:758–765
21. Yuan J, Reed A, Chen F, Stewart CN (2006) Statistical analysis of real-time PCR data. BMC Bioinformatics 7:85
22. Meacham F, Boffelli D, Dhahbi J, Martin D, Singer M, Pachter L (2011) Identification and correction of systematic error in high-throughput sequence data. BMC Bioinformatics 12:451
23. Meyer S, Pfaffl M, Ulbrich S (2010) Normalization strategies for microRNA profiling experiments: a 'normal' way to a hidden layer of complexity? Biotechnol Lett 32:1777–1788
24. Chang K, Mestdagh P, Vandesompele J, Kerin M, Miller N (2010) MicroRNA expression profiling to identify and validate reference genes for relative quantification in colorectal cancer. BMC Cancer 10:173
25. Bolstad BM, Irizarry RA, Astrand M, Speed TP (2003) A Comparison of Normalization Methods for High Density Oligonucleotide Array Data Based on Variance and Bias. Bioinformatics 19:9
26. Dudoit S, Yang YH, Callow MJ, Speed TP (2002) Statistical methods for identifying genes with differential expression in replicated cDNA microarray experiments. Stat Sin 12:111–139
27. Smyth G, Speed T (2003) Normalization of cDNA microarray data. Methods 31:265–273
28. Wirth H, von Bergen M, Binder H (2012) Mining SOM expression portraits: feature selection and integrating concepts of molecular function. BioData Min 5:18
29. Cakir V, Wirth H, Hopp L, Binder H (2013) miRNA expression landscapes in stem cells, tissues and cancer. Methods of Molecular Biology
30. Guo Y, Eichler GS, Feng Y, Ingber DE, Huang S (2006) Towards a holistic, yet gene-centered analysis of gene expression profiles: a case study of human lung cancers. J Biomed Biotechnol 2006, Article ID 69141
31. Saitou N, Nei M (1987) The neighbor-joining method: a new method for reconstructing phylogenetic trees. Mol Biol Evol 4:406–425
32. Opgen-Rhein R, Strimmer K (2007) Accurate Ranking of Differentially Expressed Genes by a Distribution-Free Shrinkage Approach. Statist. Appl Genet Mol Biol 6
33. Strimmer K (2008) A unified approach to false discovery rate estimation. BMC Bioinformatics 9:303
34. Chi SW, Zang JB, Mele A, Darnell RB (2009) Argonaute HITS-CLIP decodes microRNA-mRNA interaction maps. Nature 460: 479–486

Chapter 17

MicroRNA Expression Landscapes in Stem Cells, Tissues, and Cancer

Mehmet Volkan Çakir, Henry Wirth, Lydia Hopp, and Hans Binder

Abstract

MicroRNAs play critical roles in the regulation of gene expression with two major functions: marking mRNA for degradation in a sequence-specific manner or repressing translation. Publicly available data sets on miRNA and mRNA expression in embryonal and induced stem cells, human tissues, and solid tumors are analyzed in this case study using self-organizing maps (SOMs) to characterize miRNA expression landscapes in the context of cell fate commitment, tissue-specific differentiation, and its dysfunction in cancer. The SOM portraits of the individual samples clearly reveal groups of miRNA specifically overexpressed without the need of additional pairwise comparisons between the different systems. Sets of miRNA differentially over- and underexpressed in different systems have been detected in this study. The individual portraits of the expression landscapes enable a very intuitive, image-based perception which clearly promotes the discovery of qualitative relationships between the systems studied. We see perspectives for broad applications of this method in standard analysis to many kinds of high-throughput data of single miRNA and especially combined miRNA/mRNA data sets.

Key words Gene expression analysis, Self-organizing maps, miRNA/mRNA coexpression, Gene set enrichment analysis

1 Introduction

MicroRNAs play critical roles in the regulation of gene expression. As the set of transcription factors (TFs) in a given cell type constitutes a "code" that specifies cellular differentiation via mRNA activity, "miRNA codes" are likely to have conceptually similar roles in the regulation of gene activity [1]. Both TFs and miRNA are trans-acting factors that exert their activity through composite cis-regulatory elements. The resulting miRNA/mRNA coexpression might come true in concert, in anti-concert, or also in a more complex fashion. Here, the two major functions that miRNAs are involved in are degrading the mRNA or repressing its translation.

Mehmet Volkan Cakir and Henry Wirth have contributed equally to this work.

Malik Yousef and Jens Allmer (eds.), *miRNomics: MicroRNA Biology and Computational Analysis*, Methods in Molecular Biology, vol. 1107, DOI 10.1007/978-1-62703-748-8_17,

The situation is usually complex because a particular miRNA may have multiple mRNA targets and because multiple miRNAs could target the same mRNA constituting a complex regulatory network.

The most widely studied mechanism of regulation involves binding of a miRNA to the target mRNA. MicroRNAs regulate their targets by triggering mRNA degradation or translational repression. As a result, translation of the target mRNA is inhibited and the mRNA may be destabilized (*see* previous chapters). The inhibitory effects of miRNAs have been linked to diverse cellular processes including malignant proliferation, apoptosis, development, differentiation, and metabolic processes (*see* Chapters 2, 3, and 19). The negative relationship between miRNAs and their targets suggests that the regulatory effect of a miRNA could be determined from the expression levels of its targets. Note also that miRNAs do not need a perfect alignment with their targets to act. In consequence one miRNA may regulate several mRNAs or one mRNA may be regulated by several miRNAs. In this manner miRNAs potentially regulate approximately 60 % of all genes encoding human proteins and appear to interfere in a wide range of cellular functions, such as cell generation, differentiation, and proliferation [2].

The opposite effect, namely, direct correlations between miRNA and mRNA expression, can be caused by the fact that intronic miRNAs are usually coordinately expressed with their host gene mRNA, implying that they derive from a common transcript and that analysis of host gene expression can be used to probe the spatial and temporal localization of intronic miRNA [3]. In addition, also proximal pairs of miRNA are often coexpressed. It was found that an abrupt transition in the correlation between pairs of expressed miRNA occurs at a distance of 50 kb, implying that miRNAs separated by less than 50 kb typically derive from a common transcript [3].

Exact understanding of how miRNAs regulate gene expression is vital to the field of miRNA research. Systematic analysis of miRNA expression landscapes and also of miRNA/mRNA coexpression patterns thus constitute a basal objective in miRNA research which complements miRNA gene and target discovery. In this contribution, we analyzed miRNA expression using self-organizing maps (SOMs), a machine learning clustering technique based on neural network. The method was described in detail in the accompanying chapter [4]. In this chapter we apply the SOM portraying method to discover combined miRNA and mRNA expression landscapes in the context of cell fate decisions and stemness, fully differentiated human tissues, and diseased cancer samples of different origin in order to illustrate the performance of the method in the form of an extended case study. To the best of our knowledge, this method was applied here for the first time to analyze miRNA expression data.

2 Analyzing miRNA/mRNA Coexpression: A Brief Overview

In general, miRNA/mRNA coexpression can be studied in two different ways:

1. Correlation between the expression values of miRNA and mRNA species is directly analyzed. For example, in such studies on human brain biopsies the authors report that the distribution of correlation coefficients for all possible mRNA–miRNA pairs exceeds a random distribution at their tails at high positive and negative values of the correlation coefficient [5]. Part of the negative correlations selected tends to predict targets, and positive correlations tend to predict physically proximate pairs, as expected (*see* [4] and references cited therein). In contrast, other studies report that miRNA activity shows very weak correlation with mRNA expression which indicates more complex regulation mechanisms between miRNAs and their target genes (*see* [6] and Chapter 18).
2. The second type of coexpression studies pursues a more indirect approach based on databases which collect miRNA–mRNA target relationships (for an overview *see* [7]). These data are obtained via in silico target prediction of miRNA-binding motifs (*see* databases: PITA [8], PICTAR [9], and TargetScan [2]), via meta-analyses of experimentally validated miRNA target genes (TarBase [10], miRecords [11], and miR2Disease [11]), or via text mining of biomedical abstracts (miRSel [12]). The collected data constitute sets of mRNA target species for individual miRNA (or families of miRNA) which are subsequently used in gene expression enrichment analyses. This approach searches for significant expression changes of the target sets compared with appropriate random sets [13]. Enriched sets suggest that their expression is potentially regulated by the associated miRNA.

The computational methods predict about 40 mRNA targets per miRNA on the average, while only 2–8 targets are associated with each miRNA in the experimentally validated and text mining databases. The latter approaches usually consider less miRNA in total (from 93 in miRecords to 176 in miR2Disease) than the former ones (from 163 in PICTAR to 640 in PITA). In silico miRNA target prediction is usually not very accurate with fairly high false-positive and/or false-negative rates.

3 Methods and Data

3.1 Expression Analysis Using Self-Organizing Maps

A SOM is a neural network algorithm, widely used to categorize large, high-dimensional data sets [14]. In bioinformatics, SOMs have been successfully applied to gene expression analysis [15] and have enabled characterization of genome-wide expression landscapes in a sample-specific way [16, 17]. Our implementation of the method, called SOM-cartography or SOM-portraying, transforms large and heterogeneous sets of expression data into an atlas of sample-specific portraits which can be directly compared in terms of similarities and dissimilarities (*see* Chapter 16 for a detailed description of the method). This global view on the behavior of defined modules of correlated and differentially expressed genes is more intuitive than ranked lists of hundreds or thousands of individual genes usually obtained in standard expression analysis. Particularly, SOM analysis is featured by several important benefits:

1. It provides an individual visual identity for each sample.
2. It reduces the dimension of the original data.
3. It preserves the information richness of the molecular portraits allowing the detailed, multivariate explorative comparisons between samples.
4. Its output can be treated as a new complex object for next-level analysis in terms of visual recognition.

Here we will apply the method to different data sets to portray the miRNA expression landscapes, to characterize the similarities between the different samples studied, and to extract lists of miRNAs relevant in the context of stem cells, differentiated tissues, and cancer. MicroRNA data are complemented with associated mRNA expression landscapes.

3.2 Data Sets

We analyzed the following three data sets:

The *WILSON-data* set refers to expression data of 697 miRNAs from embryonic stem cells (ESC), fibroblasts, and derived from induced pluripotent stem (IPS) cells obtained in a microarray study [18].

The *LIANG-tissue* set contains expression values of 175 miRNAs measured in 24 human tissues by standard TaqMan qPCR assay [19]. Expression values are given in units of the threshold cycle (CT) defined as the fractional cycle number at which the fluorescence exceeds a fixed threshold. Four human miRNAs (miR-30e, miR-92, miR-92N, and miR-423) that were least variable among the tissues in this study were used to normalize the miRNA expression. SOM training was performed using log CT values after quantile normalization. The obtained miRNA expression landscapes in human tissues were compared with mRNA expression data analyzed previously [17].

The *LU-cancer* set contains miRNA and mRNA measurements from the same samples of seven healthy and tumor tissues (colon, kidney, bladder, prostate, uterus, lung, breast) [20]. MicroRNAs were measured using a bead-based profiling method: Oligonucleotide-capture probes complementary to miRNAs of interest were coupled to carboxylated 5-μm polystyrene beads impregnated with variable mixtures of two fluorescent dyes for "barcoding," each representing a single miRNA. The abundances of 217 miRNAs were measured after hybridization, amplification by PCR, and staining. Messenger RNA expression was determined using microarrays.

4 Case Studies

4.1 Stem Cells

MicroRNAs are important regulators for ESC self-renewal, pluripotency, and differentiation [21]. The SOM portraits of human fibroblasts, pluripotent stem cells induced from them (IPS), and ESC show a simple and consistent spot pattern (Fig. 1a, e). Overexpressed genes in fibroblasts and underexpressed in both types of stem cells (ESC and IPS) aggregate into one spot A. Genes with antagonistic activity cluster into spot B showing high expression in ESC and IPS and low expression in somatic cells. IPS cells reveal another moderately overexpressed spot C in addition to this "stemness" spot which differentiates between both types of stem cells. It becomes underexpressed in ESC and in fibroblasts.

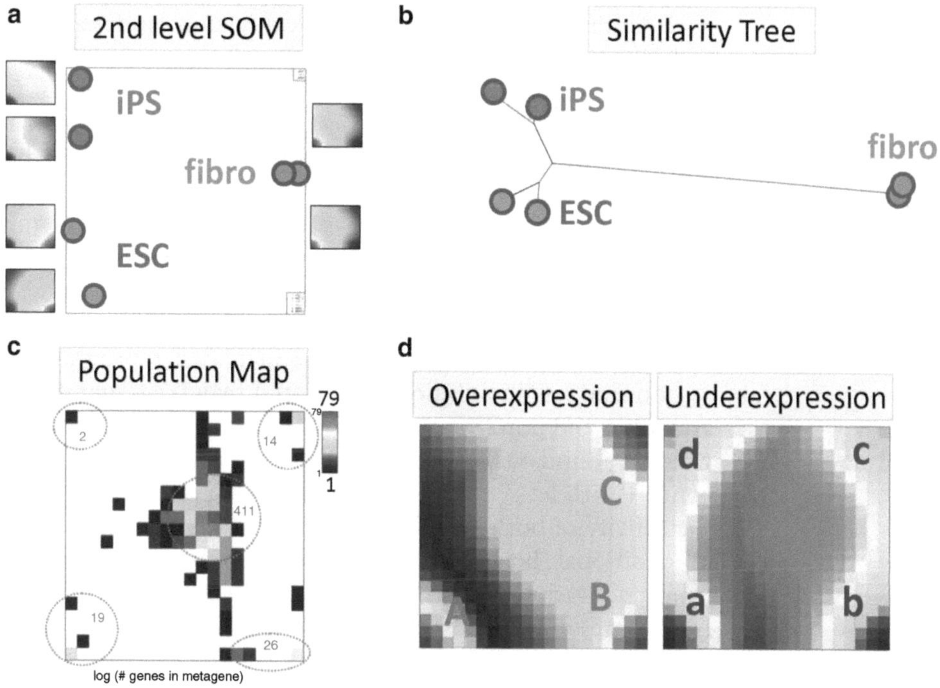

Fig. 1 miRNA expression portraying of stem cells (WILSON-data set): The second level SOM arranges the first-level SOM portraits (*see* the *small images*) of stem cells (ESC and IPS) and of differentiated cells (fibroblasts) according to their mutual similarities (panel **a**). The two mosaics per cell system refer to biological replicates [18]. The similarity tree (panel **b**) expresses the Euclidean distances between the different samples in-scale. The population map (panel **c**) illustrates the number of miRNA per tile in the first-level SOM mosaic. The over- and underexpression summary maps (panel **d**) assign the spots detected

Table 1
miRNAs differentially regulated in stem cells

Spot[a]	UP[b]	DN[b]	miRNA within the spot cluster[c]
A	Fibro	ESC, iPS	let-7d; -let-7f; -let-7e; -let-7a; -145; -100; -let-7i; -29a; -let-7c; -199a-3p; -125b; -143; -222; -23b; -23a; -24; -34c-3p; -21; -26a
B	ESC (iPS)	Fibro	mir-106a; -17; -302a; -302d; -302b; -638; -93; -106b; -20a; -302c; -19b; -25; -663; -19a; -103; -130a; -107; -20b; -16; -182; -183
C	iPS	Fibro (ESC)	mir-149; -18b; -92b; -92a; -30c; -15b; -18a; -151-5p; -30b; -923; -302c; -148a; -363; -200c; -361-5p; -335; -454; -1
d	Fibro	ESC	mir-27a; -27b; -125a-5p
	Invariant		mir-768-5p; -15b; -150; -219-2-3p; -517c; -504; -612; -629; -30a; -652; -519d; -342-5p; -330-3p; -624; -193b; -296-3p; -93; -187; -196a

[a]*See* Fig. 1e for spot assignment
[b]Systems showing up- or downregulation of expression compared with the mean expression of each miRNA
[c]mirRNAs are sorted with decreasing concordance

The sample similarity tree (Fig. 1b) reveals close similarity, but not identity, between the miRNA expression landscapes of ESC and IPS, which, in turn, clearly differs from that of the somatic fibroblasts. The cell type-specific spots A–C contain about one to two dozen miRNAs per spot. More than 400 miRNAs remain not regulated and cluster within the invariant central area of the map (*see* population map, Fig. 1c).

Analysis of the spots shows that miRNAs of the mir-302 and -17 families are upregulated in ESC and IPS (spot B) and miRNAs of the let-7 family are upregulated in fibroblasts (spot A). This is in agreement with previous results [18] by Wilson and colleagues (Table 1). It has been argued that increased reprogramming in response to let-7 inhibition is mediated by let-7 target genes such as c-Myc and Lin28 [21]. Lin28 is also repressed by miR-125, which is abundantly expressed in differentiated cells and changes in concert with let-7 according to our results, presumably inhibiting the activity of both miR-125 and let-7 miRNAs, which may result in additional beneficial effects during reprogramming due to robust activation of Lin28 expression. Another miRNA from spot A, mir-145, induces ESC differentiation by inhibiting the expression of key pluripotency/reprogramming factors, such as Sox2, Oct4, Klf4, and c-Myc. Other miRNAs from spot A (mir-24, mir-23, and mir-21) either inhibit cell proliferation by targeting important cell cycle regulators, such as c-Myc and E2F2 (mir-24), or suppress TGF-β/activin signaling. MicroRNAs from the mir-302 family are found in spot B showing antagonistic activation compared with spot A. It has been demonstrated that Sox2 and Oct4 bind the

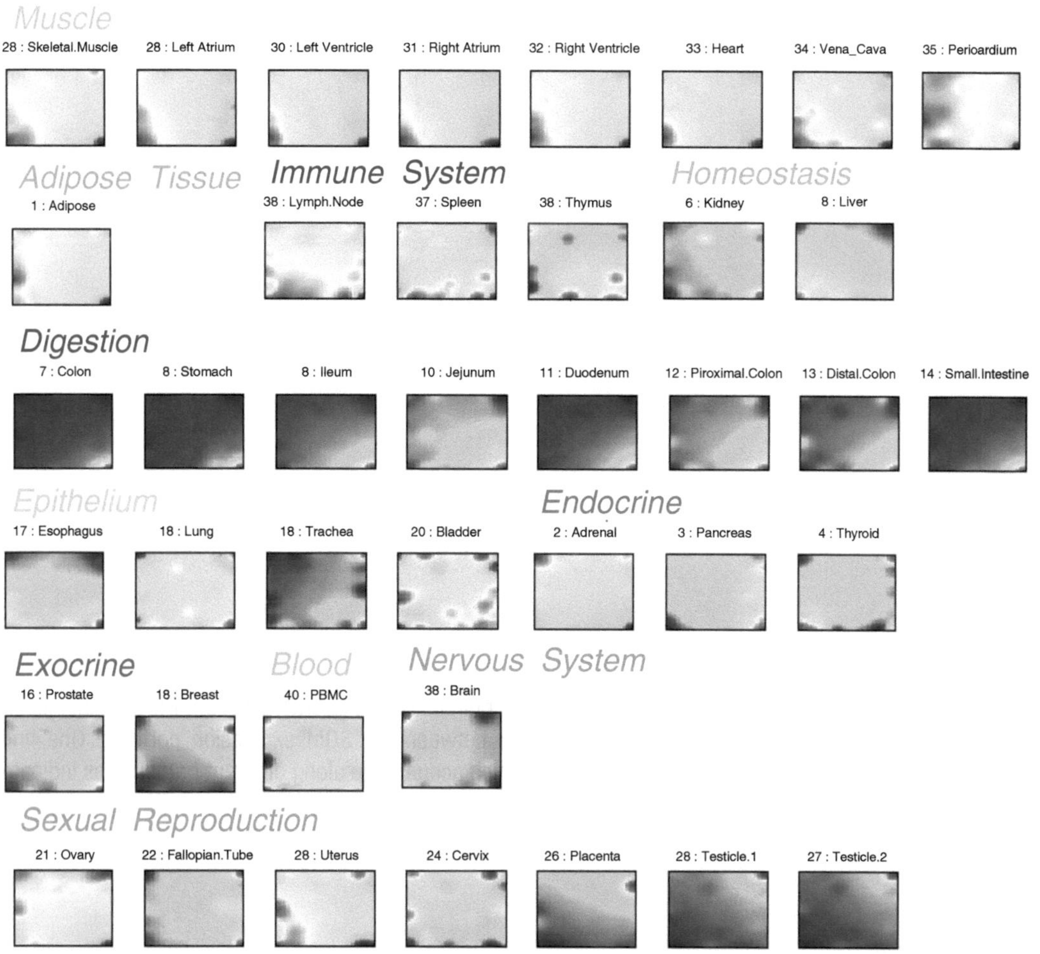

Fig. 2 miRNA expression portraits of human tissues (LIANG-data set). Tissues are grouped into 11 categories

miR-302 promoter and are essential for expression of miR-302 in human ESC.

Hence, our SOM analysis identifies signature groups of miRNA in ESC, IPS, and differentiated fibroblasts. The SOM portraits clearly assign the expression landscapes to one of the three cell types studied. A similar pattern was observed for mRNA expression [22].

4.2 Human Tissues

The expression of 345 human miRNAs was studied in a spectrum of 40 normal human tissues that included specimens derived from brain, muscle, circulatory, respiratory, lymphoid, gastrointestinal, urinary, reproductive, and endocrine systems. Their SOM portraits in Fig. 2 reveal a much more diverse spot texture compared with the simple pattern observed in stem and somatic cells discussed in the previous section. The different tissues are grouped into 11 tissue categories in analogy with the classification of human tissues in a

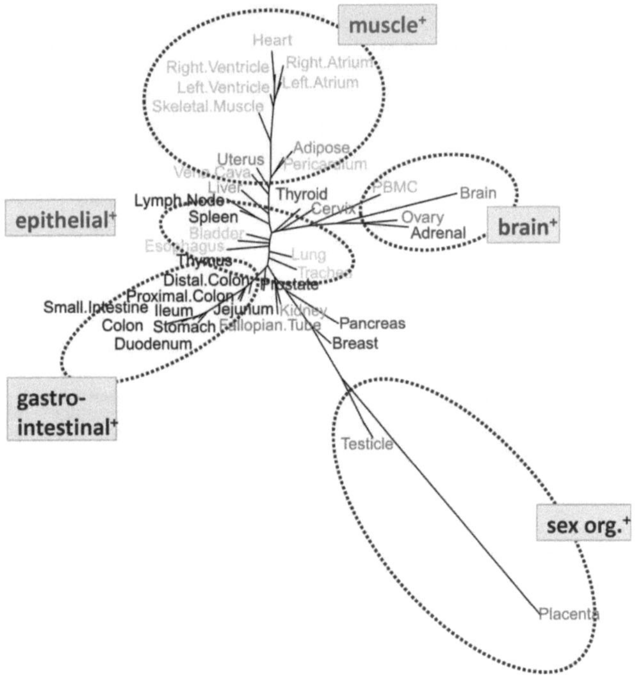

Fig. 3 miRNA expression in human tissues (LIANG-data set). The neighbor joining tree illustrates similarities between the SOM expression portraits. One finds essentially five clusters which accumulate along different branches as indicated by the *dotted ellipses*

previous study on mRNA expression [17]. Portraits of the same category mostly look very similar, hence reflecting similar miRNA expression landscapes. For example, tissues derived from different parts of heart (atrium versus ventricle) resemble those of skeletal muscle. Portraits of tissues from the gastrointestinal system ("digestion": stomach, small intestine, and colon), immune system (spleen and lymph node), female sexual organs (ovary, uterus, and cervix), and respiratory/epithelial tissues (lung and trachea) show mostly consistent spot patterns within their respective category.

Similarity tree analysis identifies five larger clusters of tissues containing muscle, gastrointestinal, epithelial, brain, and sexual organs which group along different branches of the tree (*see* Fig. 3, where the "plusses" indicate that these clusters usually include also tissues of other categories such as adipose tissue which are found in the "muscle+" branch of the tree). The similarities among samples can be

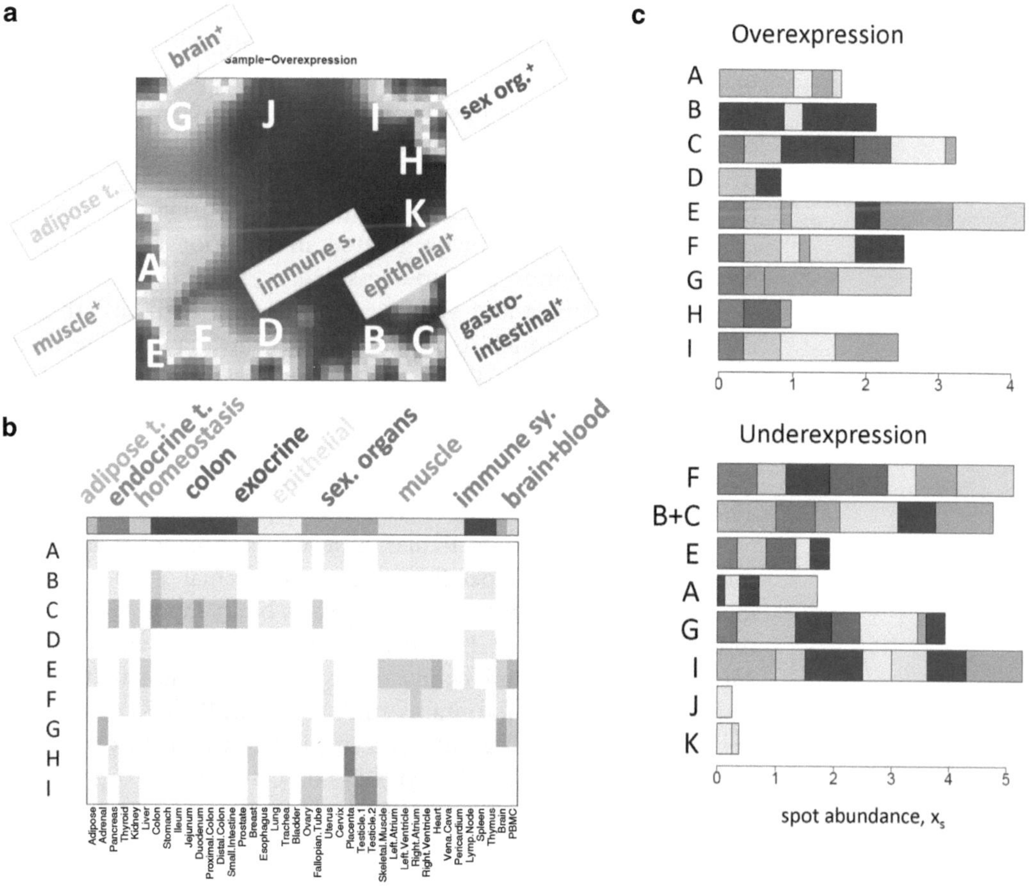

Fig. 4 Spot characteristics of miRNA expression in human tissues (LIANG-data set): The overexpression spot summary map shows that each cluster is characterized by specific spot patterns (panel **a**). The spot expression heatmap (panel **b**) shows the mean expression of the spots detected in all tissues (*red* to *white* (*darker* to *lighter grey*) indicates *high* to *low*). Over- and underexpression spot abundance in ten tissue categories (panel **c**, the colors of the *bars* are assigned in the legend on top of the heatmap). For example, spot A is especially overexpressed in adipose tissue (*orange bar*, here *light grey*) which, in turn, is characterized by underexpression of spots B, C, and I

attributed to characteristic groups of spots collected in the respective summary map (Fig. 4a). Spot analysis characterizes the miRNA expression landscape in more detail. The spot expression heatmap (Fig. 4b) shows the mean expression of each of the spots in all tissues. For example, high expression levels of spots B and C are observed in colon samples and high expression of spots A, E, and F in muscle samples. Over- and underexpression spot abundance analysis provides further details (Fig. 4c). For example, spot B collects miRNAs which are highly expressed in colon and immune system tissues, whereas spot C is populated more with miRNAs which are highly expressed in many tissues, however with slight preference for colon and epithelium. In turn, miRNAs from both spots B and C are underexpressed in adipose, muscle, and nervous tissues.

One also sees that, for example, miRNAs highly expressed in adipose tissue accumulate within spot A only, whereas endocrine tissue miRNAs show a broad-distribution high expression over different spots. The "adipose" spot A is also observed in a series of other tissues such as pericardium (muscle), bladder (epithelium), ovary, uterus, and cervix (all sexual organs) reflecting accumulation of fat in different regions of the body. Contrarily, miRNAs in spot A are strongly downregulated in fat-poor tissues such as blood, brain, thymus, and, to a less degree, kidney.

The population and variance maps in Fig. 5 reveal that the expression of 34 miRNAs remains almost invariant in the tissues studied. Most of the miRNAs (54 species) are found in spots I and H upregulated in sexual organs, endocrine and epithelial tissues. The variance of miRNA expression is maximal in placenta due to extraordinarily high expression of genes from spot H (Fig. 5c). The miRNA species found in each of the spots are listed in Table 2.

According to the particular spot abundance one can identify miRNAs expressed in specific tissues with minimal or no expression in other tissues such as miR-9/219 in brain (spot G), whereas miR-124a/124b (spot E) are also strongly expressed in muscle. MicroRNAs of the let-7 family are also found in this partly ubiquitous spot which also contains miR-1 and miR-133a/b showing high expression in different parts of the heart and skeletal muscle as well as in vena cava and thyroid. These are hollow organs composed of smooth muscle-containing wall, such as the gastrointestinal system, suggesting that expression of miR-1 and miR-133a/b might mark some features shared by different muscle types.

Note also that selected miRNAs of the mir-302 family, highly expressed in ESC, are among the invariant species in the tissue series. Our spot analysis thus provides an opportunity to extract tissue-specific expression patterns of groups of miRNAs which can be further analyzed with respect to the genomic location of their coding sequences and to their involvement in common pathways.

Figure 6 shows the clustering pattern of mRNA expression in different tissues as seen by our previous SOM analysis [17]. On the one hand, part of the tissue categories such as muscle or nervous tissues form separate branches due to category-specific mRNA expression. On the other hand, similarities between tissues and also tissue categories are different compared with the miRNA expression patterns (compare with Fig. 3). For example, miRNA expression of brain shares partial similarity with miRNA expression in muscle (spot E), whereas mRNA expression in nervous tissue is highly specific and different from mRNA expression in muscle. Note however that the tissue data sets used for mRNA and miRNA analysis are partially different with respect to the tissues included which may distort direct comparison. Figure 7 matches the tissues used in both data sets. It can be seen that nervous tissues are highly

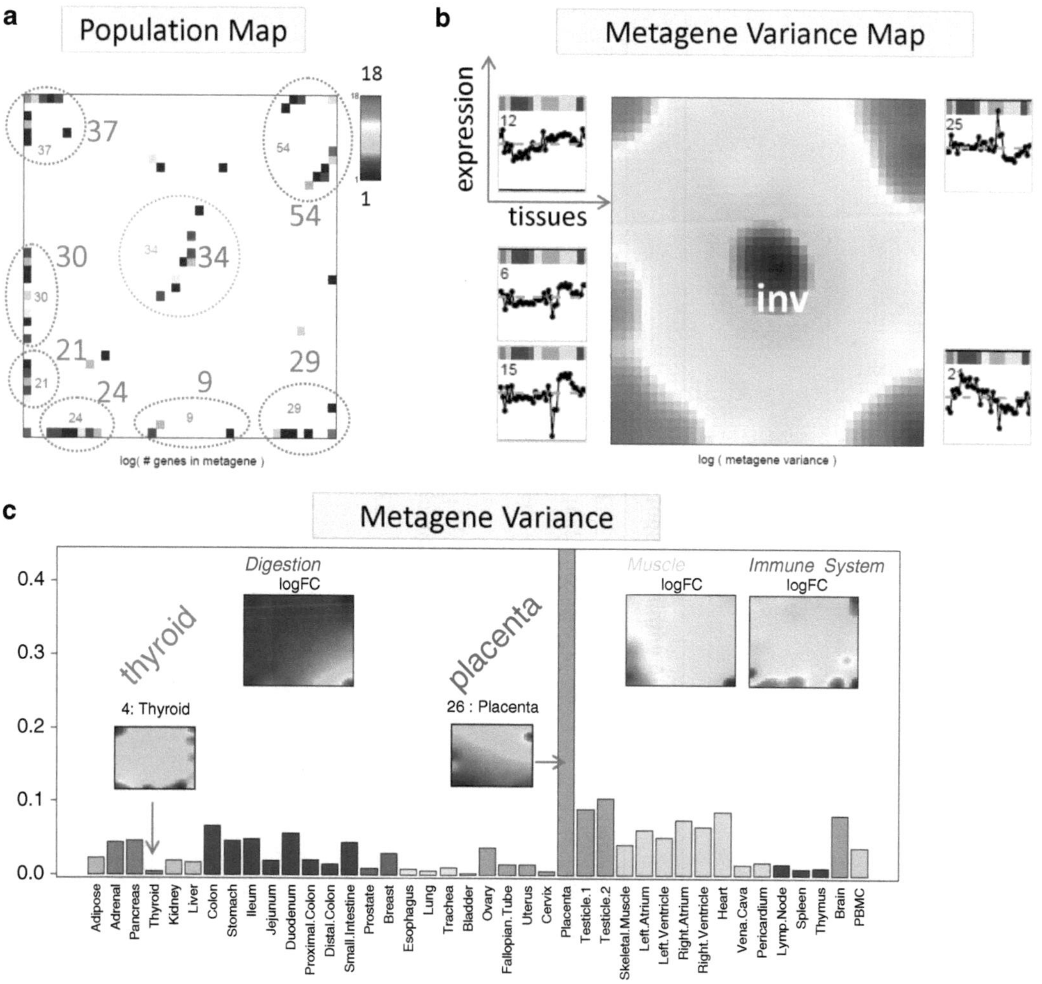

Fig. 5 Population and variance map of expression portraits of human tissues (panels **a** and **b**, respectively) and their tissue-specific metagene variance (panel **c**). MicroRNAs with highly variant expression profiles accumulate in the spots located along the border of the map (the number of miRNAs per spot is given in the population map). 34 miRNAs with almost invariant profiles form the central spot in the variance map. Selected spot profiles are shown nearby the respective spots. The metagene variance of placenta is maximal due to one strong overexpression of spot H (*see* Figs. 4 and 5)

underrepresented in the LIANG-miRNA data set. On the other hand, other tissue categories such as muscle, digestion, and immune system tissues are well represented in both data sets.

Figure 7 also compares the tissue specifics of the miRNA and mRNA expression landscapes in terms of their metagene entropies. It reveals interesting differences such that miRNA expression is more specific for epithelial and immune system tissues (low entropy) than the respective mRNA expression patterns. For endocrine and reproductive organs this relation reverses.

Table 2
miRNAs differentially regulated in human tissues

Spot[a]	Group of tissues[b]	UP[c]	DN[c]	miRNA within the spot[d]
A	Muscle+, adipose	Adipose, cervix	Blood, thymus, placenta	mir-224; -452; -452; -193b; -335; -365; -362; -199b; -27aN; -425; let-7a_control; -23a; -126; -214; -27a; -126; -152; -374; -27b
B	Epithelial+, gastrointestinal+	Immune system, digestion	Trachea, thyroid	mir-142-3p; -338; -363; -146; -146b; -20b; -301
C	Gastrointestinal+	Epithelial, digestion, kidney, prostate	Muscle, adipose, nervous system	mir-215; -375; -192.1; -194; -192; -141.1; -141 N; -141; -200a; -200c; -200b; -200cN; -200bN; -31.1; -31; -429
D	Epithelial+, immune system	Homeostasis, immune system	Thyroid, trachea, prostate, cervix	mir-92 N; -17-5p; -183; -92; -18; -15b; -106a; -15a; -25; -106b
E	Muscle+	Muscle, homeostasis, blood, brain	Immune system, sexual repr., exocrine	mir-122a; -124b; -124a; cel-124; -30e-3p; -133a; -302d; -302b; -1; -302a; -422a; -128a; -133b; -197; -491; -378; -328; -181b.1; -138; -128b; -340; -190; -423; -490; -129; -107; -422b; -425.1; -22; -296, cel-2, cel-lin-4; -159a; -104; -108; -105; -136
F	Muscle+	Muscle, immune system, homeostasis	Blood, digestion, sexual repr., exocrine, kidney	mir-188.1; -95; -150; -155; -181c; -511; -139.1; -505; -342; -101; -193; -345; let-7i; let-7e; let-7iN; -17-3p; -324-3p; let-7f
G	Brain+	Blood, brain	Thyroid, homeostasis, digestion, breast	mir-9; -137; -323; -433; -219; -204; -153; -485-3p; -432; -127; -382; -370; -383; -149; -134; -379; -132; -299; -125b

H	Sex. organs+	Immune system, placenta, breast	Thymus, trachea, prostate, bladder	mir-517c; -516-5p; -518c; -519d; -512-3p; -520 h; -518e; -525; -520 g; -517b; -517a; -519c; -518b; -518a; -518f; -515-3p; -515-5p; -525; -520a; -512-5p; -519e; -520d; -372
I	Sex. organs+	Sexual repr., endothelial	Adipose, digestion, liver, spleen, nervous tissues	mir-514; -508; -206; -509; -449; -506; -34c; -202; -202; -34cN; -34bN; -34b.1; -510; -196a; -513; -196b; -424; -135a; -432; -10b; -199a; -34aN; -199-s
J		Bladder	Digestion, trachea	mir-451
K		Bladder	Trachea, thyroid	mir-29a.1
inv				mir-448; -453; -488; -492; -496; -498; -504; -518f; -519a; -520d; -410; -412; -524; -526c; -527; -384; -371; -377; -517; -326; -329; -33; -337; -368; -373; -376b; -380-3p; -380-5p; -450; -503; -526b; -325; -325 N; -381; -302c; -96; -519b; -302b; -220; -299-3p; -302a; -208

[a] *See* Fig. 3 for spot assignment
[b] Tissue categories
[c] Systems showing up- or downregulation of expression compared with the mean expression of each miRNA
[d] mirRNAs are sorted with decreasing Pearson's correlation between their expression profile and that of the respective metagene

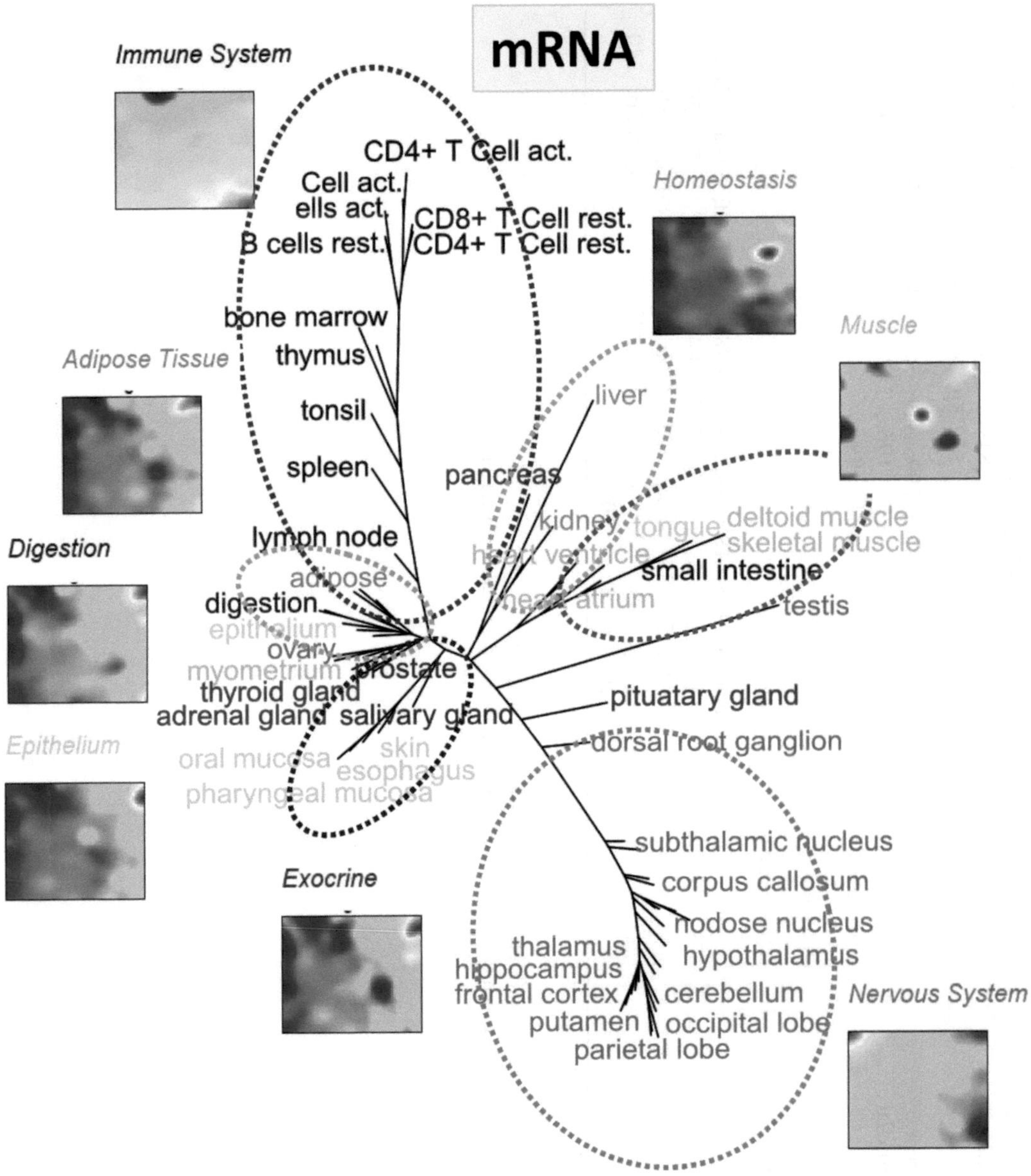

Fig. 6 mRNA expression in human tissues. The neighbor joining tree illustrates similarities between the SOM metagene expression of 67 different tissues grouped into eight categories. They preferentially accumulate along specific branches as indicated by the *ellipses*. The respective mean portraits are shown in the figure (for the full gallery of all single-tissue portraits *see* ref. 17). Compare with the miRNA expression tree of human tissues in Fig. 4

4.3 Cancer

Cancer is a disease of gene function in most respects caused by genetic and epigenetic alterations affecting many molecular pathways involving both canonical protein-coding "mRNA" genes as well as noncoding "miRNA" genes. Aberrant miRNA expression signatures can serve as a hallmark of cancer where miRNA genes can function as oncogenes and tumor suppressors. Global up- [23] as well as

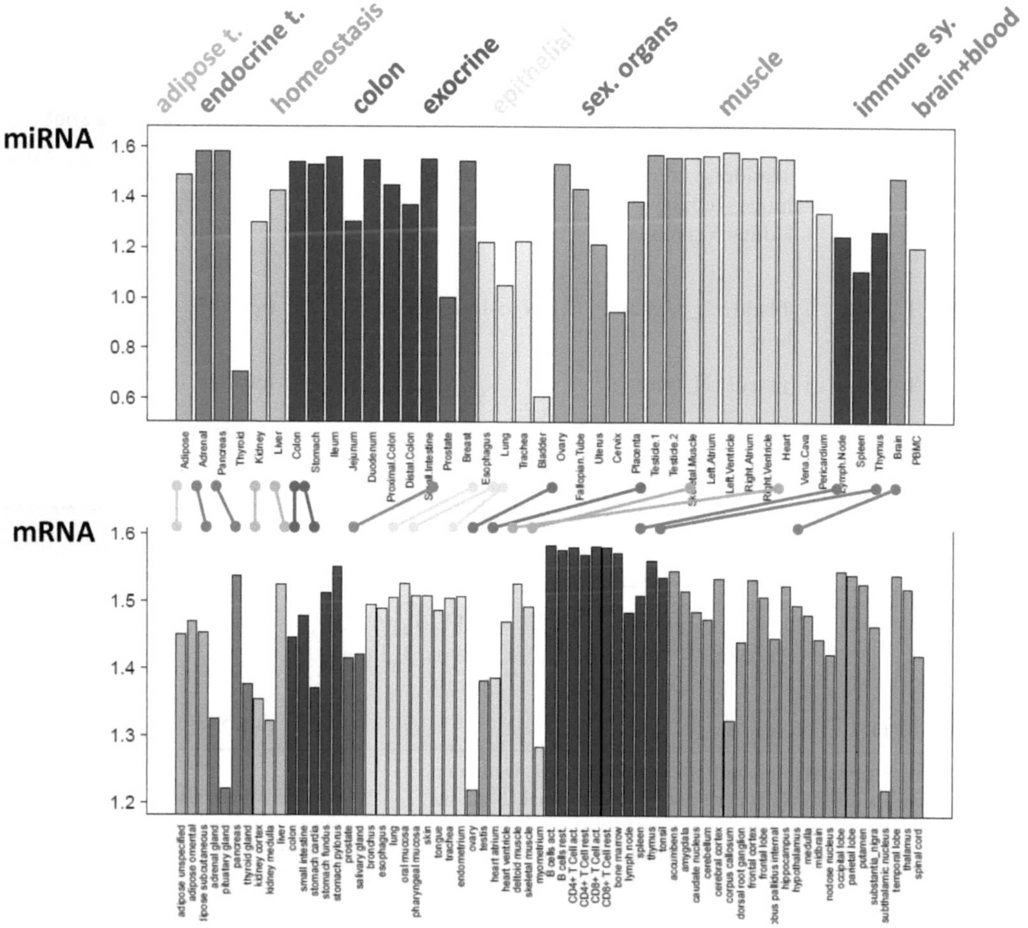

Fig. 7 Metagene entropies of the miRNA (*top*) and mRNA (*bottom*) expression portraits of human tissues. Identical tissues in both parts of the figure are connected by the *dumbbell* lines

downregulation [20] of miRNA activity in cancer compared with normal tissues have been reported previously. Thus, an evaluation of changes in miRNA expression could provide an insight into mechanisms of cancer genesis and progression.

In ref. 4 we identified groups of miRNAs differently expressed in healthy and tumor tissues. Table 3 lists the miRNAs taken from the spot clusters extracted from the miRNA-SOM images. Many of them are previously reported as differentially expressed in different cancers (compare with Table 1 in [24], Table 1 in [25], and also refs. 26, 27). Upregulated miRNAs in cancer act as OncoMiRs, whereas miRNAs acting as tumor suppressors are often downregulated. For example, miRNAs of the let-7, mir-126, and mir-130 families are known as tumor suppressors. They accumulate in spot A which is downregulated in colon, bladder, lung, and breast cancer. Spot F contains mir-34 family with miR-34a inhibiting the expression of multiple oncogenes (e.g., c-Met, Notch-1/Notch-2,

Table 3
miRNAs differentially regulated in normal and tumor samples (LU-data set)

Spot[a]	[b]	UP[c]	DN[c]	miRNA within the spot[d]
H	T_UP	T_bladder, _uterus; _lung; _breast; _kidney; (_colon)	All N_tissues	mir-296; -183; -153; -339; -324-3p; -181c; -182; -200b; -208; -146a; -200a; -141; -210; -328
E	T_UP	T_bladder, _prostate, _lung	N_colon, _kidney	mir-205; -144; -323
F	T_UP	T_uterus; N_lung		mir-34c; -34b
C	T_UP	T_kidney		mir-10a, -10b, -196a, -128b
A	T_DN	N_uterus, N_lung	T_colon, _bladder, _lung, _breast	mir-130a, -126, let-7e, -140, -133a, let-7d, -99a, -1, -126, -189
B	T_DN	N_prostate, N_breast	T_kidney,	mir-214, -199b, -136
D	T_DN	N_colon, N_kidney, T_colon	T_bladder, T_prostate, T_uterus, T_lung, T_breast	mir-194, -215, -192
G	T_DN	N_bladder		mir-216, -217

[a]Spot letters are assigned in the miRNA summary maps given in ref. 4
[b]Up- or downregulated mainly in tumor (T) and normal (N) tissue. T_DN spots are usually also N_UP spots and vice versa
[c]Particular samples showing this spot with up- (UP) or down- (DN) regulated genes
[d]miRNAs are ordered with decreasing significance according to the concordance *t*-score

and CDK6) by binding to their 3′-UTR and suppressing, e.g., tumor growth in human gliomas. Many miRNAs found in the T_UP (tumor up) spot H are known as OncoMiRs (e.g., mir-181, -200, -146). Other OncoMiRs such as mir-10 and -196 are specifically upregulated in the tumor-specific spot C. Another upregulated miRNA in spot H, mir-296, is related to angiogenesis commonly activated in many solid tumors [25].

The analogous analysis provides groups of mRNAs together with their functional context using gene set enrichment analysis. Table 4 lists the top-enriched gene sets taken from the gene ontology category "biological process" in the overexpression spots of the mRNA SOM portraits together with the top ten concordant genes in each spot.

In the next step we combined both data sets in terms of spot–spot correlation analysis as described in the methodical section of Chapter 18. The pairwise correlation heatmap in Fig. 8 visualizes positive and negative correlations. Each negative correlation might refer to downregulation of miRNAs and upregulation of mRNAs in cancer or vice versa. The dark tiles thus include miRNAs acting

Table 4
mRNAs differentially regulated in cancer (LU-data set)

Spot[a]	[b]	UP[c]	DN[c]	Enriched gene sets in the spot[d]	Top-10 concordant genes in the spot[e]
F	T_UP	T_kidney, T_bladder, T_uterus, T_lung, T_breast	T_prostate, N_prostate	Receptor activity (−6); keratinization (−5); cation transport (−4);	CDH15, LTK, SYN1, CCL22, DDX11, ARFRP1, HLA-DOA, NHLH1, PSMB6, PTPRN
L	T_UP	T_bladder	T_colon	Glucose homeostasis (−4); defense response to Gram-negative bacterium (−4)	KIF14, RP4-669P10.16, RBL1, CDC7, PSG7, ADD2, HIST1H4E, SAG, CLCN5, CD44
J	T_UP	T_lung		Wnt receptor signaling pathway (−5); negative regulation of cyclin-dependent protein kinase activity (−4); humoral immune response (−4)	ZBTB33, FMO3, HBQ1, MMP10, PIM2, SERPINC1, HTR2C, POU2AF1, CCR6, KIAA0368
M	T_UP	T_uterus		Apoptosis (−4)	ATIC, WFDC2, AIMP2, MFSD10, IRAK1, LTBR, NPIPL3, TUBG1, FLAD1, IER3
C	T_DN	T_lung, N_lung	T_colon, T_prostate, N_prostate	Respiratory gaseous exchange (−8); response to hyperoxia (−7); complement activation, classical pathway (−5)	MNDA; SFTPA1; PRR4; SFTPC; SFTPA2; SFTPB; NRGN; IL1RL1; SCGB1A1; ITGB2
a1	T_DN	T_colon	T_lung, N_colon	Translation (−10); viral transcription (−6); mRNA metabolic process (−5)	NFKB1, RPS16, ATP6V1G2-DDX39B, RP11-40H20.2, EEF1A1P11, SRSF3, HMGN2P17, SERINC3, PTGES3, ACTN1

(continued)

Table 4 (continued)

Spot[a]	[b]	UP[c]	DN[c]	Enriched gene sets in the spot[d]	Top-10 concordant genes in the spot[e]
a2	T_DN	N_uterus, N_lung N_breast	T_colon, T_kidney, T_bladder, T_uterus, T_breast	Platelet degranulation (−9); muscle contraction (−6); complement activation (−6)	SPARCL1, CNN3, PURA, UQCRB, NFE2L2, LEPROT, SAFB, TJP1, FNTA, DSTN, ATP2A2
D	T_DN			Proteolysis (−11); digestion (−5); lipid catabolic process (−4)	PNLIPRP2, CTRB2, REG3A, PRSS1, CEL, CELA2A, CPA1, PNLIP, FGL1, AMYP1
E	T_DN	N_kidney, N_bladder	T_prostate	Muscle cell homeostasis (−4); intracellular receptor-mediated signaling pathway (−4)	MCAM; SPEG; MFAP5; CAMK2G; TLE2; MCAM; PDLIM7; MATN2; MYOC; LEFTY2
B	T_DN	T_prostate, T_breast	T_colon, T_kidney, T_breast	Pituitary gland development (−5); nucleosome assembly (−5); cholesterol biosynthetic process (−4)	CD38; ALDH1A3; SC5DL; IDI1; ACPP; CCK; PPP3CA; KLK2:MSMB; TGFB3
G	T_DN	T_bladder	T_colon, T_prostate	Excretion (−7); oxidation–reduction process (−6); metabolic process (−6)	CLCNKB; CYP4A11;RIPK1; SLC1A6; MST1P9; SERPINF2; PLG CYFIP2; HRG; FGB
H	T_DN	T_lung	T_prostate	Cellular response to organic cyclic compound (−5); excretion (−4); midgut development (−4)	STIL; TTLL12; DRD3; TOMM34; CKMT1B; EEA1; GUCA2B; MEP1A; HNF1A; CCBL1

[a]Spot letters are assigned in the mRNA summary maps given in [4]

[b]Up- or downregulated mainly in tumor (T). T_DN spots are usually N_UP spots and vice versa

[c]Particular samples showing this spot with up- (UP) or down- (DN) regulated genes

[d]Top enriched gene sets of the gene ontology category "Biological Process." The number in the brackets is the logged *p*-value according to Fishers exact test, e.g., −6 means $p \sim 10^{-6}$

[e]Top genes in the list of most concordant mRNAs according to the concordance *t*-score

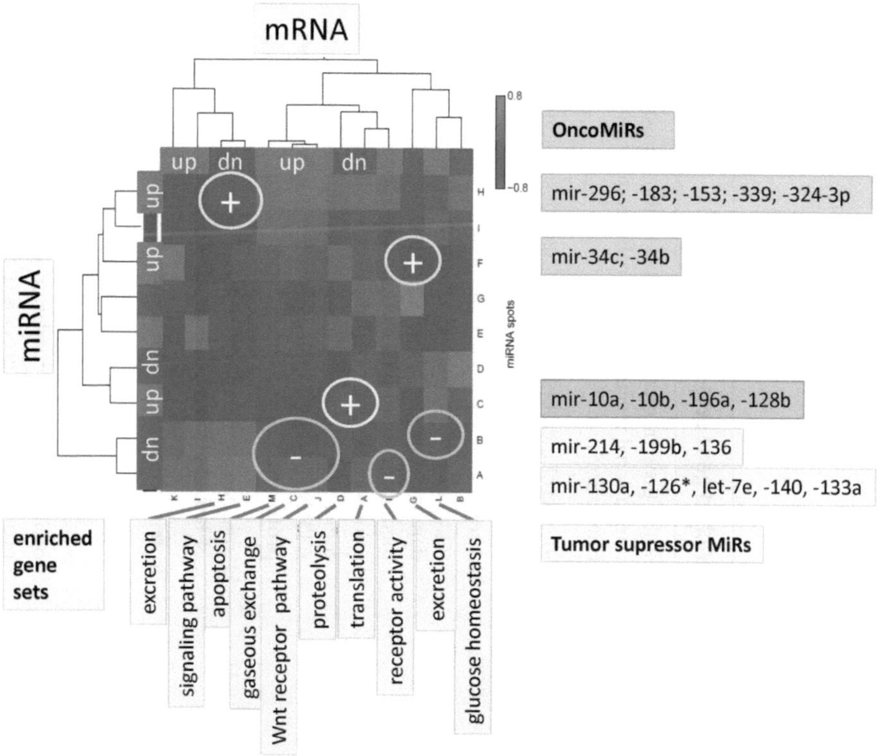

Fig. 8 Pairwise miRNA/mRNA spot covariance map of the LU-data set. The spots are taken from the respective spot summary maps of miRNA and mRNA overexpression shown in [4]. The map color codes the covariance of all pairwise combinations of spots using their mean metagene expression profiles ($-0.8 < cov < +0.8$). Their preferential up- or downregulation in cancer is indicated by the *bars* at the *left* and *top* borders of the map. *Minuses* and *plusses* indicate potential tumor suppressors (miRNA_DN and mRNA_UP) and OncoMiRs (miRNA_UP and mRNA_DN), respectively. The respective top-ranked miRNAs and enriched gene sets are taken from Tables 3 and 4, respectively

as tumor suppressors (indicated by minuses in Fig. 8) and as OncoMiRs (plusses) as well. The miRNAs regulated in the respective spots of the SOM portraits and the top-enriched gene sets in the respective mRNA spots are given in Fig. 8. For example, tumor-suppressor miRNAs of the let-7, mir-126, and mir-130 families are clearly identified together with upregulated mRNAs showing enrichment of processes such as "translation" and "receptor activity." OncoMiRs such as mir-10, -196, and -128 are related to mRNA downregulation in the context of "translation" and "proteolysis".

As a second option of joint miRNA and mRNA analysis we trained the SOM based on the combined covariance features (*see* Chapter 18 for details). It quantifies the degree of concerted miRNA and mRNA expression in each sample. Each spot in the obtained portraits thus refers to miRNA/mRNA metagene pairs which, in turn, are each associated with lists of single miRNA and mRNA interactions. Table 5 provides the particular miRNAs

Table 5
Combined miRNA/mRNA expression in tumor and normal tissues (LU-data set)

Spot[a]	N_DN[b]	T_DN[b]	miRNA in the spot[c]	Enriched gene sets in the spot[d]
a	N_bladder		{mir-196a; -128b; -30a-3p; -30b; -30c; -10b; -10a; -190}; -204; -135a; -135b; -124a; -187	Cellular protein metabolic process; viral transcription; translational termination; aerobic respiration; translation
b		T_prostate	{mir-199b; -136; -214; -199a; -17-3p	Cholesterol metabolic process; hemidesmosome assembly
c		T_lung	{mir-1; -133a; -99a}; {-199b; -136; -214; -199a; -199a}; -338; -100; -125b; -195; -130a; -189; {let-7d; let-7e; -126; -140; -126; -101}; -215; -194; -192; -106b; -142-3p; -142-5p	Neuromuscular synaptic transmission; positive regulation of glycogen biosynthetic process; DNA damage response, signal transduction by p53 class mediator resulting in induction of apoptosis; ion transport; synaptic transmission, cholinergic; transferrin transport; smoothened signaling pathway; ATP hydrolysis-coupled proton transport; transport; proton transport
d	N_breast		{mir-205; -323; -144; -10a; -215; -194; -192}	Hemidesmosome assembly; RNA splicing; protein N-linked glycosylation via asparagine
e	N_prostate		{mir-196a; -128b; -10b; -10a; -190}	Protein N-linked glycosylation via asparagine
f	N_lung		{mir-1; -133a; -99a}; {-100; -125b; -195; -130a; -189}	Protein N-linked glycosylation via asparagine
g	N_uterus		{mir-1; -133a; -99a}; -205; {-100; -125b; -195}; -323; -144; -200a; -200b; -141	Protein N-linked glycosylation via asparagines; cellular protein metabolic process; Leydig cell differentiation
h	N_lung	T_breast	{mir-205; -323; -144; -215; -194; -192}; -200a; -200b; -141	Aerobic respiration; response to reactive oxygen species

i		T_uterus	{mir-1; -133a; -99a}; {-100; -125b; -195: -130a; -189}; -197; {let-7d; let-7e; -126; -140; -126; -101}; -182; -200a; -200b; -141	Tissue development; epidermis development; rRNA processing; RNA metabolic process
j	N_colon, N_uterus	T_colon,	mir-199a; -17-3p; -335; -181b; -181c	Methylation; single fertilization
k		T_bladder, T_breast	mir-7; -106b; -142-3p; -142-5p	Hemidesmosome assembly; cholesterol transport; cholesterol metabolic process; cell junction assembly
l	N_prostate, N_breast		{mir-1; -133a; -99a}; {-100; -125b; -195; -130a; -189}; -197; -335; {let-7d; let-7e; -126; -140; -126; -101}	Fibrinolysis; complement activation; collagen fibril organization; complement activation; classical pathway

[a]"Underexpression" spots of the combined map shown in ref. 4

[b]Systems with downregulated spots

[c]miRNA extracted from the miRNA/mRNA metafeature combinations taken from the respective spot. The brackets collect miRNAs referring to one miRNA metagene (see text)

[d]Enriched gene sets taken of the gene ontology category "biological process" in the list of mRNA extracted from the miRNA/mRNA metafeature combinations taken from the respective spot

collected in each of the "antiexpression" spots identified in the combined cov-portraits shown in ref. 4. The mRNA species collected in each of the spots are characterized using gene set enrichment analysis using gene sets of the gene ontology category "biological process".

Note that spots of the cov-map refer to combined miRNA and mRNA metagenes, each representing lists of single miRNA and mRNA genes. One and the same metagene-related list of miRNAs (and of mRNAs) can be found in different spots: For example, mir-1, -133a, and -99a appear together in antiexpression spots c, f, g, i, and l (as indicated by {...} in Table 5) however in combination with different mRNA metagenes and thus with different functional themes such as "glycogen biosynthesis" (spot c), "protein glycosylation via asparagine" (f and g), "tissue development" (i), and "fibrinolysis" (f). On the other hand, similar lists of mRNAs giving rise to enrichment of genes of the set "protein glycosylation via asparagine" appear in combination with different miRNA lists such as {mir-196a; -128b; -10b; -10a; -190} (spot e), {mir-1, -133a, and -99a} (f and g), and {-100; -125b; -195} (g). Finally, the cov-spots of antiexpression are predominantly observed either in cancer (e.g., spots c and i) or in healthy (f and g) tissues. Hence, the combined SOM provides very detailed and diverse view on interrelations between miRNA and mRNA expression patterns. Its resolution clearly exceeds that of the spot–spot correlation analysis discussed above.

5 Summary and Conclusions

Case studies illustrating the potency of the portraying approach using SOMs were presented. Sample portraits of ESC, IPS cells, and differentiated fibroblasts clearly reveal groups of miRNAs specifically overexpressed without the need of additional pairwise comparisons between the different systems. More than one dozen miRNAs are found to be differentially expressed between ESC and IPS reflecting the difference between the two kinds of stem cells with respect to miRNA activity.

The much more heterogeneous series of human tissues splits roughly into five groups (brain, muscle, epithelial, gastrointestinal, sexual organs) according to their miRNA expression landscapes. They are characterized by about one dozen distinguishable expression modules. This diversity of miRNA expression is comparable with the diversity of mRNA expression patterns in human tissues despite the lower number of miRNA species available. Mixed samples with miRNA signatures of different tissues can be clearly identified. For example pericardium combines signatures of muscle and adipose tissues. Interestingly, immune system tissues are clearly separated from other tissue types according to their mRNA expression signature in contrast to the miRNA expression patterns.

The joint analysis of miRNAs and mRNAs in healthy and tumor samples allows identifying potential OncoMiRs and tumor-suppressor MiRs, their potential mRNA targets, and functional annotations according to positive and negative correlations between both entities and enrichment of sets of genes of known function.

In summary, these analyses demonstrate that the individual portraying of the expression landscape of each sample is highly sophisticated because it provides unmistakable fingerprints of the underlying expression phenotypes. It enables a very intuitive, image-based perception which clearly promotes the discovery of qualitative relationships between the samples in the absence of existing hypotheses. We see perspectives for broad applications of this method in standard analysis of single miRNA and especially combined miRNA/mRNA data set analysis.

Acknowledgements

This publication is supported by LIFE Center for Civilization Diseases, University of Leipzig, Leipzig, Germany. LIFE is funded by the European ERDF fund and by the Free State of Saxony.

References

1. Hobert O (2004) Common logic of transcription factor and microRNA action. Trends Biochem Sci 29:462–468
2. Friedman RC, Farh KK-H, Burge CB, Bartel DP (2009) Most mammalian mRNAs are conserved targets of microRNAs. Genome Res 19:92–105
3. Baskerville S, Bartel DP (2005) Microarray profiling of microRNAs reveals frequent coexpression with neighboring miRNAs and host genes. RNA 11:241–247
4. Wirth H, Cakir V, Hopp L, Binder H (2013) Analysis of miRNA expression using machine learning. Methods of Molecular Biology. Springer, New York, NY
5. Liu T et al (2007) Detection of a microRNA signal in an *in vivo* expression set of mRNAs. PLoS One 2:e804
6. Liang Z, Zhou H, Zheng H, Wu J (2011) Expression levels of microRNAs are not associated with their regulatory activities. Biol Direct 6:43
7. Alexiou P, Maragkakis M, Hatzigeorgiou AG (2010) Online resources for microRNA analysis. J Nucleic Acids Investig 2:e4
8. Kertesz M, Iovino N, Unnerstall U, Gaul U, Segal E (2007) The role of site accessibility in microRNA target recognition. Nat Genet 39: 1278–1284
9. Krek A et al (2005) Combinatorial microRNA target predictions. Nat Genet 37:495–500
10. Papadopoulos GL, Reczko M, Simossis VA, Sethupathy P, Hatzigeorgiou AG (2009) The database of experimentally supported targets: a functional update of TarBase. Nucleic Acids Res 37:D155–D158
11. Xiao F, Zuo Z, Cai G, Kang S, Gao X, Li T (2009) miRecords: an integrated resource for microRNA–target interactions. Nucleic Acids Res 37:D105–D110
12. Naeem H, Kuffner R, Csaba G, Zimmer R (2010) miRSel: automated extraction of associations between microRNAs and genes from the biomedical literature. BMC Bioinformatics 11:135
13. Cheng C, Li LM (2008) Inferring microRNA activities by combining gene expression with microRNA target prediction. PLoS One 3:e1989
14. Kohonen T (1982) Self-organized formation of topologically correct feature maps. Biol Cybern 43:59–69
15. Törönen P, Kolehmainen M, Wong G, Castrén E (1999) Analysis of gene expression data using self-organizing maps. FEBS Lett 451:142–146
16. Eichler GS, Huang S, Ingber DE (2003) Gene expression dynamics inspector (GEDI): for integrative analysis of expression profiles. Bioinformatics 19:2321–2322
17. Wirth H, Loeffler M, von Bergen M, Binder H (2011) Expression cartography of human tissues using self organizing maps. BMC Bioinformatics 12:306

18. Wilson KD, Venkatasubrahmanyam S, Jia F, Sun N, Butte AJ, Wu JC (2009) MicroRNA profiling of human-induced pluripotent stem cells. Stem Cells Dev 18:749–757
19. Liang Y, Ridzon D, Wong L, Chen C (2007) Characterization of microRNA expression profiles in normal human tissues. BMC Genomics 8:166
20. Lu J et al (2005) MicroRNA expression profiles classify human cancers. Nature 435: 834–838
21. Mallanna SK, Rizzino A (2010) Emerging roles of microRNAs in the control of embryonic stem cells and the generation of induced pluripotent stem cells. Dev Biol 344: 16–25
22. Binder H, Hopp L, Cakir V, Fasold M, von Bergen M, Wirth H (2011) Genomic and molecular phenotypic portraits – exploring the 'OMes' with individual resolution, Proceedings: Health Informatics and Bioinformatics (HIBIT), 6th International Symposium Izmir; IEEE Xplore: 99–107
23. Israel A, Sharan R, Ruppin E, Galun E (2009) Increased microRNA activity in human cancers. PLoS One 4:e6045
24. Visone R, Croce CM (2009) MiRNAs and cancer. Am J Pathol 174:1131–1138
25. Hamano R, Ishii H, Miyata H, Doki Y, Mori M (2010) Role of microRNAs in solid tumors. J Nucleic Acids Investig 2:e2
26. Osada H, Takahashi T (2007) MicroRNAs in biological processes and carcinogenesis. Carcinogenesis 28:2–12
27. Raia R, Calin GA (2010) Non-coding RNAs and cancer: microRNAs and beyond. J Nucleic Acids Investig 2:e5

Chapter 18

Master Regulators of Posttranscriptional Gene Expression Are Subject to Regulation

Syed Muhammad Hamid and Bünyamin Akgül

Abstract

MicroRNAs (miRNAs) are small noncoding RNAs of 17–25 nt in length that control gene expression posttranscriptionally. As master regulators of posttranscriptional gene expression, miRNAs themselves are subject to tight regulation at multiple steps. The most common mechanisms include miRNA transcription, processing, and localization. Additionally, intricate feedback loops between miRNAs and transcription factors result in unidirectional, reciprocal, or self-directed elegant control mechanisms. In this chapter, we focus on the posttranscriptional regulatory mechanisms that generate miRNAs whose sequence might be slightly different from the miRNA-coding sequences. Hopefully, this information will be helpful in the discovery of novel miRNAs as well as in the analysis of deep-sequencing data and ab initio prediction of miRNAs.

Key words miRNA, miRNA regulation, Small RNAs, Posttranscription

1 Introduction

Regulation of the genetic information has been the key to the evolution of complex, multicellular organisms and the generation of biological diversity. Intricate molecular regulatory programs that span over DNA, RNA, and protein levels dictate the biological activity of any gene product. In 1993, a novel mechanism of posttranscriptional regulation of gene expression by small noncoding RNAs was discovered in animals [1]. Over time, the number of these regulatory RNAs has enormously increased, and they have been shown to be active from development to disease [2–4].

The estimated number of all microRNAs (miRNAs) reaches out to nearly 1–5 % of all the predicted genes in nematodes, flies, and mammals [5–7]. A large number of these miRNA genes are dispersed throughout the genome. Some miRNAs are found in clusters that are co-expressed as polycistronic units showing their functional relationships. More than half of miRNAs reside in introns of their host genes and are co-expressed with their neighboring

Malik Yousef and Jens Allmer (eds.), *miRNomics: MicroRNA Biology and Computational Analysis*, Methods in Molecular Biology, vol. 1107, DOI 10.1007/978-1-62703-748-8_18, © Springer Science+Business Media New York 2014

protein-coding sequences. [7–9]. Recent developments in sequencing technologies (e.g., deep sequencing) have accelerated novel miRNA discoveries as it is now possible to identify rarely expressed miRNAs owing to the high coverage rate of deep sequencing technologies. This technology also allows for extensive and detailed comparison of miRNA expression under various physiological and pathological cellular states. However, single-nucleotide changes introduced into miRNAs during biogenesis or variations at 5′ and 3′ termini require careful bioinformatics analyses so as not to lose any valuable information that may be associated with the phenotype of interest. As an introduction to miRNAs in biological systems and the role of miRNAs in human diseases are covered in the first two chapters, we discuss the transcriptional regulation of miRNAs followed by a focus on the posttranscriptional regulatory mechanisms.

2 Transcriptional Regulation of MicroRNA Genes

Despite the initial predictions about the intergenic localization of miRNA genes, subsequent findings showed that the majority of mammalian miRNA genes are organized within transcription units [8, 10]. Based on their genomic location, miRNAs can be contained within intronic sequences in protein-coding or noncoding transcription units or exonic sequences in noncoding transcription units. Since Drosha processing precedes splicing, a single transcript can give rise to both miRNAs and an mRNA [9, 11, 12]. Thus, the transcriptional regulation of miRNAs depends upon the genomic localization of miRNA genes.

In most of the cases, miRNAs are transcribed by RNA polymerase II which initially generates a primary miRNA (pri-miRNA) transcript consisting of one or more hairpin structures. These hairpins are then processed both in the nucleus and the cytoplasm to generate the mature miRNA. Pri-miRNAs have 5′ cap, undergo splicing, may have poly A tail, and often generate more than one functional miRNA [13].

A tightly controlled multi-step process of transcription provides the first line of regulation of miRNA expression, RNA polymerase II being the major player (Fig. 1). The promoter structure of miRNA genes has been shown to be largely similar to that of protein-coding genes [14, 15]. Consequently, the members of the transcription machinery also largely overlap between protein and miRNA genes. Some transcription factors have been shown to regulate the expression of both proteins and miRNAs, c-myc, p53, and hypoxia-inducible factor (HIF) being potent examples. C-myc has been shown to bind to the E-boxes in the promoter and activate the transcription of the miR 17-92 cluster [16]. STAT1 is another transcription factor that has recently been reported to regulate about 9 % of the total 1,105

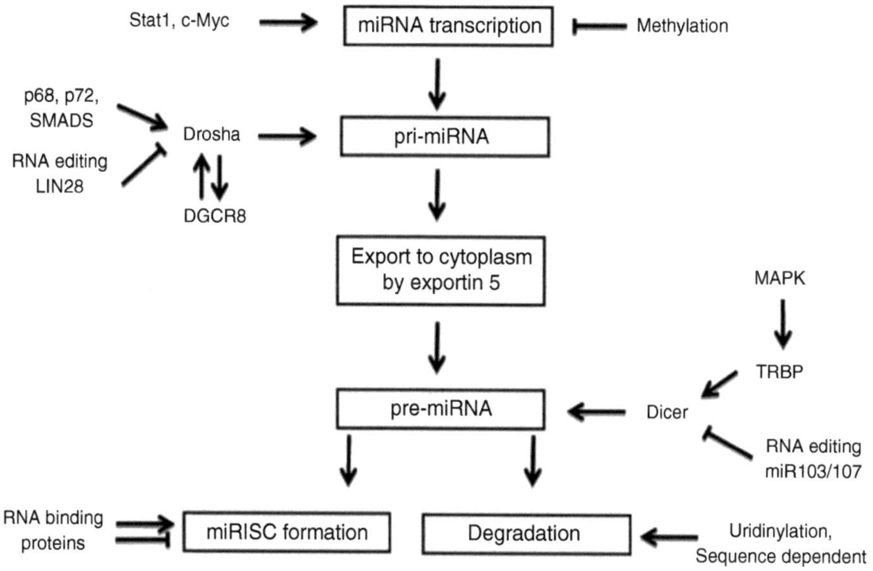

Fig. 1 A brief overview of transcriptional and posttranscriptional regulation of miRNAs. miRNA transcription itself might be subject to regulation by transcription factors and/or modulators of chromatin structure, e.g., methylation. The nuclear processing by Drosha might generate miRNAs with heterogeneous termini in addition to internal editing events. Similar terminal heterogeneity and internal editing events might be introduced in the cytoplasm during processing by Dicer. Additionally, modulation of intracytoplasmic RISC location and 3′-uridinylation adds further complexity to the cytoplasmic regulatory events that determine the fate of a miRNA

miRNAs in response to interferon γ stimulation in melanoma cells [17]. There may also be unilateral, reciprocal, or double-negative feedback loops between miRNAs and transcription factors [18]. For example, PITX3, a transcription factor involved in dopaminergic neuron differentiation, activates miR-133b transcription, which results in suppression of PITX3 expression through a negative autoregulatory mechanism [19].

Another mechanism of transcriptional control of miRNA biogenesis lies in their epigenetic regulation. Epigenetic modifications of miRNA loci may result in altered transcription of these genes. In several human cancers, promoter regions of the genes encoding miR-9-1, -193a, -137, -342, -203, and -34b/c have been shown to be hypermethylated ([20], Chapter 19). Histone deacetylase (HDAC) inhibitors have been reported to increase the expression of some miRNAs including miR-1 in cancer cells [21, 22].

3 Regulation of Drosha Activity

In the nucleus, a multiprotein complex, called microprocessor complex, cuts the pri-miRNAs into about 70 nt long hairpin structures called pre-miRNAs. Drosha, an RNase III enzyme, and

DGCR8 (Pasha), a double-strand RNA (dsRNA)-binding protein, are the vital components of the microprocessor complex along with cofactors including the DEAD box RNA helicases p68, p72, and heterogeneous nuclear proteins [23]. Drosha cleaves the pri-miRNA co-transcriptionally to generate a product with 2 nt overhang at the 3′ end [9].

There are many factors that modulate Drosha activity both positively and negatively. Two members of the DEAD box RNA helicase family, p68 and p72, have been reported as components of Drosha and DGCR8 [23]. Both single- and double-knockout p68 and p72 are lethal in mice [24]. These factors probably serve as scaffold proteins that recruit other modulatory proteins to the microprocessor complex to enhance pri-miRNA processing. LIN-28 and SMADs are other well-known examples of accessory proteins that modulate the processing efficiency of Drosha [18].

Regulation of the total protein levels of Drosha and DGCR8 also plays a vital role in the regulation of miRNA processing. DGCR8 has a stabilizing effect on the Drosha protein level [25]. In turn, Drosha determines the DGCR8 levels by processing hairpins present in the DGCR8 transcript. Consequently, the ratio of Drosha to DGCR8 appears to be crucial for the activity of the microprocessor complex. In addition to cleaving pre-miRNA hairpins, the microprocessor complex can also promote cleavage of hairpin structures within annotated protein-coding genes [25, 26], a more direct form of gene regulation than the indirect miRNA approach.

An important issue to consider while analyzing miRNA data generated from SAGE or deep sequencing is the heterogeneity at the miRNA termini. Not only Drosha but also Dicer generates nonuniform miRNA termini. Certainly this heterogeneity could have a dramatic effect on the duplex stability and strand selection [27]. As a result, miRNA:miRNA* ratio would change if extensive heterogeneity is introduced. The terminal heterogeneity also influences the target mRNA selection, particularly the heterogeneity at the 5′ end that affects the seed register of miRNAs.

4 Export of Pre-miRNA to the Cytoplasm

Exportin 5 is responsible for the transfer of pre-miRNA from nucleus to the cytoplasm in a ran-GTP-dependent manner [28–30]. A 16–18 base pair long stem of pre-miRNA and modifications of 3′ overhang affect its binding to and transport efficiency by exportin 5 [31]. In the cytoplasm, pre-miRNA is released from exportin 5 by the hydrolysis of GTP and is processed further. This nuclear-to-cytoplasmic transfer may be differentially regulated under certain conditions. The precursor hairpins of miR-105, -108, and -31 are found at high levels in many cells, but mature miRNAs are not detectable [32]. A specific example of this phenomenon is miR-31. The nuclear form of this miRNA is almost equally expressed in both MCF7 and HS766T cell lines. In HS766T cells, high levels of mature miR-31 are found but are not detectable in MCF7. In HS766T cells, pre-miR31 shows cytoplasmic localization while it is accumulated in the nucleolus in MCF7 cells. This shows that the cytoplasmic export of miR-31 is cell type dependent and is regulated by some factors that are still to be unraveled.

5 Regulation of Dicer Activity

Once in the cytoplasm, another RNase III enzyme, Dicer, cuts the pre-miRNA into a ~22 nt long miRNA duplex with the help of dsRBD proteins TRBP/PACT [33–35]. The two miRNA strands are then separated, and one of the strands associates with Argonaute protein to form the RNA-induced silencing complex (RISC).

Regulation of miRNA processing by Dicer involves inhibition of Dicer activity. In colorectal tumor samples, miR-143 and miR-145 show very low expression as compared to the normal tissues despite the equal pre-miRNA levels. This suggests that the nuclear transcription and processing are similar, but the cytoplasmic processing by Dicer may be altered in colorectal tumor samples [36]. Another example includes developmental regulation of miR-138 processing. The mature form of this miRNA is only detectable in adult mouse brain and foetal liver, whereas pre-miR138 is expressed in all tissues [37].

Dicer activity is known to be modulated by several factors. Perhaps the best-characterized protein is TRBP, which is required for Dicer stability as mutations in TRBP lead to the impairment of Dicer function. TRBP in turn is stabilized by MAPK-mediated phosphorylation of serine residues [38], establishing a link between signalling pathways and Dicer-mediated modulation of cellular functions. Accessory proteins may regulate Dicer activity, resulting in the modulation of specific miRNAs, rather than modulating global Dicer activity. LIN-28 binds to the terminal loop of pri-let-7 in embryonic stem cells and blocks its processing by Dicer [39]. KSRP is another protein that promotes processing by Dicer [18].

The expression of Dicer may also be subject to regulation by miRNAs. The miR-103/107 family has been shown to target the expression of Dicer and down-regulate global miRNA biogenesis [40, 41]. The miR-103/107-mediated regulation of Dicer expression is associated with increased epithelial-mesenchymal transition (EMT) and metastasis. The amino terminal helicase domain of Dicer may have an autoinhibitory function, as removal of this domain increases the catalytic activity of Dicer processing [42].

6 RNA Editing

RNA editing is catalyzed by members of the adenosine deaminase acting on RNA (ADAR) protein family, found in most metazoans [43]. A-to-I editing may happen in all types of RNA where adenosine is converted to inosine through hydrolytic deamination. The stem–loop structures of both pri- and pre-miRNAs are potential targets for editing where editing of pri-miRNAs may occur at more than one adenosine (residue) [44–46]. Pri-miR-22 was the first miRNA to be reported as a miRNA edited at six different positions including sites within the mature miRNA. Editing of miRNAs may decrease their stability or inhibit their processing by Drosha or Dicer. For example, editing of pri-miR-142 at two adenosines close to the Drosha cleavage site not only inhibits its processing by Drosha but also makes it less stable. Editing of miR-151 inhibits its processing by Dicer but not by Drosha [47].

Editing events not only influence the processing efficiency of miRNAs, but it also affects the target mRNA selection. Because there is partial complementarity between the miRNAs and the 3′ untranslated region of target mRNAs, even the change in a single-nucleotide sequence may have a dramatic effect on the stability of a miRNA:mRNA pairing. Thus, it is quite important, during bioinformatics analyses, to consider potential editing events to obtain information at the maximum level. It is quite challenging, however, to identify the edited miRNAs as it is possible for the sequences with single-nucleotide mismatches to originate from other genomic loci. Thus, prior to calling a sequence an edited miRNA, it is imperative that the genomic origin of the sequences be identified unequivocally. Additionally, functional reporter assays would be required to illustrate the functional significance of these editing events.

7 Conclusion

Since the discovery of miRNAs way back in 1993, the number and scope of their function have tremendously increased over the years. They not only fulfil the duty of a carrier of information between DNA and protein but also fine-tune the various steps of posttranscriptional regulation to maintain the tight balance in protein expression. To achieve this goal, miRNAs are themselves very tightly regulated both at the levels of transcription and posttranscription. An extreme example of miRNA gene regulation comes from the chromatin studies. Cis-regulatory elements of gene expression, scaffold/matrix attachment regions (MARs), have been shown to define the cell-specific expression of let-7b, miR-93, miR-17, and miR-221 by tethering the chromatin to the nuclear matrix [48]. A minority of miRNAs has been found to respond to circadian rhythm regulatory mechanisms. For example miR-219 is targeted by the CLOCK and BMAL1 complexes [49]. So the cell- and context-type specificity of miRNAs appear to involve a coordinated activity of cis-regulation, transcription factor binding, and chromatin modulation resulting in specific gene expression. This also involves the tightly regulated miRNA processing factors both in the nucleus and cytoplasm, loading of mature miRNA to the RISC complex, editing, and degradation to ensure a precise balance in the expression of miRNAs.

Acknowledgement

This work was supported by the Scientific and Technical Research Council of Turkey (104T144, 107T475, and 210T006 to BA).

References

1. Lee Y, Feinbaum RL, Ambros V (1993) The C. Elegans heterochronic gene lin-4 encodes small RNAs with antisense complementarity to lin-14. Cell 75:843–854
2. Bushati N, Cohen SM (2007) Micro RNA functions. Annu Rev Cell Dev Biol 23:175–205
3. Kim VN, Han J, Siomi MC (2009) Biogenesis of small RNAs in animals. Nat Rev Mol Cell Biol 10:126–139
4. Fabian MR, Sonenberg N (2012) The mechanics of miRNA-mediated gene silencing: a look under the hood of miRISC. Nat Struct Mol Biol 19:586–593
5. Lai EC, Tomancak P, Williams RW et al (2003) Computational identification of *Drosophila* microRNA genes. Genome Biol 4:R42
6. Lim LP, Glasner ME, Yekta S et al (2003) Vertebrate microRNA genes. Science 299:1540
7. Baskerville S, Bartel DP (2005) Microarray profiling of microRNAs reveals frequent coexpression with neighbouring miRNAs and host genes. RNA 11:241–247
8. Rodriguez A, Griffiths JS, Ashurst JL et al (2004) Identification of mammalian microRNA host genes and transcription units. Genome Res 14:1902–1910
9. Kim YK, Kim VN (2007) Processing of intronic microRNAs. EMBO J 26:775–783
10. Lau NC, Lim LP, Weinstein EG et al (2001) An abundant class of tiny RNAs with probable regulatory roles in *Caenorhabditis elegans*. Science 294:858–862
11. Cai X, Hagedorn CH, Cullen BR (2004) Human microRNAs are processed from capped, polyadenylated transcripts that can also function as mRNAs. RNA 10:1957–1966
12. Morlando M, Ballarino M, Gromak N et al (2008) Primary microRNA transcripts are processed co-transcriptionally. Nat Struct Mol Biol 15:902–909
13. Carthew RW, Sontheimer EJ (2009) Origins and mechanisms of miRNAs and siRNAs. Cell 136:642–655
14. Ozsolak F, Poling LL, Wang Z et al (2008) Chromatin structure analyses identify miRNA promoters. Genes Dev 22:3172–3183
15. Corcoran DL, Pandit KV, Gordon B et al (2009) Features of mammalian microRNA promoters emerge from polymerase II chromatin immunoprecipitation data. PLoS One 4:e5279
16. O'Donnell KA, Wentzel EA, Zeller KI et al (2005) c-Myc regulated microRNAs modulate E2F1 expression. Nature 435:839–843
17. Susanne ER, Petr VN, Demetra P et al (2012) RNA Biol 9:978–989
18. Krol J, Loedige I, Filipowicz W (2010) The widespread regulation of microRNA biogenesis, function and decay. Nat Rev Genet 11:597–610
19. Kim J, Inoue K, Ishii J et al (2007) A microRNA feedback circuit in midbrain dopamine neurons. Science 317:1220–1224
20. Lujambio A, Calin GA, Villanueva A et al (2008) A microRNA DNA methylation signature for human cancer metastasis. Proc Natl Acad Sci U S A 105:13556–13561
21. Nasser MW, Datta J, Nuovo G et al (2008) Suppression of tumorigenic property of lung cancer cells and their sensitization to doxorubicin

induced apoptosis by miR-1. J Biol Chem 283:33394–33405
22. Saito Y, Liang G, Egger G et al (2006) Specific activation of microRNA-127 with downregulation of the proto-oncogene BCL6 by chromatin-modifying drugs in human cancer cells. Cancer Cell 9:435–443
23. Gregory RI, Yan KP, Amuthan G et al (2004) The microprocessor complex mediates the genesis of microRNAs. Nature 432:235–240
24. Fukuda T, Yamagata K, Fujiyama S et al (2007) DEAD-box RNA helicase subunits of the Drosha complex are required for processing of rRNA and a subset of microRNAs. Nat Cell Biol 9:604–611
25. Han J, Lee Y, Yeom KH et al (2006) Molecular basis for the recognition of primary microRNAs by the Drosha-DGCR8 complex. Cell 125:887–901
26. Han JJ, Lee Y, Yeom KH et al (2004) The Drosha-DGCR8 complex in primary microRNA processing. Genes Dev 18:3016–3027
27. Chiang HR, Schoenfeld LW, Ruby JG et al (2010) Mammalian microRNAs: experimental evaluation of novel and previously annotated genes. Genes Dev 24:992–1009
28. Yi R, Qin Y, Macara IG et al (2003) Exportin-5 mediates the nuclear export of pre-microRNAs and short hair pin RNAs. Genes Dev 17:3011–3016
29. Bohnsack MT, Czaplinski K, Gorlich D (2005) Exportin 5 is a RanGPT-dependent dsRNA binding protein that mediates nuclear export of pre-microRNAs. RNA 10:185–191
30. Lund E, Guttinger S, Calado A et al (2004) Nuclear export of microRNA precursors. Science 303:1959
31. Zeng Y, Cullen BR (2004) Structural requirements for pre-microRNA binding and nuclear export by Exportin 5. Nucleic Acids Res 32:4776–4785
32. Lee EJ, Baek M, Gusev Y et al (2008) Systematic evaluation of microRNA processing patterns in tissues, cell lines and tumors. RNA 14:35–42
33. Chendrimada TP, Gregory RI, Kumaraswamy E et al (2005) TRBP recruits the Dicer complex to Ago2 for microRNA processing and gene silencing. Nature 436:740–744
34. Haase JP, Piskounova E, Gregory RI (2009) Lin28 recruits the TUTase Zcchc11 to inhibit let-7 maturation in mouse embryonic stem cells. Nat Struct Mol Biol 16:1021–1025
35. Lee Y, Hur I, Park SY et al (2006) The role of PACT in the RNA silencing pathway. EMBO J 25:522–532
36. Michael MZ, O'Connor SM, van Holst Pellekaan NG et al (2003) Reduced accumulation of specific microRNAs in colorectal neoplasia. Mol Cancer Res 1:882–891
37. Obernosterer G, Leuschner PJ, Alenius M et al (2006) Post-transcriptional regulation of microRNA expression. RNA 12:1161–1167
38. Paroo Z, Ye X, Chen S et al (2009) Phosphorylation of the human microRNA generating complex mediates MAPK/Erk signalling. Cell 139:112–122
39. Viswanathan SR, Daley GQ (2010) Lin28: a microRNA regulator with a macro role. Cell 140:445–459
40. Grelier G, Voirin N, Ay S et al (2009) Prognostic value of Dicer expression in human breast cancers and association with the mesenchymal phenotype. Br J Cancer 101:673–683
41. Martello G, Rosato A, Ferrari F et al (2010) A microRNA targeting dicer for metastasis control. Cell 141:1195–1207
42. Ma E, MacRae IJ, Kirsch JF et al (2008) Autoinhibition of human dicer by its internal helicase domain. J Mol Biol 380:237–243
43. Jin Y, Zhang W, Li Q (2009) Origins and evolution of ADAR-mediated RNA editing. IUBMB Life 61:572–578
44. Luciano DJ, Mirsky H, Vendetti NJ et al (2004) RNA editing of a miRNA precursor. RNA 10:1174–1177
45. Yang W, Chendrimada TP, Wang Q et al (2006) Modulation of microRNA processing and expression through RNA editing by ADAR deaminases. Nat Struct Mol Biol 13:13–21
46. Kawahara Y, Zinshteyn B, Chenrimada TP et al (2007) RNA editing of the microRNA-151 precursor blocks cleavage by the Dicer-TRBP complex. EMBO Rep 8:763–769
47. Kawahara Y, Megraw M, Krieder E et al (2008) Frequency and fate of microRNA editing in human brain. Nucleic Acids Res 36:5270–5280
48. Chavali PL, Funa K, Chavali S (2011) Cis-regulation of microRNA expression by scaffold/matrix-attachment regions. Nucleic Acids Res 39:6908–6918
49. Cheng HY (2007) MicroRNA modulation of circadian-clock period and entrainment. Neuron 54:813–829

Chapter 19

Use of MicroRNAs in Personalized Medicine

Çiğir Biray Avci and Yusuf Baran

Abstract

Personalized medicine comprises the genetic information together with the phenotypic and environmental factors to yield healthcare tailored to an individual and removes the limitations of the "one-size-fits-all" therapy approach. This provides the opportunity to translate therapies from bench to clinic, to diagnose and predict disease, and to improve patient-tailored treatments based on the unique signatures of a patient's disease and further to identify novel treatment schedules.

Nowadays, tiny noncoding RNAs, called microRNAs, have captured the spotlight in molecular biology with highlights like their involvement in DNA translational control, their impression on mRNA and protein expression levels, and their ability to reprogram molecular signaling pathways in cancer. Realizing their pivotal roles in drug resistance, they emerged as diagnostic targets orchestrating drug response in individualized therapy examples.

It is not premature to think that researchers could have the US Food and Drug Administration (FDA)-approved kit-based assays for miRNA analysis in the near future. We think that miRNAs are ready for prime time.

Key words miRNAs, Personalized medicine, Pharmacogenomics

1 Introduction

Pharmacogenomics investigations have emphasized genes that contribute to an individual patient's drug sensitivity, resistance, and toxicity. It has also designated the causes of interindividual variations in the expression and function of many of the genes, including the roles of microRNAs, DNA methylation, copy number variations, and single-nucleotide polymorphisms.

In this chapter, we focus on miRNAs, 19–24-nucleotide noncoding RNAs that function as gene regulators and have roles in countless cellular processes. We discuss how miRNAs, arranging the expression of pharmacogenomic-relevant genes, play a considerable role in drug efficacy and toxicity and have potential clinical reflections for personalized medicine.

Malik Yousef and Jens Allmer (eds.), *miRNomics: MicroRNA Biology and Computational Analysis*, Methods in Molecular Biology, vol. 1107, DOI 10.1007/978-1-62703-748-8_19, © Springer Science+Business Media New York 2014

2 MicroRNA Pharmacogenomics

The National Cancer Institute of the National Institutes of Health, USA, defined "personalized medicine" in 2011 as a form of healthcare that considers information about a person's genes, proteins, and environment to prevent, diagnose, and treat disease [1]. Developments in the field of pharmacogenomics are not only due to better sequencing technology but also stem from gene expression alterations on the account of regulatory elements and epigenetic variations that occur in response to the environment. In the realm of RNA, these modulations usually result from two comprehensive components: first, the direct regulatory impact of miscellaneous forms of noncoding RNA (hnRNA, microRNA, etc.) that have the capacity to modulate expression by interacting with DNA regulatory sequences in promoters or in different regions of target genes and, second, the environmentally stimulated chemical modulation of DNA nucleotides, principally by cytosine methylation.

The application of genomics in personalized medicine includes its potential to improve risk assessment, diagnosis, prognosis, and treatment. Classical examples of current genomic investigation in personalized medicine are (1) BRCA1/2 testing for risk assessment of breast cancer, (2) gene expression profiles to diagnose breast cancer subtypes [2], (3) number of trinucleotide repeats predicting the seriousness of Fragile X syndrome [3], and (4) Herceptin® (trastuzumab) management that is restricted using a companion diagnostics to women protecting from HER2-positive breast cancer [4, 5] and CYP2C19 variants associated with minimized clopidogrel response [6].

Metabolism of xenobiotics (drugs, environmental chemicals, carcinogens) and endobiotics (steroids, bile acids, and fatty acids) are catalyzed by essential enzymes called P450s that are transcriptionally regulated by nuclear receptors.

Cytochrome P450s and nuclear receptors, such as CYP1B1, CYP2A3 (rat), CYP2E1, CYP3A4, CYP24A1, pregnane X receptor (PXR), vitamin D receptor (VDR), peroxisome proliferator-activated receptor γ (PPARγ), RXRα (rat), hepatocyte nuclear factor 4α (HNF4α), estrogen receptor α (ERα), and human glucocorticoid receptors (GR) that are regulated by miRNAs, were studied in a wide range of studies [7].

3 Diagnostic and Prognostic Implementations: MicroRNA Signatures in Solid Tumors

Potential links to the etiology of a disease have been obtained from sets of differentially expressed mRNAs and miRNAs (expression signatures) acquired via expression profiling of whole tumor samples (Table 1).

Table 1
Diagnostic and prognostic applications of miRNA signatures in solid tumors

Signature in breast cancer	Clinical application(s)	References
miR-7, -128a, -210, -516-3p	Disease progression (distant metastasis) of estrogen receptor+ lymph node– cases	[8]
miR-30a-3p, -30c, -182	Response to adjuvant tamoxifen in advanced ER+ cases	[9]
miR-128a, -135a, -767-3p, -769-3p	Recurrence-free survival in ER+ cases	[10]
miR-27b, -30c, -144, -150, -210, -342	Recurrence-free survival in ER– cases	[10]
miR-21, -181a	Recurrence-free and overall survival in all-comers	[11]
miR-21, -210, -221, -222	Recurrence-free and overall survival in ER–, progesterone receptor–, human EGF receptor 2– cases	[11]
mir-21	Response to neoadjuvant trastuzumab treatment, recurrence-free survival in stage I/II all-comers, overall survival in all-participants	[12–14]
miR-205	Disease recurrence and overall survival in ER–, progesterone receptor–, HER2– cases	[15]
miR-210	Disease progression (distant metastasis) in ER– lymph node– cases, disease progression (distant metastasis) in ER– PR– HER2– LN– cases, recurrence-free and overall survival in all attendants, response to tamoxifen in ER+ cases	[8, 16, 17]

In colorectal cancer, 18 miRNA signatures (*see* **Note 1**) are responsible for subtype classification like microsatellite stable (MSS) vs. microsatellite instable (MSI-high) [18, 19]. Additionally, miR-320 and -498 signatures are accountable for recurrence-free survival in stage II MSS cases [18].

The prevailing oncomir in colorectal cancer is miR-21 involved in diagnostic and prognostic applications. This worthy miRNA plays a key role in response to neoadjuvant chemoradiotherapy in advanced rectal cancer [20], recurrence-free survival in stage II colon cases [21], recurrence-free survival in all-comers [22], recurrence-free and overall in all-contributors [23], and overall survival in all-contributors [24].

In lung cancer, miRNA signatures are miscellaneous. There are 34 miRNA signatures (*see* **Note 2**) responsible for prediction of subtype classification into adenocarcinoma (AdCa) vs. squamous cell carcinoma (SCC) in male smokers [25]. Signatures of the miRNAs let-7a and miR-221, 137, -182, and -372 are essential for recurrence-free and overall survival in non-small-cell lung cancer (NSCLC) cases [26].

In the literature, it was shown that nineteen miRNA signature (*see* **Note 3**) can play a major role in prediction of overall survival in SCC cases [27], and let-7e and miR-34a, -34c-5p, -25, and -191 signatures have a role in estimation of overall survival in male smoker SCC cases [25].

According to the studies concerning lung cancer let-7a has prognostic factors alone like overall survival in NSCLC [28] and AdCa cases [29], miR-21 has prognostic value in overall survival in NSCLC [30] and SCC cases [31], and miR-34a is an important marker for recurrence-free and overall survival in NSCLC cases [32]. The miRNA miR-155 has crucial role in overall survival in AdCa cases [29], and miR-205 distinguishes AdCa from SCC [33–35].

These abovementioned issues contain important data for applications of miRNA signatures especially in solid tumors, but they are just preclinical data, and further validation and understanding of what these signatures reflect are required for converting them to approved approaches.

4 Correlation of Drug Resistance with Deregulation of MicroRNA Expression

Aberration of miRNA expression in malignant tumor cells is considerably observed and can be triggered by three distinct mechanisms:

1. Location of miRNAs in cancer-associated tissue.
2. Genomic region or at fragile sites, epigenetic regulation of miRNA genes.
3. Abnormalities in miRNA processing genes and proteins [36].

Changes in expression levels of miRNAs affect numerous target mRNAs and for this reason multiple proteins pioneering the diversity in the chemosensitivity of cancer cells through miscellaneous cellular processes. Cellular response to anticancer agents shifted by most of the miRNAs by means of survival signaling pathways and programmed cell death response modulation [37]. Additionally, there are reports about miRNAs affecting mechanisms such as drug targets and DNA repair systems [38, 39]. Ultimately, miRNAs also have duties in the regulation of drug metabolism by regulating the expression of drug-metabolizing enzymes and drug transporters [40].

Numerous studies have shown the effect of specific microRNAs on factors encompassed in apoptosis and survival pathways impressing anticancer drug sensitivity and resistance (Table 2).

Table 2
Some important miRNAs and drug resistance/effect

Expression change	miRNA	Resistance/ sensitivity	Drug	Target gene	Cancer or cell line	Reference
Upregulation of expression	Let-7a	Resistance	IFNγ	Caspase-3	Hepatocellular carcinoma	[41]
			Doxorobicin Paclitaxel		Human squamous cell carcinoma	
Overexpression	Let-7a/b	Radio-sensitization	-	RAS	Lung cancer	[42]
Overexpression	Let-7g	Resistance	-	RAS	Lung cancer	[42]
Increased expression	miR-1	Sensitivity	Doxorobicin	Mcl-1	NSCLC cells	[43]
Overexpression	miR-15b/16	Sensitivity	Several drugs	Bcl-2	Gastric cancer	[44]
Increased expression	miR-27a	Resistance	Paclitaxel	HIPK2	Ovarian cancer	[45]
Overexpression	miR-27a/451	Resistance	Vinblastine Doxorubicin	MDR1	Ovarian cancer Cervical cancer	[46]

Examples could be multiplied according to recent studies in literature due to the loss of, increased, or decreased expressions of 22 specific miRNAs (*see* **Note 4**) targeting genes (*see* **Note 4**) involved in apoptosis and survival pathways affecting anticancer drug sensitivity in various cancer types including breast, cervical, prostate, colorectal, colon, cholangiocarcinoma, T-cell leukemia, and osteosarcoma [7].

The mRNA, miR-21, is important in chemoresistance. Increased expression of miR-21 prevents gemcitabine-induced apoptosis targeting PTEN in cholangiocarcinoma [47] and unknown molecules in pancreatic cancer [48]. Overexpression of miR-21 triggers increased resistance to VM-26 in glioblastoma [49] and topotecan in breast cancer [50] by targeting LRRFIP1 and Bcl-2, respectively. Contrariwise, the expression inhibition of miR-21 increased the sensitivity to TRAIL-induced apoptosis in glioma [51]. Interestingly, it increased the expression protected against temozolomide-induced apoptosis via targeting Bax in glioblastoma [52] and arsenic trioxide-induced apoptosis by targeting PDCD4 in chronic myeloid leukemia [53] and also prevented apoptosis induced by arabinosylcytosine by targeting PDCD4 in acute myeloid leukemia [54].

Table 3
MicroRNAs and drug relationships

miRNA	Change	Target gene	Cancer or cell line	Result	Reference
miR-24	C to T	DHFR 3′UTR	DG44 cells	Resistance to methotrexate	[39]
miR-206b	C to T	ER-alpha	Breast cancer	Sensitivity to endocrine therapy	[58]
miR-519c	Deletion	ABCG2 3′UTR	Colon cancer	Inhibition of repression/ multidrug resistance	[59]
pri-miR-26a1/ pri-miR-100	Several SNPs	–	Colon cancer	Longer time to progression after 5-fluorouracil or irinotecan treatment	[60]

5 Drug Resistance miRNA Pharmacogenetics Association

The expression levels of mature miRNAs may be affected by two pathways: sequence variations in miRNA regions and miRNA-processing pathways [55, 56]. This issue is discussed in more detail in Chapter 18. The pharmacogenetic analysis of miRNAs may represent a revolutionary area of investigation for predicting treatment response or chemoresistance [57]. These alterations are owing to single-nucleotide polymorphisms (SNPs [56]). SNPs can occur in three different ways: (1) polymorphisms affecting miRNA biogenesis, by modulating the transcription of pri-miRNA or pre-miRNA processing and maturation; (2) polymorphisms in miRNA target sites; and (3) polymorphisms altering epigenetic regulation of miRNA genes [56] (Table 3).

6 MicroRNAs in Cancer Stem Cells: Association with Drug Resistance

Recently, it has become clear that miRNAs can play a pivotal role in the drug resistance of cancer cells and cancer stem cells. Drug resistance in cancer stem cells inserts another viewpoint to the challenge of drug resistance in cancer. Cancer stem cell hypothesis states that when not all cancer stem cells from a tumor are extirpated, the tumor will always reoccur [61] (Table 4).

The fact that cancer stem cells are often found to be resistant to one or multiplexed drugs increases the complexity for successful cancer cures. For this reason, it is extremely important to explore occasions to selectively target the cancer stem cell population of a tumor. Selective killing of cancer stem cells would properly enhance patient outcome by preventing metastasis and recurrence of the

Table 4
miRNAs in cancer stem cells: Association with drug resistance

miRNA	Characteristics	Targets	Tumor model	Reference
miR-34	Regulated by p53, downregulated in most tumors/drug resistance	Notch, HMGA2,Bcl-2	Cancer stem cells lacking p53 gene expression	[62]
miR-125b	Decreased sensitivity of ATRA-induced apoptosis via upregulation	Unknown	Glioma	[63, 64]
miR-140	Methotrexate and 5-FU resistance increased via upregulation	HDAC4	Colon and osteosarcoma stem cells	[65]
miR-215	Methotrexate and tomudex resistance increased by downregulation of DHFR and TS	DHFR, TS	Osteosarcoma and colon cancer stem cells	[66]

primary tumor. In the last years, some encouraging studies have been succeeded in this research field, with distinct molecules developed to specifically kill cancer stem cells, such as monoclonal antibodies targeting leukemic stem cell marker CD44 [67] or drugs like nigericin and abamectin that inhibit cancer stem cell growth [68] or gene therapy [69, 70].

There are a lot of studies concerning with this issue. The miRNA, miR-34, is directly regulated by p53 and downregulated in most tumors, which suggests a common drug resistance via targeting Notch, HMGA2, and Bcl-2 in cancer stem cells that lack p53 expression [62].

Upregulation of miR-125b increases the rate of proliferation and decreases sensitivity to ATRA-induced apoptosis by targeting Bmf in glioma stem cells [102].

Overexpression of miR-140 elevates the resistance against methotrexate and 5-FU HDAC4 osteosarcoma and colon cancer stem cells [65].

Upregulation of miR-215 downregulates dihydrofolate reductase (DHFR) and thymidylate synthase (TS) which in turn leads to increased resistance against methotrexate and tomudex in osteosarcoma and colon cancer stem cells, respectively [65, 71].

7 MicroRNAs as Drugs

Rukov et al. reported that attempts are in progress to improve miRNA-based drugs, either in the form of miRNA mimics, amplifying the effect of a miRNA, or miRNA inhibitors, fundamentally suppressing the effect of a miRNA [7]. MicroRNA drugs have the benefit that one miRNA may target and modify the expression of several genes with different roles in the same pathway. The most developed miRNA drug to date is a miRNA inhibitor targeting miR-122 in liver to treat hepatitis C virus (HCV [72]). One issue as such drugs approach clinical use is testing for interactions between the novel miRNA drugs and traditional drugs already in the market. The connection from miRNAs to drugs allows Pharmaco-miR to predict default interactions between novel miRNA drugs and more traditional drugs, which can then be tested experimentally [7]. For example miR-122 is estimated to target estrogen receptor 1 (ESR1), whose gene product is necessary for the important drug families of estrogens (e.g., estradiol) and antiestrogens (e.g., tamoxifen). If this predicted target is functional, treating patients for HCV with miR-122 may cause adverse drug effects if the patient is also undergoing treatment with estrogens or antiestrogens.

In recent years, an increasing number of papers describe a link between miRNAs and drug function through deregulation of pharmacogenomic-relevant genes. These studies are mainly performed in cancer cell lines and mainly describe chemoresistance. Drug toxicity studies and studies on drug metabolizers are remarkably rare. Also, many studies report miRNA deregulation in drug-resistant cells but fail to identify the miRNA target effector genes. The lack of such studies highlights how elusive it can be to link miRNA expression with the connection between genes and drug efficacy/toxicity. Pharmaco-miR is a web server designed to help in defining such interactions between miRNAs, target genes, and associated drugs by complementary of the pioneering resources on miRNA targeting and pharmacogenomics. The outcome usually comprises a miRNA pharmacogenomic set consisting of a miRNA, a target gene, and a drug commented in the literature as being linked with the target gene. Pharmaco-miR is thus a useful tool when predicting the effect of miRNAs on drug efficacy and toxicity or when developing hypothesis within miRNA pharmacogenomics. Identification of miRNA pharmacogenomic sets makes it possible to outline potential mechanisms for miRNA–drug interactions when planning experiments and assists in the interpretation of results. As the field of miRNA pharmacogenomics matures, Pharmaco-miR can be extended to collect relevant information within the field and allow searches specifically for miRNA pharmacogenomic sets, where the full set has been investigated in a pharmacogenomic context [7].

According to the National Center for Toxicological Research's 2011–2012 annual report, new biomarkers of liver damage are needed to improve detection of injury in animals and humans. A genomics approach was used on urine samples of rats treated with drugs and chemicals that cause liver injury. Several urinary miRNAs (also called epigenetic biomarkers) were identified that may serve as predictive biomarkers of hepatotoxicity.

MicroRNAs are differently expressed in diseases, and they have crucial responsibilities in diverse biological pathways. Notwithstanding alteration of DNA methylation or histone modifications, deregulated miRNA expression patterns of tumor cells have been described as colliding with drug response. Approaches to arrange the expression of chosen miRNAs have partly led to intriguing developments of chemotherapy response.

MicroRNAs are potential targets of therapeutics. The arrangement of miRNA levels covers a wide spectrum of technologies from gene therapy to antisense therapy. Targeting miRNAs for therapy could be an emerging field, although there are many obstacles to be overcome: stability, convenient in vivo delivery systems, and selectivity. An individual miRNA could regulate several genes and pathways simultaneously suggesting that miRNA modulation could be powerful. However, attention must be paid to the possibility that miRNA manipulation may cause adverse influences, for the reason that each miRNA target has not been identified. Chiefly there are two strategies to target miRNA expression: by blocking the expression of an oncomiR or by re-expression of a tumor-suppressor miRNA, or by targeting the genes involved in their transcription and processing. Anti-miRNA oligonucleotides (AMOs) [73–76], locked nucleic acids (LNA) [77–81], small-molecule inhibitors (SMIRs), miRNA sponges or decoys [71, 82, 83], nanoparticles [84–86], and miRNA mimics and adenovirus-associated vectors (AAV) [87–90] are the current ways to target miRNAs as drugs.

There is growing evidence that diagnostic and prognostic miRNA signatures are indispensable in tumor classification and treatment protocols. Differences in individual gene expressions stimulated the interest of global miRNA expression studies in human diseases that exposed modest variations between normal and tumor tissues [64, 91–93]. For instance, in tumor tissues, miR-125b and miR-145 are mostly determined at lower levels while miR-21, miR-155, and miR-210 at higher levels [50, 94–96]. But of course, there are slight differences in miRNA expressions, and further studies are needed to detect alterations in the signaling pathways and enlighten treatment response [95, 97]. There are high numbers of studies demonstrating miRNA signatures and their prognostic value correlation [53, 91, 98, 99].

Studies in CRC cell lines declared that mRNA and miRNA signatures could improve prediction of treatment response over K-Ras mutation status alone and thus inform patient eligibility to anti-EGFR-based treatments [100, 101]. Ragusa et al. reported on using cetuximab-sensitive and -resistant CRC cell lines to profile miRNA expression, and they suggested three miRNA signatures for estimating treatment response to cetuximab [101]. Two of these miRNAs, let-7b and let 7e, were found to be negative regulators of K-Ras expression. Several studies showed the importance of let-7's role in a poor outcome of cetuximab-treated metastatic

CRC patients [102, 103]. Also, let-7a-mediated regulation of K-Ras expression was first described in lung cancer [104].

The most important problem in the majority of the cancer cases is a lack of targeted therapy. MicroRNA signature analyses are indispensable in every step of therapy approach like tumor aggressiveness and resistance to treatment in almost every type of cancer affecting different gene expressions like EGFR and K-Ras and in several signaling pathways [105–107]. These studies also identified the organ site of carcinomas of unknown primary origin [108, 109].

Additionally, miRNA signatures are crucial in blood samples of patients to detect diseases earlier and monitor progression of diseases in a noninvasive way after treatment [110–113].

Global miRNA profiling pioneered to improve disease management from bench to clinical applications.

8 Conclusion

MicroRNAs are potential providers of drug response prediction. MicroRNAs are also the headliners of drug resistance and pharmacological drug response dilemma and conducives of drug dosage tailoring and inhibitors of adverse drug reactions that procreate personalized therapy.

Complicated dynamic natured networks between diagnostic, prognostic, and therapeutic miRNAs and their targets are complex, and in vitro studies alone may be improper to estimate miRNA significance related with early detection, progression, recurrence, and treatment response in vivo.

In the light of these novel findings, more studies should be conducted in order to demonstrate the utilization of these tiny but indispensable molecules in diagnosis and treatment of various diseases: a dream could be realized. Individualizing current anticancer regimens by predicting the potential intrinsic/acquired resistance and future therapeutic strategies to get over resistance, including specific targeting of miRNA (via mimics or antagomirs) is very important. Also these regimens target synergistic interaction with anticancer agents. These regimens use the modulation of expression of key proteins in different molecular mechanisms involved in drug activity.

9 Notes

1. In colorectal cancer, miR-142-3p, -144, -151, -212, -17, -20, -25, -32, -92, -93, -106a, -125a, -155, -191, -192, -203, -215, and -223 are responsible for subtype classification like MSS vs. MSI-high.

2. MicroRNA signatures such as let-7a, -7b, -7c, -7d, -7e, -7f, -7g, and -7i and miR-16, -17, -19b, -20a, -26a, -26b, -29a, -29b, -29c, -30b, -30d, -98, -103, -106a, -106b, 107, -146b-5p, -181a, -191, -195, -453, -491-5p, -498, -509-3p, -654-5p, and -663 are responsible for the subtype classification of AdCa vs. SCC.
3. MicroRNA signatures, such as let-7e and miR-17-5p, -20a, -20b, -21, -93, -106a, -106b, -126, -146b, -155, -182, -183, -191, 200a, -200c, -203, -210, and -224, have a major role in the prediction of the overall survival in SCC cases.
4. Specific miRNAs such as miR-27b, -29a/181a/221, -34a, -98, -122, -125b, -140, -143, -148a, -155, -192/215, -199a-3p, -200c, -204, -205, -212, -214, -221/222, -320, -328, -451, and -512 targeting genes (such as CYPIB1, SIRT1, HMGA2, Cyclin G1, BAK1, HDAC4, ERK5, MSK1, PXR, FOXO3a, TS, mTOR, C-Met, TUBB3, Mcl-1, HER3, PED, PTEN, P27, ERα, Bcl-2, ABCG2, and CSA) are involved in apoptosis and survival pathways affecting anticancer drug sensitivity in various cancer types, including breast, cervical, prostate, colorectal, colon, cholangiocarcinoma, T-cell leukemia, and osteosarcoma.

References

1. The Case for Personalized Medicine, 3rd Edition (2011) Personalized medicine coalition. www.personalizedmedicinecoalition.org/sites/default/files/files/Case_for_PM_3rd_edition.pdf
2. Sorlie T, Perou CM, Tibshirani R et al (2001) Gene expression patterns of breast carcinomas distinguish tumor subclasses with clinical implications. Proc Natl Acad Sci USA 98: 10869–10874
3. Sherman S, Pletcher BA, Driscoll DA (2005) Fragile X syndrome: diagnostic and carrier testing. Genet Med 7:584–587
4. Piccart-Gebhart MJ, Procter M, Leyland-Jones B et al (2005) Trastuzumab after adjuvant chemotherapy in HER2-positive breast cancer. N Engl J Med 353:1659–1672
5. Romond EH, Perez EA, Bryant J et al (2005) Trastuzumab plus adjuvant chemotherapy for operable HER2-positive breast cancer. N Engl J Med 353:1673–1684
6. Shuldiner AR, O'Connell JR, Bliden KP et al (2009) Association of cytochrome P450 2C19 genotype with the antiplatelet effect and clinical efficacy of clopidogrel therapy. JAMA 302:849–857
7. Rukov JL, Wilentzik R, Jaffe I, et al (2013) Pharmaco-miR: linking microRNAs and drug effects. Brief Bioinform Jan 31. [Epub ahead of print]
8. Foekens JA, Sieuwerts AM, Smid M et al (2008) Four miRNAs associated with aggressiveness of lymph node-negative, estrogen receptor-positive human breast cancer. Proc Natl Acad Sci USA 105:13021–13026
9. Rodriguez-Gonzalez FG, Sieuwerts AM, Smid M et al (2011) MicroRNA-30c expression level is an independent predictor of clinical benefit of endocrine therapy in advanced estrogen receptor positive breast cancer. Breast Cancer Res Treat 127:43–51
10. Buffa FM, Camps C, Winchester L et al (2011) microRNA associated progression pathways and potential therapeutic targets identified by integrated mRNA and microRNA expression profiling in breast cancer. Cancer Res 71(17):5635–5645
11. Radojicic J, Zaravinos A, Vrekoussis T, Kafousi M, Spandidos DA, Stathopoulos EN (2011) MicroRNA expression analysis in triple-negative (ER, PR and Her2/neu) breast cancer. Cell Cycle 10:507–517
12. Gong C, Yao Y, Wang Y et al (2011) Up-regulation of miR-21 mediates resistance to trastuzumab therapy for breast cancer. J Biol Chem 286:19127–19137

13. Qian B, Katsaros D, Lu L et al (2009) High miR-21 expression in breast cancer associated with poor disease-free survival in early stage disease and high TGF-β1. Breast Cancer Res Treat 117:131–140
14. Yan LX, Huang XF, Shao Q et al (2008) MicroRNA miR-21 overexpression in human breast cancer is associated with advanced clinical stage, lymph node metastasis and patient poor prognosis. RNA 14:2348–2360
15. Sempere LF, Christensen M, Silahtaroglu A et al (2007) Altered microRNA expression confined to specific epithelial cell subpopulations in breast cancer. Cancer Res 67:11612–11620
16. Rothe F, Ignatiadis M, Chaboteaux C et al (2011) Global MicroRNA expression profiling identifies miR-210 associated with tumor proliferation, invasion and poor clinical outcome in breast cancer. PLoS One 6:e20980
17. Camps C, Buffa FM, Colella S et al (2008) hsa-miR-210 is induced by hypoxia and is an independent prognostic factor in breast cancer. Clin Cancer Res 14:1340–1348
18. Schepeler T, Reinert JT, Ostenfeld MS et al (2008) Diagnostic and prognostic microRNAs in stage II colon cancer. Cancer Res 68:6416–6424
19. Lanza G, Ferracin M, Gafa R et al (2007) mRNA/microRNA gene expression profile in microsatellite unstable colorectal cancer. Mol Cancer 6:54
20. Drebber U, Lay M, Wedemeyer I et al (2011) Altered levels of the onco-microRNA 21 and the tumor suppressor microRNAs 143 and 145 in advanced rectal cancer indicate successful neoadjuvant chemoradiotherapy. Int J Oncol 39:409–415
21. Nielsen BS, Jorgensen S, Fog JU et al (2011) High levels of microRNA-21 in the stroma of colorectal cancers predict short disease-free survival in stage II colon cancer patients. Clin Exp Metastasis 28:27–38
22. Kulda V, Pesta M, Topolcan O et al (2010) Relevance of miR-21 and miR-143 expression in tissue samples of colorectal carcinoma and its liver metastases. Cancer Genet Cytogenet 200:154–160
23. Shibuya H, Iinuma H, Shimada R, Horiuchi A, Watanabe T (2010) Clinicopathological and prognostic value of microRNA-21 and microRNA-155 in colorectal cancer. Oncology 79:313–320
24. Schetter AJ, Leung SY, Sohn JJ et al (2008) MicroRNA expression profiles associated with prognosis and therapeutic outcome in colon adenocarcinoma. JAMA 299:425–436
25. Landi MT, Zhao Y, Rotunno M et al (2010) MicroRNA expression differentiates histology and predicts survival of lung cancer. Clin Cancer Res 16:430–441
26. Yu SL, Chen HY, Chang GC et al (2008) MicroRNA signature predicts survival and relapse in lung cancer. Cancer Cell 13:48–57
27. Raponi M, Dossey L, Jatkoe T et al (2009) MicroRNA classifiers for predicting prognosis of squamous cell lung cancer. Cancer Res 69:5776–5783
28. Takamizawa J, Konishi H, Yanagisawa K et al (2004) Reduced expression of the let-7 microRNAs in human lung cancers in association with shortened postoperative survival. Cancer Res 64:3753–3756
29. Yanaihara N, Caplen N, Bowman E et al (2006) Unique microRNA molecular profiles in lung cancer diagnosis and prognosis. Cancer Cell 9:189–198
30. Markou A, Tsaroucha EG, Kaklamanis L, Fotinou M, Georgoulias V, Lianidou ES (2008) Prognostic value of mature microRNA-21 and microRNA-205 overexpression in non-small cell lung cancer by quantitative real-time RT-PCR. Clin Chem 54:1696–1704
31. Gao W, Shen H, Liu L, Xu J, Xu J, Shu Y (2011) miR-21 overexpression in human primary squamous cell lung carcinoma is associated with poor patient prognosis. J. Cancer Res. Clin Oncol 137:557–566
32. Gallardo E, Navarro A, Vinolas N et al (2009) miR-34a as a prognostic marker of relapse in surgically resected non-small-cell lung cancer. Carcinogenesis 30:1903–1909
33. Lebanony D, Benjamin H, Gilad S et al (2009) Diagnostic assay based on hsa-miR-205 expression distinguishes squamous from nonsquamous non-small-cell lung carcinoma. J Clin Oncol 27:2030–2037
34. Fassina A, Cappellesso R, Fassan M (2011) Classification of non-small cell lung carcinoma in transthoracic needle specimens using microRNA expression profiling. Chest 140:1305–1311. doi:10.1378/chest.11-0708
35. Bishop JA, Benjamin H, Cholakh H, Chajut A, Clark DP, Westra WH (2010) Accurate classification of non-small cell lung carcinoma using a novel microRNA-based approach. Clin Cancer Res 16:610–619
36. Calin GA, Croce CM (2006) microRNA signatures in human cancers. Nature 6(11):857–866
37. Zheng T, Wang J, Chen X, Liu L (2009) Role of microRNA in anticancer drug resistance. Int J Cancer 126(1):2–10
38. van Jaarsveld MT, Helleman J, Berns EM, Wiemer EA (2010) MicroRNAs in ovarian cancer biology and therapy resistance. Int J Biochem Cell Biol 42(8):1282–1290

39. Mishra PJ, Bertino JR (2009) microRNA polymorphisms: the future of pharmacogenomics, molecular epidemiology and individualized medicine. Pharmacogenomics 10(3):399–416
40. Fojo T (2007) Multiple paths to a drug resistance phenotype: mutations, translocations, deletions and amplification of coding genes or promoter regions, epigenetic changes and microRNAs. Drug Resist Updat 10(1–2):59–67
41. Tsang WP, Kwok TT (2008) Let-7a microRNA suppresses therapeutics induced cancer cell death by targeting caspase-3. Apoptosis 13(10): 1215–1222
42. Pothof J, Verkaik NS, van IJcken W et al (2009) MicroRNA-mediated gene silencing modulates the UV-induced DNA-damage response. EMBO J 28(14):2090–2099
43. Nasser MW, Datta J, Nuovo G et al (2008) Down-regulation of microRNA-1 (mir-1) in lung cancer: suppression of tumorigenic property of lung cancer cells and their sensitization to doxorubicin-induced apoptosis by mir-1. J Biol Chem 283(48):33394–33405
44. Wang F, Sun GP, Zou YF, Hao JQ et al (2012) MicroRNAs as promising biomarkers for gastric cancer. Cancer Biomark 11:259–267
45. Li Z, Hu S, Wang J et al (2010) MiR-27a modulates MDR1/P-glycoprotein expression by targeting HIPK2 in ovarian cancer cells. Gynecol Oncol 119(1):125–130
46. Zhu H, Wu H, Liu X et al (2008) Role of MicroRNA miR-27a and miR- 451 in the regulation of MDR1/P-glycoprotein expression in human cancer cells. Biochem Pharmacol 76(5):582–588
47. Meng F, Henson R, Lang M et al (2006) Involvement of human micro-RNA in growth and response to chemotherapy in human cholangiocarcinoma cell lines. Gastroenterology 130(7):2113–2129
48. Giovannetti E, Funel N, Peters GJ et al (2010) microRNA-21 in pancreatic cancer: correlation with clinical outcome and pharmacologic aspects underlying its role in the modulation of gemcitabine activity. Cancer Res 70(11):4528–4538
49. Li J, Huang H, Sun L et al (2009) MiR-21 indicates poor prognosis in tongue squamous cell carcinomas as an apoptosis inhibitor. Clin Cancer Res 15(12):3998–4008
50. Si ML, Zhu S, Wu H et al (2007) miR-21-mediated tumor growth. Oncogene 26(9): 2799–2803
51. Corsten MF, Miranda R, Kasmieh R et al (2007) microRNA-21 knockdown disrupts glioma in vivo and displays synergistic cytotoxicity with neural precursor cell-delivered S-TRAIL in human gliomas. Cancer Res 67(19):8994–9000
52. Shi L, Chen C, Yang J et al (2010) MiR-21 protected human glioblastoma U87MG cells from chemotherapeutic drug temozolomide induced apoptosis by decreasing Bax/Bcl-2 ratio and caspase-3 activity. Brain Res 1352:255–264
53. Gu J, Zhu X, Li Y et al (2011) miRNA-21 regulates arsenic-induced anti-leukemia activity in myelogenous cell lines. Med Oncol 28(1):211–218
54. Li Y, Zhu X, Gu J et al (2010) Anti-miR-21 oligonucleotide enhances chemosensitivity of leukemic HL60 cells to arabinosylcytosine by inducing apoptosis. Hematology 15(4):215–221
55. Ryan BM, Robles AI, Harris CC (2010) Genetic variation in microRNA networks: the implications for cancer research. Nat Rev Cancer 10(6):389–402
56. Duan R, Pak C, Jin P (2007) Single nucleotide polymorphism associated with mature miR-125a alters the processing of pri-miRNA. Hum Mol Genet 16(9):1124–1131
57. Passetti F, Ferreira CG, Costa FF (2009) The impact of microRNAs and alternative splicing in pharmacogenomics. Pharmacogenomics J 9(1):1–13
58. Tchatchou S, Jung A, Hemminki K et al (2009) A variant affecting a putative miRNA target site in estrogen receptor (ESR) 1 is associated with breast cancer risk in premenopausal women. Carcinogenesis 30(1):59–64
59. To KK, Zhan Z, Litman T, Bates SE (2008) Regulation of ABCG2 expression at the 3' untranslated region of its mRNA through modulation of transcript stability and protein translation by a putative microRNA in the S1 colon cancer cell line. Mol Cell Biol 28(17):5147–5161
60. Boni V, Zarate R, Villa JC et al (2011) Role of primary miRNA polymorphic variants in metastatic colon cancer patients treated with 5-fluorouracil and irinotecan. Pharmacogenomics J 11:429–436
61. Reya T, Morrison SJ, Clarke MF et al (2001) Stem cells, cancer, and cancer stem cells. Nature 414(6859):105–111
62. Bommer GT, Gerin I, Feng Y et al (2007) p53-mediated activation of miRNA34 candidate tumor-suppressor genes. Curr Biol 17(15):1298–1307
63. Xia HF, He TZ, Liu CM et al (2009) MiR-125b expression affects the proliferation and apoptosis of human glioma cells by targeting Bmf. Cell Physiol Biochem 23(4–6):347–358

64. Du L, Pertsemlidis A (2010) MicroRNAs and lung cancer: tumors and 22-mers. Cancer Metastasis Rev 29:109–122
65. Song B, Wang Y, Xi Y et al (2009) Mechanism of chemoresistance mediated by mir-140 in human osteosarcoma and colon cancer cells. Oncogene 28(46):4065–4074
66. Zou GM (2008) Cancer initiating cells or cancer stem cells in the gastrointestinal tract and liver. J Cell Physiol 217(3):598–604
67. Jin L, Hope KJ, Zhai Q et al (2006) Targeting of CD44 eradicates human acute myeloid leukemic stem cells. Nat Med 12(10):1167–1174
68. Riccioni R, Dupuis ML, Bernabei M et al (2010) The cancer stem cell selective inhibitor salinomycin is a p-glycoprotein inhibitor. Blood Cells Mol Dis 45(1):86–92
69. Wang Z, Li Y, Ahmad A et al (2010) Targeting miRNAs involved in cancer stem cell and EMT regulation: an emerging concept in overcoming drug resistance. Drug Resist Updat 13(4–5):109–118
70. Aboody KS, Najbauer J, Danks MK (2008) Stem and progenitor cell-mediated tumor selective gene therapy. Gene Ther 15(10): 739–752
71. Valastyan S, Reinhardt F, Benaich N, Calogrias D, Szasz AM, Wang ZC, Brock JE, Richardson AL, Weinberg RA (2009) A pleiotropically acting microRNA, miR-31, inhibits breast cancer metastasis. Cell 137:1032–1046
72. Lanford RE, Hildebrandt-Eriksen ES, Petri A et al (2010) Therapeutic silencing of microRNA-122 in primates with chronic hepatitis C virus infection. Science 327:198–201
73. Krutzfeldt J, Rajewsky N, Braich R, Rajeev KG, Tuschl T, Manoharan M, Stoffel M (2005) Silencing of microRNAs in vivo with antagomirs. Nature 438:685–689
74. Krutzfeldt J, Kuwajima S, Braich R, Rajeev KG, Pena J, Tuschl T, Manoharan M, Stoffel M (2007) Specificity, duplex degradation and subcellular localization of anatagomirs. Nucleic Acids Res 35:2885–2892
75. Fabani MM, Gait MJ (2008) miR-122 targeting with LNA/20-O-methyl oligonucleotide mixers, peptide nucleotides (PNA) and PNA-peptide conjugates. RNA 14:336–346
76. Fabani MM, Abreu-Goodger C, Williams D, Lyons PA, Torres AG, Smith KGC, Enright AJ, Gait MJ, Vigorito E (2010) Efficient inhibition of miR-155 function in vivo by peptide nucleic acids. Nucleic Acids Res 38:4466–4475
77. Koshkin AA, Rajwanshi VK, Wengel J (1998) Novel convenient synthesys of LNA [2.2.1] bicycle nucleotides. Tetrahedron Lett 39:4381–4384
78. Orom UA, Kaupinnen S, Lund AH (2006) LNA-modified oligonucleotides mediate specific inhibition of microRNA function. Gene 372:137–141
79. Thomas JR, Hergenrother PJ (2008) Targeting RNA with small molecules. Chem Rev 108:1171–1224
80. Gumireddy K, Young DD, Xiong X, Hogenesch JB, Huang Q, Deiters A (2008) Small-molecule inhibitors of microRNA miR-21 function. Angew Chem Int Ed 47:7482–7484
81. Zhang S, Chen L, Jung EJ, Calin GA (2010) Targeting microRNAs with small molecules: from dream to reality. Clin Pharmacol Ther 87:754–758
82. Ebert MS, Sharp PA (2010) MicroRNA sponges: progress and possibilities. RNA 16:2043–2050
83. Ma L, Reinhardt F, Pan E, Soutschek J, Bhat B, Marcusson E, Bell GW, Teruya-Feldstein J, Weinberg RA (2010) Therapeutic silencing of miR-10b inhibits metastasis in a mouse mammary tumor model. Nat Biotechnol 28:341–347
84. Chen Y, Zhu X, Zhang X, Liu B, Huang L (2010) Nanoparticles modified with tumor targeting scFv deliver siRNA and miRNA for cancer therapy. Mol Ther 18:1650–1656
85. Shi SJ, Zhong ZR, Liu J, Zhang ZR, Sun X, Gong T (2011) Solid lipid nanoparticles loaded with anti-microRNA oligonuleotides (AMOs) for suppression of microRNA-21 functions in human lung cancer cells. Pharm Res 29:97–109
86. Liu XQ, Song WJ, Sun TM, Zhang PZ, Wang J (2010) Targeted delivery of antisense inhibitor of miRNA for antiangiogenesis therapy using Crgd functionalized nanoparticles. Mol Pharm 8:250–259
87. Bader AG, Brown D, Winkler M (2010) The promise of microRNA replacement therapy. Cancer Res 70:7027–7030
88. Takeshita F, Patrawala L, Osaki M, Takahashi RU, Yamamoto Y, Kosaka N, Kawamata M, Kelnar K, Bader AG, Brown D, Ochiya T (2010) Systemic delivery of synthetic microRNA-16 inhibits the growth of metastatic prostate tumors via downregulation of multiple cell-cycle genes. Mol Ther 18:181–187
89. Wu Z, Asokan A, Samulski RJ (2006) Adeno-associated virus serotypes: vector toolkit for human gene therapy. Mol Ther 14:316–327
90. Kota J, Chivukula RR, O'Donnell KA (2009) Therapeutic microRNA delivery suppresses tumorigenesis in a murine liver cancer model. Cell 137:1005–1017

91. Volinia S, Calin GA, Liu CG et al (2006) A microRNA expression signature of human solid tumors defines cancer gene targets. Proc Natl Acad Sci USA 103:2257–2261
92. Lin PY, Yu SL, Yang PC (2010) MicroRNA in lung cancer. Br J Cancer 103:1144–1148
93. Yendamuri S, Kratzke R (2011) MicroRNA biomarkers in lung cancer: MiRacle or quagMiRe? Transl Res 157:209–215
94. Iorio MV, Ferracin M, Liu CG et al (2005) MicroRNA gene expression deregulation in human breast cancer. Cancer Res 65:7065–7070
95. Andorfer CA, Necela BM, Thompson EA, Perez EA (2011) MicroRNA signatures: clinical biomarkers for the diagnosis and treatment of breast cancer. Trends Mol Med 17:313–319
96. Le QJ, Caldas C (2010) Micro-RNAs and breast cancer. Mol Oncol 4:230–241
97. Fu SW, Chen L, Man YG (2011) miRNA biomarkers in breast cancer detection and management. J Cancer 2:116–122
98. Luo X, Burwinkel B, Tao S, Brenner H (2011) MicroRNA signatures: novel biomarker for colorectal cancer? Cancer Epidemiol Biomarkers Prev 20:1272–1286
99. Ma Y, Zhang P, Yang J, Liu Z, Yang Z, Qin H (2012) Candidate microRNA biomarkers in human colorectal cancer: systematic review profiling studies and experimental validation. Int J Cancer 130:2077–2087. doi:10.1002/ijc.26232
100. Solmi R, Lauriola M, Francesconi M et al (2008) Displayed correlation between gene expression profiles and submicroscopic alterations in response to cetuximab, gefitinib and EGF in human colon cancer cell lines. BMC Cancer 8:227
101. Ragusa M, Majorana A, Statello L et al (2010) Specific alterations of microRNA transcriptome and global network structure in colorectal carcinoma after cetuximab treatment. Mol Cancer Ther 9:3396–3409
102. Zhang W, Winder T, Ning Y et al (2011) A let-7 microRNA-binding site polymorphism in 3′-untranslated region of KRAS gene predicts response in wild-type KRAS patients with metastatic colorectal cancer treated with cetuximab monotherapy. Ann Oncol 22:104–109
103. Graziano F, Canestrari E, Loupakis F et al (2010) Genetic modulation of the let-7 microRNA binding to KRAS 3′-untranslated region and survival of metastatic colorectal cancer patients treated with salvage cetuximab–irinotecan. Pharmacogenomics J 10:458–464
104. Read ML, Spice R, Parker AL, Mir S, Logan A (2005) 12th annual congress of the European society of gene therapy. Expert Opin Biol Ther 5:137–141
105. Weiss GJ, Bemis LT, Nakajima E et al (2008) EGFR regulation by microRNA in lung cancer: correlation with clinical response and survival to gefitinib and EGFR expression in cell lines. Ann Oncol 19:1053–1059
106. Rai K, Takigawa N, Ito S et al (2011) Liposomal delivery of microRNA-7-expressing plasmid overcomes epidermal growth factor receptor-tyrosine kinase inhibitor-resistance in lung cancer cells. Mol Cancer Ther 10(9):1720–1727
107. Webster RJ, Giles KM, Price KJ, Zhang PM, Mattick JS, Leedman PJ (2009) Regulation of epidermal growth factor receptor signaling in human cancer cells by microRNA-7. J Biol Chem 284:5731–5741
108. Ferracin M, Pedriali M, Veronese A et al (2011) MicroRNA profiling for the identification of cancers with unknown primary tissue-of-origin. J Pathol 225(1):43–45
109. Varadhachary GR, Spector Y, Abbruzzese JL et al (2011) Prospective gene signature study using microRNA to identify the tissue of origin in patients with carcinoma of unknown primary. Clin Cancer Res 17:4063–4070
110. Schwarzenbach H, Hoon DS, Pantel K (2011) Cell-free nucleic acids as biomarkers in cancer patients. Nat Rev Cancer 11:426–437
111. Mostert B, Sieuwerts AM, Martens JW, Sleijfer S (2011) Diagnostic applications of cell-free and circulating tumor cell-associated miRNAs in cancer patients. Expert Rev Mol Diagn 11:259–275
112. Brase JC, Wuttig D, Kuner R, Sultmann H (2010) Serum microRNAs as non-invasive biomarkers for cancer. Mol Cancer 9:306
113. Scholer N, Langer C, Dohner H, Buske C, Kuchenbauer F (2010) Serum microRNAs as a novel class of biomarkers: a comprehensive review of the literature. Exp Hematol 38:1126–1130

INDEX

Malik Yousef and Jens Allmer (eds.), *miRNomics: MicroRNA Biology and Computational Analysis*, Methods in Molecular Biology, vol. 1107, DOI 10.1007/978-1-62703-748-8, © Springer Science+Business Media New York 2014

MIX
Papier aus verantwortungsvollen Quellen
Paper from responsible sources
FSC® C105338

If you have any concerns about our products,
you can contact us on
ProductSafety@springernature.com

In case Publisher is established outside the EU,
the EU authorized representative is:
Springer Nature Customer Service Center GmbH
Europaplatz 3, 69115 Heidelberg, Germany

Printed by Libri Plureos GmbH
in Hamburg, Germany